Submarine Warfare in the 20th & 21st Centuries:
A Bibliography

**Compiled by
Michaele Lee Huygen**

Dudley Knox Library
Naval Postgraduate School
December 2003

Report Documentation Page

Form Approved
OMB No. 0704-0188

Public reporting burden for the collection of information is estimated to average 1 hour per response, including the time for reviewing instructions, searching existing data sources, gathering and maintaining the data needed, and completing and reviewing the collection of information. Send comments regarding this burden estimate or any other aspect of this collection of information, including suggestions for reducing this burden, to Washington Headquarters Services, Directorate for Information Operations and Reports, 1215 Jefferson Davis Highway, Suite 1204, Arlington VA 22202-4302. Respondents should be aware that notwithstanding any other provision of law, no person shall be subject to a penalty for failing to comply with a collection of information if it does not display a currently valid OMB control number.

1. REPORT DATE 00 DEC 2003	2. REPORT TYPE N/A	3. DATES COVERED -
4. TITLE AND SUBTITLE **Submarine Warfare in the 20th & 21st Centuries: A Bibliography**		5a. CONTRACT NUMBER
		5b. GRANT NUMBER
		5c. PROGRAM ELEMENT NUMBER
6. AUTHOR(S)		5d. PROJECT NUMBER
		5e. TASK NUMBER
		5f. WORK UNIT NUMBER
7. PERFORMING ORGANIZATION NAME(S) AND ADDRESS(ES) **Naval Postgraduate School Dudley Knox Library Monterey, CA 93943**		8. PERFORMING ORGANIZATION REPORT NUMBER
9. SPONSORING/MONITORING AGENCY NAME(S) AND ADDRESS(ES)		10. SPONSOR/MONITOR'S ACRONYM(S)
		11. SPONSOR/MONITOR'S REPORT NUMBER(S)

12. DISTRIBUTION/AVAILABILITY STATEMENT
Approved for public release, distribution unlimited

13. SUPPLEMENTARY NOTES
The original document contains color images.

14. ABSTRACT

15. SUBJECT TERMS

16. SECURITY CLASSIFICATION OF:			17. LIMITATION OF ABSTRACT	18. NUMBER OF PAGES	19a. NAME OF RESPONSIBLE PERSON
a. REPORT **unclassified**	b. ABSTRACT **unclassified**	c. THIS PAGE **unclassified**	**UU**	**375**	

Standard Form 298 (Rev. 8-98)
Prescribed by ANSI Std Z39-18

PAGE INTENTIONALLY LEFT BLANK

Submarine Warfare in the 20th & 21st Centuries: A Bibliography

Compiled by
Michaele Lee Huygen

Dudley Knox Library
Naval Postgraduate School
December 2003

INTRODUCTION

There are constant motions in the sea caused by atmospheric and seabed activities, volcanic disruptions, marine animals, ships, and submarines -- all of which create what is called the ambient noise level of the oceans.

Today acoustics is the basis of both submarine and antisubmarine warfare: the single most significant element upon which all undersea warfare activity depends. When the submarine was first developed, its security lay beneath the surface of the ocean where it would not be seen. Today its security lies in its ability to avoid being heard.

-- Thomas S. Burns. **The secret war for the ocean depths: Soviet-American rivalry for mastery of the seas**. 1st ed. New York: Rawson Associates Publishers, c1978. p. 66, 68.

This bibliography is a revised and expanded version of **Submarine Warfare in the 20th Century**, 2002. It is a selected bibliography listing books, periodical articles, and web sites related to submarine and antisubmarine warfare in the 20th century. Some entries have brief annotations taken directly from library cataloger's notes. To ensure a compact citation format, the series statement(s), when present, are in parentheses following the publisher statement and before the pagination. The letters "NPS/DKL," followed by location and call number information, identify the books held by the Naval Postgraduate School's Dudley Knox Library. Many of the journal articles are also accessible in the Dudley Knox Library. Consult with the Reference or Interlibrary Loan Department of your local library for advice on obtaining materials of interest to you.

This Bibliography will also available online at
http://library.nps.navy.mil/home/bibs/submarine/

December 2003

TABLE OF CONTENTS

BOOKS & TECHNICAL REPORTS 6

HISTORY OF NAVAL STRATEGY / WARFARE 6

GENERAL   6
- -- AMERICAN   11
- -- AUSTRALIAN   14
- -- AUSTRIAN   15
- -- BRAZILIAN   15
- -- BRITISH   15
- -- CANADIAN   17
- -- CHINESE   17
- -- FRENCH   18
- -- GERMAN   19
- -- INDIAN   20
- -- IRANIAN   21
- -- ITALIAN   21
- -- JAPANESE   22
- -- KOREAN   22
- -- NEW ZEALAND   23
- -- NORWEGIAN   23
- -- POLISH   23
- -- RUSSIAN/SOVIET   23
- -- SWEDISH   25
- -- TURKISH   25

-- WWI
- -- GENERAL   25
- -- AMERICAN   27
- -- AUSTRALIAN   27
- -- AUSTRIAN   28
- -- BRITISH   28
- -- GERMAN   30
- -- ITALIAN   30

-- WWII
- -- GENERAL   31
- -- AMERICAN   33
- -- AUSTRALIAN   33
- -- BRITISH   33
- -- GERMAN   35
- -- ITALIAN   38
- -- JAPANESE   38
- -- POLISH   39
- -- RUSSIAN/SOVIET   40

SPECIAL TOPICS
-- ANTISUBMARINE WARFARE (ASW)
- -- GENERAL 40
- -- WWI 42
- -- WWII 42
- -- NPS THESES & TECHNICAL REPORTS 44
- -- OTHER THESES & TECHNICAL REPORTS 95

-- CONFERENCES ON LIMITATION OF ARMAMENT
- -- WASHINGTON (1921-1922) 117
- -- LONDON NAVAL CONFERENCE. (1930) 120

-- CONVOYS
- -- GENERAL 122
- -- WWI 122
- -- WWII 123
  - -- ARCTIC 124
  - -- ATLANTIC 128
  - -- MALTA/MEDITERRANEAN 131

-- PACIFIC	132
-- FALKLAND ISLANDS WAR, 1982	132
-- JUTLAND	133
-- MIDGET SUBMARINES, KAITEN & FROGMEN	138
-- Q-SHIPS	146

SUBMARINES

-- GENERAL	147		-- JAPANESE	
-- BIBLIOGRAPHIES	155		-- GENERAL	256
-- FICTION	155		-- WWI	257
-- NUCLEAR	160		-- WWII	257
-- WWI	165		-- KOREAN	261
-- WWII	166		-- PERUVIAN	262
-- AMERICAN			-- POLISH – WWII	262
-- GENERAL	169		-- RUSSIAN/SOVIET	
-- KOREAN WAR	179		– GENERAL	262
-- NUCLEAR	180		-- NUCLEAR	270
-- PERSIAN GULF WAR	185		-- SWEDISH	274
-- WWI	186		-- TURKISH	275
-- WWII	187		-- YUGOSLAV	275
-- AUSTRALIAN				
– GENERAL	197			
-- WWI	197			
-- AUSTRIAN	198			
-- BRAZILIAN	198			
-- BRITISH				
-- GENERAL	198			
-- WWI	200			
-- WWII	204			
-- CANADIAN	208			
-- CHINESE	209			
-- DANISH	209			
-- DUTCH	209			
-- FRENCH	209			
-- GERMAN				
-- GENERAL	211			
-- WWI	214			
-- WWII	223			
-- GREEK	255			
-- INDIAN	255			
-- ISRAELI	255			
-- ITALIAN				
-- GENERAL	255			
-- WWI	255			
-- WWII	256			

PERIODICAL ARTICLES

SUBMARINES – GENERAL   276

SUBMARINES BY COUNTRY

ARGENTINA	329	IRANIAN	340
AUSTRALIAN	329	ISRAELI	341
AUSTRIAN	331	ITALIAN	342
BRITISH	332	JAPANESE	342
CANADIAN	332	KOREAN	342
CHINESE	333	MALAYSIAN	343
DUTCH	333	PAKISTANI	344
EGYPTIAN	334	POLISH	344
FRENCH	334	RUSSIAN/SOVIET	344
GERMAN	334	SINGAPORE	347
GREEK	339	SOUTH AFRICAN	347
INDIAN	339	SWEDISH	347
INDONESIAN	340	TAIWANESE	348
		TURKISH	348

ANTISUBMARINE WARFARE (ASW)   349

SUBMARINE CHASERS AND PATROL BOATS   368

SUBMARINE WEBSITES

GENERAL   370

BY COUNTRY

AMERICAN	370	GERMAN	371
AUSTRALIAN	370	ITALIAN	372
BRITISH	371	JAPANESE	372
CANADIAN	371	POLISH	372
DUTCH	371	SINGAPORE	372
FRENCH	371	TURKISH	372

MIDGET SUBMARINES   372

WWII SUBMARINE HISTORY   372

WWII SUBMARINE STATISTICS   373

HISTORY OF NAVAL STRATEGY/STRATEGY/WARFARE -- GENERAL

Balano, Randy Carol and Craig L. Symonds, eds. **New interpretations in naval history: selected papers from the Fourteenth Naval History Symposium, held at Annapolis, Maryland, 23-25 September 1999.** Annapolis, MD: Naval Institute Press, 2001. 430 p.
NPS/DKL Location: GENERAL V47 .U55 2001

Black, Jeremy, ed. **European warfare, 1815-2000.** New York: Palgrave, c2002. (Problems in focus series). 247 p.
Contents: Europe's way of war, 1815-1864 / Dennis Showalter -- European warfare, 1864-1914 / Jeremy Black -- The First World War / Spencer Tucker -- The European civil war: Reds versus Whites in Russia and Spain, 1917-1939 / Francisco J. Romero Salvadó -- The Second World War / S.P. MacKenzie -- Colonial wars / Bruce Vandervort -- Naval power and warfare / Lawrence Sondhaus -- The transformation of war in Europe, 1945-2000 / Warren Chin.
NPS/DKL Location: GENERAL D361 .E97 2001

Cable, James. **Diplomacia de cañoneras: empleo político de fuerzas navales limitadas [traducción, Benjamín Oscar Cosentino y Mariette D. de Cosentino].** Buenos Aires, República Argentina: Instituto de Publicaciones Navales, Centro Naval, c1977. (Ediciones del Instituto de Publicaciones Navales. Colección Relaciones internacionales; 5o libro). 200 p.
Spanish translation of Gunboat diplomacy.

_____. **Gunboat diplomacy 1919-1979: political applications of limited naval force.** 2d ed. New York, NY: St. Martin's Press, c1981. (International Institute for Strategic Studies. Studies in international security; 16). 288 p.
Originally published: London, Chatto and Windus for the Institute for Strategic Studies; New York ,Praeger for the Institute for Strategic Studies, 1971, as Gunboat diplomacy; political applications of limited naval force.
NPS/DKL Location: RESERVE V25 .C22

_____. **Gunboat diplomacy 1919-1991: political applications of limited naval force.** 3rd ed. New York: St. Martin's Press, c1994. 246 p.
NPS/DKL Location: GENERAL V25 .C327 1994

_____. **Navies in violent peace.** New York: St. Martin's Press, 1989. 155 p.

_____. **The political influence of naval force in history.** New York: St. Martin's Press; London: Macmillan, 1998. 213 p.
NPS/DKL Location: GENERAL V25 .C3324 1998

Cline, Ray S. and William M. Carpenter, eds. **Report of the Seventh International Conference on the Security of Sea Lines of Communication in the Western Pacific and Indian Oceans** (7th: 1990: Airlie House) Warrenton, Virginia, May 7-10, 1990 /

[sponsored by] United States Global Strategy Council, SRI International. Washington, DC: The Council, 1991. 236 p.
NPS/DKL Location: GENERAL HE327 .I54 1990

Collins, John M. **American and Soviet military trends since the Cuban missile crisis.** Washington, DC: Center for Strategic and International Studies, Georgetown University; [New Brunswick, NJ: Distributed by Transaction Books c1978. 496 p.
NPS/DKL Location: GENERAL UA23 .C645

Cowman, Ian. **Dominion or decline: Anglo-American naval relations on the Pacific, 1937-1941.** Oxford, UK; Washington, DC: Berg Publishers, 1996. 327 p.

Collier, Basil. **The lion and the eagle; British and Anglo-American strategy, 1900-1950.** New York, Putnam, [1972]. 499 p.
Originally published: London: Macdonald and Co., 1972.

Diwald, Hellmut. **Die Erben Poseidons: Seemachtpolitik im 20. Jahrhundert.** 1. Aufl. München: Droemer-Knaur, c1984. 511 p.

Dunaway, William Michael. **Gunboat diplomacy in a new world order: strategic considerations for U.S. naval intervention in the twenty-first century.** Springfield, VA: Available from the National Technical Information Service, 1991. (ADA256442). 1 v.
Thesis (M.A. in International Relations), The Fletcher School of Law and Diplomacy, April 1991.
NPS/DKL Location: THESIS D78873

Duppler, Jörg and Werner Rahn. **Seemacht und Seestrategie im 19. und 20. Jahrhundert.** Hamburg; Berlin; Bonn: Mittler, 1999. (Vorträge zur Militärgeschichte; Bd. 18). 282 p.

Ellinger, Werner B. & Herbert Rosinski. **Sea power in the Pacific, 1936-1941: a selected bibliography of books, periodical articles, and maps from the end of the London naval conference to the beginning of the war in the Pacific.** London: H. Milford, Oxford university press, 1942. 80 p.
NPS/DKL Location: BUCKLEY V27 .A2

Gartner, Scott Sigmund. **Strategic Assessment in War.** New Haven, CT: Yale Univ. Press, 1997. 177 p.

George, James L. **History of warships: from ancient times to the twenty-first century.** Annapolis, MD: Naval Institute Press, 1998. 353 p.
NPS/DKL Location: GENERAL V750 .G46 1998

Hough, Richard Alexander. **Naval battles of the twentieth century.** Woodstock, NY: Overlook Press, 2001. 304 p.

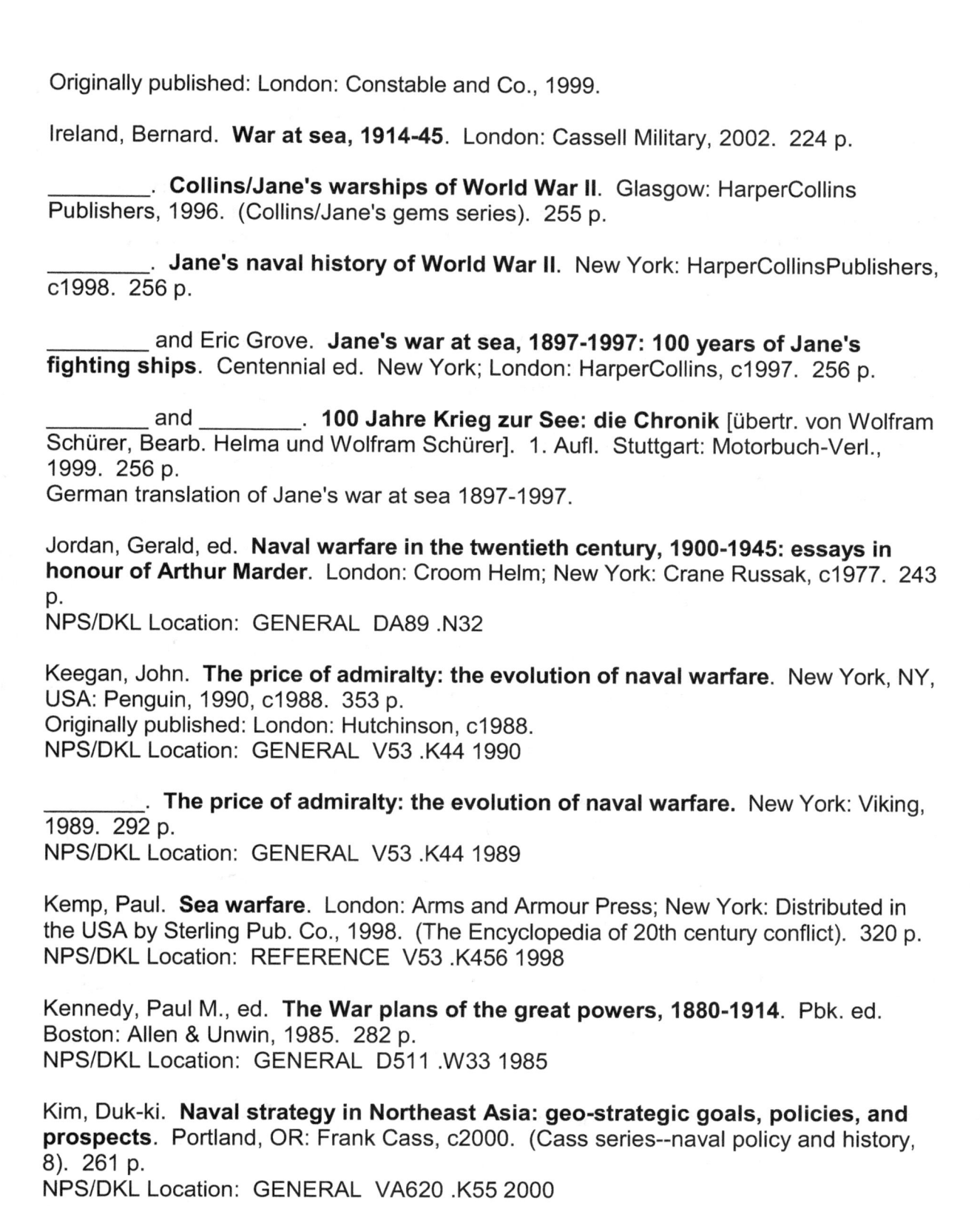

Originally published: London: Constable and Co., 1999.

Ireland, Bernard. **War at sea, 1914-45**. London: Cassell Military, 2002. 224 p.

_____. **Collins/Jane's warships of World War II**. Glasgow: HarperCollins Publishers, 1996. (Collins/Jane's gems series). 255 p.

_____. **Jane's naval history of World War II**. New York: HarperCollinsPublishers, c1998. 256 p.

_____ and Eric Grove. **Jane's war at sea, 1897-1997: 100 years of Jane's fighting ships**. Centennial ed. New York; London: HarperCollins, c1997. 256 p.

_____ and _____. **100 Jahre Krieg zur See: die Chronik** [übertr. von Wolfram Schürer, Bearb. Helma und Wolfram Schürer]. 1. Aufl. Stuttgart: Motorbuch-Verl., 1999. 256 p.
German translation of Jane's war at sea 1897-1997.

Jordan, Gerald, ed. **Naval warfare in the twentieth century, 1900-1945: essays in honour of Arthur Marder**. London: Croom Helm; New York: Crane Russak, c1977. 243 p.
NPS/DKL Location: GENERAL DA89 .N32

Keegan, John. **The price of admiralty: the evolution of naval warfare**. New York, NY, USA: Penguin, 1990, c1988. 353 p.
Originally published: London: Hutchinson, c1988.
NPS/DKL Location: GENERAL V53 .K44 1990

_____. **The price of admiralty: the evolution of naval warfare**. New York: Viking, 1989. 292 p.
NPS/DKL Location: GENERAL V53 .K44 1989

Kemp, Paul. **Sea warfare**. London: Arms and Armour Press; New York: Distributed in the USA by Sterling Pub. Co., 1998. (The Encyclopedia of 20th century conflict). 320 p.
NPS/DKL Location: REFERENCE V53 .K456 1998

Kennedy, Paul M., ed. **The War plans of the great powers, 1880-1914**. Pbk. ed. Boston: Allen & Unwin, 1985. 282 p.
NPS/DKL Location: GENERAL D511 .W33 1985

Kim, Duk-ki. **Naval strategy in Northeast Asia: geo-strategic goals, policies, and prospects**. Portland, OR: Frank Cass, c2000. (Cass series--naval policy and history, 8). 261 p.
NPS/DKL Location: GENERAL VA620 .K55 2000

Kitamura, Kenichi. **Ima naze shi ren boei ka: Higashi Ajia Nishi Taiheiyo no chiseigakuteki senryakuteki bunseki.** Shohan. Tokyo: Shingaku Shuppan: Hatsubaimoto Seiunsha, Showa 63 [1988]. 140 p.

Library of Congress. Foreign Affairs Division. **Means of measuring naval power with special reference to U.S. and Soviet activities in the Indian Ocean; prepared for the Subcommittee on the Near East and South Asia of the Committee on Foreign Affairs.** Washington: U.S. Govt. Print. Off., 1974. 16 p.
Committee print. 93d Congress, 2d session.

Lutz, Dieter S., Erwin Müller, Andreas Pott. **Seemacht und Sicherheit: Beitr. zur Diskussion maritimer Rüstung u. Rüstungskontrolle.** 1. Aufl. Baden-Baden: Nomos-Verlagsgesellschaft, 1986. (Militär, Rüstung, Sicherheit; Bd. 28). 213 p.

Macintyre, Donald G. F. W. **Sea power in the Pacific; a history from the sixteenth century to the present day.** New York, Crane, Russak ,[1972]. 281 p.
Originally published: London, A. Barker, [1972].

McKee, Alexander. **Against the odds: battles at sea, 1591-1949.** Annapolis, MD: Naval Institute Press, c1991. 272 p.
NPS/DKL Location: GENERAL D215 .M35 1991

Miller, David and Chris Miller. **Modern naval combat.** London; New York: Salamander Books, c1986. 208 p.
NPS/DKL Location: GENERAL V53 .M553 1986

Modelski, George and William R. Thompson. **Seapower in global politics, 1494-1993.** Seattle: University of Washington Press, c1988. 380 p.
Originally published: Houndmills, Basingstoke, Hampshire: Macmillan Press, 1988.
NPS/DKL Location: GENERAL V25 .M63 1988

Moineville, Hubert. **La guerre navale: réflexions sur les affrontements navals et leur avenir.** Paris Presses universitaires de France, c1982. (Perspectives internationales). 199 p.

_____. **Naval warfare today and tomorrow [translation and picture captions by P.R. Compton-Hall].** Oxford, England: B. Blackwell, 1983. 141 p.
English translation of: La guerre navale.
NPS/DKL Location: GENERAL V163 .M6413 1983

Murfett, Malcolm H. **Fool-proof relations: the search for Anglo-American naval cooperation during the Chamberlain years, 1937-1940.** Singapore: Singapore University Press: National University of Singapore, c1984. 324 p.

Murray, Williamson, and Allan R. Millett, eds. **Military Innovation in the Interwar Period.** Cambridge, MA: Cambridge Univ. Press, 1996. 415 p.

NPS/DKL Location: GENERAL U42 .M556 1996

O'Brien, Phillips Payson. **British and American naval power: politics and policy, 1900-1936.** Westport, CT: Praeger, 1998. (Praeger studies in diplomacy and strategic thought). 274 p.
NPS/DKL Location: GENERAL E183.8.G7 O37 1998

_____, ed. **Technology and naval combat in the twentieth century and beyond.** Portland, OR: Frank Cass, 2001. (Cass series--naval policy and history; 13). 272 p.
NPS/DKL Location: GENERAL V53 .T43 2001

Polmar, Norman et al. **Chronology of the Cold War at sea, 1945-1991.** Annapolis, MD: Naval Institute Press, c1998. 241 p.
NPS/DKL Location: GENERAL D842.3 .C48 1998

Reynolds, Clark G. **History and the sea: essays on maritime strategies.** Columbia, SC: University of South Carolina Press, c1989. 232 p.

Roskill, Stephen Wentworth. **Naval policy between the wars.** New York, Walker [c1968-76]. 2 v.
NPS/DKL Location: GENERAL D436 .R7

Spector, Ronald H. **At war, at sea: sailors and naval warfare in the twentieth century.** New York, NY: Viking, 2001. 463 p.
NPS/DKL Location: GENERAL V53 .S66 2001

Till, Geoffrey. **Maritime strategy and the nuclear age.** New York: St. Martin's Press, 1982. 274 p.
NPS/DKL Location: GENERAL V163 .T56

Till, Geoffrey. **Maritime strategy and the nuclear age.** 2nd ed. New York: St. Martin's Press, c1984. 295 p.
NPS/DKL Location: GENERAL V163 .T56 1984

Tucker, Spencer, et al., eds. **Naval warfare: an international encyclopedia.** Santa Barbara, CA: ABC-CLIO, c2002. 3 v.
NPS/DKL Location: REFERENCE V23 .N32 2002

United States Naval Institute. **Combat fleets of the world.** [Annapolis, MD: United States Naval Institute], 1976-1989. v.
Continued by: Naval Institute guide to combat fleets of the world.
VA40 .C8

_____. **The Naval Institute guide to combat fleets of the world.** Annapolis, MD: Naval Institute Press, c1990-. v.
Continues: Combat fleets of the world

NPS/DKL Location: REFERENCE VA40 .C8

Willmott, H. P. **Sea warfare: weapons, tactics and strategy**. Strettington, Chichester: A. Bird, 1981. 165 p.
NPS/DKL Location: GENERAL VA10 .V68

Zhai, Qiang. **The dragon, the lion & the eagle: Chinese-British-American relations, 1949-1958**. Kent, OH: Kent State University Press, c1994. (American diplomatic history). 284 p.
Revision of thesis (Ph.D.) -- Ohio University, 1991.
NPS/DKL Location: GENERAL E183.8.C5 Z42 1994

HISTORY OF NAVAL STRATEGY/WARFARE -- GENERAL -- AMERICAN

Baer, George W. **One hundred years of sea power: the U.S. Navy, 1890-1990**. Stanford, CA: Stanford University Press, 1994. 553 p.
NPS/DKL Location: GENERAL VA58 .B283 1994

Breemer, Jan S. **U.S. naval developments**. Annapolis, MD: Nautical & Aviation Pub. Co., 1983. 194 p.
NPS/DKL Location: GENERAL VA58.4 .B73 1983

Davis, Jacquelyn K., Michael J. Sweeney, Charles M. Perry. **The submarine and U.S. national security strategy into the twenty-first century**. Cambridge, MA: A publication of the Institute for Foreign Policy Analysis in association with the Fletcher School of Law and Diplomacy, Tufts University, 1997. (National security paper; no. 19). 85 p.
NPS/DKL Location: GENERAL V858 .D38 1997

Dorwart, Jeffery M. **Conflict of duty: the U.S. Navy's intelligence dilemma, 1919-1945**. Annapolis, MD: Naval Institute Press, c1983. 262 p.
NPS/DKL Location: GENERAL VB231 .U54 D667 1983

Duncan, Francis. **Rickover: the struggle for excellence**. Annapolis, MD: Naval Institute Press, 2001. 364 p.
NPS/DKL Location: GENERAL V63.R63 D86 2001

Friedman, Norman. **The US maritime strategy**. London; New York: Jane's, 1988. 246 p.
NPS/DKL Location: GENERAL VA50 .F74 1988

George, James L. **The U.S. Navy in the 1990s: alternatives for action**. Annapolis, MD: Naval Institute Press, c1992. 246 p.
NPS/DKL Location: GENERAL VA58.4 .G46 1992

Goure, Daniel, ed. **The role of sea power in U.S. national security in the twenty-first century: a consensus report of the CSIS Working Group on Undersea warfare**.

Washington: Center for Strategic and International Studies, 1998. (CSIS panel reports). 42 p.
NPS/DKL Location: GENERAL VA50 .G68 1998

Harris, Brayton. **The Navy times book of submarines: a political, social, and military history**. 1st ed. New York: Berkley Books, 1997. 398 p.
NPS/DKL Location: BUCKLEY V857 .H37 1997

Hattendorf, John B., ed. **Naval policy and strategy in the Mediterranean: past, present, and future**. London; Portland, OR: Frank Cass, 2000. (Cass series--naval policy and history; 10). 445 p.
NPS/DKL Location: GENERAL UA646.55 .N38 2000

Hooper, Edwin Bickford. **United States naval power in a changing world**. New York: Praeger, 1988. 294 p.
NPS/DKL Location: GENERAL VA55 .H66 1988

Isenberg, Michael T. **Shield of the Republic: the United States Navy in an era of Cold War and violent peace**. 1st ed. New York: St. Martin's Press, 1993- . v.
Contents: v. 1. 1945-1962.

Levine, Robert H. **The politics of American naval rearmament, 1930-1938**. New York: Garland Pub., 1988. (Harvard dissertations in American history and political science). 498 p.
Thesis (Ph.D.), Harvard University, 1972.

Lott, Arnold S. **A long line of ships; Mare Island's century of naval activity in California**. Annapolis, United States Naval Institute, [1954]. 268 p.

McBride, William M. **Technological change and the United States Navy, 1865-1945**. Baltimore: Johns Hopkins University Press, c2000. (Johns Hopkins studies in the history of technology). 336 p.
Contents include "Anomalous technologies of the great war: airplanes, submarines, and the professional status quo." p. 111-138.
NPS/DKL Location: GENERAL VA55 .M33 2000

Merrill, John, and Lionel D. Wyld. **Meeting the Submarine Challenge: A Short History of the Naval Underwater Systems Center**. U.S. Gov't. Print. Office, 1997. 329 p.
NPS/DKL Location: FEDDOCS D 201.2:SU 1

Miller, Edward S. **War Plan Orange: the U.S. strategy to defeat Japan, 1897-1945**. Annapolis, MD: Naval Institute Press, c1991. 509 p.
NPS/DKL Location: GENERAL VA50 .M53 1991

Muir, Malcolm, Jr. **Black shoes and blue water: surface warfare in the United States Navy, 1945-1975**. Washington, DC: Naval Historical Center, Dept. of the Navy: For sale

by the U.S. G.P.O., Superintendent of Documents, 1996. (Contributions to naval history; no. 6). 348 p.
NPS/DKL Location: FEDDOCS D 207.10/4:6

Naval Studies Board and National Research Council. **Technology for the United States Navy and Marine Corps, 2000-2035: becoming a 21st-century force** / Naval Studies Board, Committee on Technology for Future Naval Forces; National Research Council, Commission on Physical Sciences, Mathematics, and Applications. Washington, DC: National Academy Press, 1997. 9 v.
Incomplete contents: v.1. Overview -- v.2. Technology -- v.3. Information in warfare -- v.4. Human resources -- v.5. Weapons -- v.6. Platforms -- v.7. Undersea warfare -- v.8. Logistics
NPS/DKL Location: GENERAL VA55 .T42 1997
Electronic access: http://www.nap.edu/html/tech_21st/tfnf.htm

O'Connell, Robert L. **Sacred vessels: the cult of the battleship and the rise of the U.S. Navy**. Boulder: Westview Press, 1991. 409 p.
NPS/DKL Location: GENERAL VA50 .O36 1991

Owens, William A. **High seas: the naval passage to an uncharted world**. Annapolis, MD: Naval Institute Press, c1995. 184 p.
NPS/DKL Location: GENERAL VA58.4 .O94 1995

Packard, Wyman H. **A century of U.S. naval intelligence**. Washington, DC: Office of Naval Intelligence; Naval Historical Center: For sale by U.S. G.P.O., Supt. of Docs., 1996. 498 p.
NPS/DKL Location: FEDDOCS D 221.2:IN 8

Palmer, Michael A. **Origins of the maritime strategy: American naval strategy in the first postwar decade.** Washington, DC: Naval Historical Center, Dept. of the Navy, 1988. (Contributions to naval history; no. 1). 129 p.
NPS/DKL Location: GENERAL VA58 .P28 1988

Rodger, N.A.M., ed. **Naval Power in the Twentieth Century**. Annapolis, MD: Naval Institute Press, 1994. 273 p.
Proceedings of a conference held at the University of Exeter, in July 1994.
NPS/DKL Location: BUCKLEY VA25 .N38 1996

Ryan, Paul B. **First line of defense: the U.S. Navy since 1945**. Stanford, CA: Hoover Institution Press, c1981. 224 p.
NPS/DKL Location: GENERAL VA58.4 .R97

Shultz, Richard H. and Robert L. Pfaltzgraff, eds. **The role of naval forces in 21st century operations**. 1st ed. Washington, DC: Brassey's, c2000. 284 p.

Sheehy, Edward John. **The U.S. Navy, the Mediterranean, and the cold war, 1945-1947**. Westport, CT: Greenwood Press, 1992. (Contributions in military studies; no. 126). 191 p.

Tuleja, Thaddeus V. **Statesmen and admirals; quest for a Far Eastern naval policy**. [1st ed.]. New York, Norton, [1963]. 256 p.
NPS/DKL Location: GENERAL E746 .T9

Uhlig, Frank, Jr. **How navies fight: the U.S. Navy and its allies**. Annapolis, MD: Naval Institute Press, c1994. 455 p.
NPS/DKL Location: GENERAL VA58 V103 .U37 1994

Utz, Curtis A. **Cordon of steel: the U.S. Navy and the Cuban missile crisis**. Washington: Naval Historical Center, Dept. of the Navy: For sale by the Supt. of Docs, U.S. G.P.O., 1993. (The U.S. Navy in the modern world series; no. 1). 48 p.
NPS/DKL Location: GENERAL VA58 E841 .U89 1993

HISTORY OF NAVAL STRATEGY/WARFARE -- GENERAL --AUSTRALIAN

Bruce, Robert H., ed. **Australia and the Indian Ocean: strategic dimensions of increasing naval involvement**. Perth, Australia: Centre for Indian Ocean Regional Studies, Curtin University of Technology, 1988. (Studies in Indian Ocean maritime affairs; no. 1). 138 p.

McCaffrie, Jack and Alan Hinge, eds. **Sea power in the new century: maritime operations in Asia-Pacific beyond 2000**. [Canberra, ACT]: Australian Defence Studies Centre, c1998. 225 p.

Stevens, David. **The Impact of the submarine threat on Australia's maritime defence 1915-1954**. [Canberra, ACT]: Australian Defence Force Academy, 2000. 1 v. Thesis (Ph.D.), Australian Defence Force Academy, School of History, 2000.

_____. **Maritime power in the 20th century: the Australian experience**. St. Leonards, NSW, Australia: Allen & Unwin, 1998. 329 p.

_____. **U-boat far from home: the epic voyage of U 862 to Australia and New Zealand**. St. Leonards, NSW, Australia: Allen & Unwin, 1997. 282 p.

_____ and John Reeve, eds. **Southern trident: strategy, history, and the rise of Australian naval power**. Crows Nest, N.S.W.: Allen & Unwin, 2001. 363 p.
NPS/DKL Location: GENERAL UA870 .S68 2001

Wilson, David, ed. **Maritime war in the 21st century: the medium and small navy perspective**. Jervis Bay, N.S.W.: RAN Sea Power Centre, c2001. (Papers in australian maritime affairs, no. 8). 298 p.

Papers from "Maritime War 21" conference.

Wilson, David and Richard John Sherwood, eds. **Oceans governance and maritime strategy**. St. Leonards, N.S.W.: Allen & Unwin, 2000. 244 p.

HISTORY OF NAVAL STRATEGY/WARFARE -- GENERAL -- AUSTRIAN

Vego, Milan N. **Austro-Hungarian naval policy, 1904-14**. London; Portland, OR: Frank Cass, 1996. 213 p.
NPS/DKL Location: GENERAL VA473 .V54 1996

HISTORY OF NAVAL STRATEGY/WARFARE -- GENERAL -- BRAZILIAN

Bonturi, Orlando. **Brazil and the vital South Atlantic**. Washington, DC: National Defense University: Supt. Of Docs., U.S. G.P.O. [distributor], 1988. 96 p.
NPS/DKL Location: GENERAL VA422 .B65 1988

Brazil. Marinha. **Poder naval**. [Brazil]: Marinha do Brasil, [between 1983 and 1986]. 54 p.

Vianna Filho, Arlindo. **Estratégia naval brasileira: abordagem à história da evolução dos conceitos estratégicos navais brasileiros**. Rio de Janeiro: Biblioteca do Exército Editora, 1995. 170 p.

HISTORY OF NAVAL STRATEGY/WARFARE -- GENERAL -- BRITISH

Alch, Mark Lee. **Germany's naval resurgence, British appeasement, and the Anglo-German Naval Agreement of 1935**. UCLA, 1977. 2 v.
Thesis (Ph.D.), UCLA, 1977.

Arthur, Max. **The Navy: 1939 to the present day**. London: Hodder & Stoughton, 1997. 416 p.

Bell, Christopher M. **The Royal Navy, seapower and strategy between the wars**. Stanford, CA: Stanford University Press, 2000. 232 p.

Cable, James. **Britain's naval future**. London: Macmillan, 1983. 220 p.

Dolby, James. **The steel Navy: a history in silhouette 1860-1962**. London: Macdonald, 1962. 168 p.

Domville-Fife, Charles William, ed. **Evolution of sea power: studies of modern naval warfare and the effect of evolution on the basis and employment of sea power**. London: Rich & Cowan, ltd., [1939]. 258 p.
NPS/DKL Location: BUCKLEY V103 .D6

Garbutt, Paul Elford. **Naval challenge, 1945-1961; the story of Britain's post-war fleet**. London, Macdonald, [1961]. 119 p.

Gordon, Gilbert Andrew Hugh. **British seapower and procurement between the wars: a reappraisal of rearmament**. Annapolis, MD: Naval Institute Press, 1988. 321 p.
NPS/DKL Location: GENERAL VC265.G7 G67 1988

Kemp, Peter Kemp, ed. **History of the Royal Navy**. [1st American ed.]. New York, Putnam [1969]. 304 p.
Originally published: London: Barker, 1969.
NPS/DKL Location: GENERAL VA454 .K3

Kennedy, Paul M. **Aufstieg und Verfall der britischen Seemacht [Übertr. aus dem Engl. von Hans und Hanne Meckel]**. 1. Aufl. Herford; Bonn: Mittler, c1978. 448 p.
German translation of The rise and fall of British naval mastery.

_____. **The rise and fall of British naval mastery**. New York: Scribner, c1976. 405 p.
NPS/DKL Location: GENERAL DA85 .K3

Lambert, Nicholas. **Sir John Fisher's Naval Revolution**. Columbia: Univ. of South Carolina Press, 1999. 364 p.

Morris, Jan. **Fisher's face, or, Getting to know the admiral**. 1st American ed. New York: Random House, c1995. 300 p.
NPS/DKL Location: BUCKLEY CT6.I74 M67 1995

Murfett, Malcolm H., ed. **From Fisher to Mountbatten**. Westport, CT: Praeger, 1995. 313 p.
NPS/DKL Location: GENERAL CT32 .F57 1995

_____. **In jeopardy: the Royal Navy and British Far Eastern defence policy, 1945-1951**. Kuala Lumpur; New York: Oxford University Press, 1995. (South-East Asian historical monographs). 178 p.
NPS/DKL Location: GENERAL VA454 .M873 1995

Preston, Antony. **Dreadnought to nuclear submarine**. London: H.M. Stationery Off.; [Palo Alto, CA: obtainable in the United States from Pendragon House], 1980. (The Ship; 9). 60 p.

_____, ed. **History of the Royal Navy in the 20th century**. Novato, CA: Presidio, c1987. 224 p.
NPS/DKL Location: FOLIO VA454 .H53 1987

Rasor, Eugene L. **British naval history since 1815: a guide to the literature**. New York: Garland Pub., 1990. (Military history bibliographies; vol. 13); (Garland reference library of the humanities; vol. 1069). 841 p.
NPS/DKL Location: GENERAL DA70 .R37 1990

Roskill, Stephen Wentworth. **The strategy of sea power; its development and application**. London, Collins, 1962. 287 p.
Based on the Lees-Knowles lectures delivered in the University of Cambridge, 1961.
NPS/DKL Location: GENERAL V163 .R7

Van der Vat, Dan. **Standard of power: the Royal Navy in the twentieth century**. London: Hutchinson, 2000. 460 p.

Warner, Oliver. **Cunningham of Hyndhope, Admiral of the Fleet: a memoir**. London, Murray, [1967]. 301 p.
American ed. (Athens, Ohio University Press) entitled: Admiral of the Fleet: Cunningham of Hyndhope.
NPS/DKL Location: GENERAL DA89 .W2

Wettern, Desmond. **The decline of British seapower**. London: Jane's; Boston, MA: Distributed by Science Books International, 1982. 452 p.
NPS/DKL Location: GENERAL VA454 .W43 1982

HISTORY OF NAVAL STRATEGY/WARFARE – GENERAL -- CANADIAN

Haydon, Peter T. and Ann L. Griffiths, eds. **Canada's Pacific naval presence: purposeful or peripheral**. Halifax, N.S.: Centre for Foreign Policy Studies, Dalhousie. Univ.: Maritime Affairs, [1999]. 181 p.

Sarty, Roger F. **The maritime defence of Canada**. [Toronto]: Canadian Institute of Strategic Studies, c1996. 223 p.

HISTORY OF NAVAL STRATEGY/WARFARE -- GENERAL -- CHINESE

Austin, Greg. **China's ocean frontier: international law, military force, and national development**. St. Leonards, NSW, Australia: Allen & Unwin in association with the Dept. of International Relations and the Northeast Asia Program, Research School of Pacific and Asian Studies, Australian National University, Canberra, ACT, 1998. (Studies in world affairs; 17). 415 p.

Breyer, Siegfried and Jürg Meister. **Die Marine der Volksrepublik China**. München: Bernard & Graefe, c1982. 343 p.

Cao, Baojian and Guo Fuwen. **Mian dui Taiping yang di chen si: hai yang yi shi yu guo fang**. Di 1 ban. [Peking]: Guo fang da xue chu ban she: Beijing shi xin hua shu dian jing xiao, 1989. (Guo fang jiao yu cong shu). 293 p.

Cole, Bernard D.. **The great wall at sea: China's Navy enters the 21st century**. Annapolis, MD: Naval Institute Press, 2001. 288 p.

Hiramatsu, Shigeo. **Chugoku no Kaiyo senryaku**. Dai 1-han. Tokyo Keiso Shobo, 1993. 189 p.

Huang, Alexander Chieh-cheng. **Chinese maritime modernization and its security implications: the Deng Xiaoping era & beyond**. Washington, DC: George Washington University, 1994. 2 v.
Thesis (Ph.D.), George Washington University, 1994.

Kane, Thomas M. **Chinese grand strategy and maritime power**. London; Portland, OR: Frank Cass, 2002. (Naval policy and history; 16). 158 p.
NPS/DKL Location: GENERAL VA633 .K36 2002

Kondapalli, Srikanth. **China's naval power**. New Delhi: Knowledge World in association with Institute for Defence Studies and Analyses, 2001. 252 p.
Cover title: Zhong guo hai jun li liang/Chung-kuo hai chün li liang

Stepanov, E. D. **Ekspansiia Kitaia na more**. Moskva: Mezhdunar. otnosheniia, 1980. 158 p.

Wu, Chunguang. **Taiping Yang shang de jiao liang: dang dai Zhongguo de hai yang zhan le wen ti**. Di 1 ban. Beijing: Jin ri Zhongguo chu ban she, 1998. (Zhongguo wen ti bao gao). 427 p.

HISTORY OF NAVAL STRATEGY/WARFARE -- GENERAL -- FRENCH

Labouérie, Guy. **Défense et océans: propos de marin**, 1969-1994. Paris: ADDIM, c1994. (Collecton Esprit de défense.) 293 p.

Laurens, Adolphe. **Histoire de la guerre sous-marine allemande (1914-1918)**. Mesnil (Eure), impr. Firmin-Didot et Cie, Paris, Société d'éditions géographiques, maritimes et coloniales, 184, boulevard Saint-Germain: 1930. 461 p.

_____. **Précis d'histoire de la guerre navale, 1914-1918**. Paris: Payot, 1929. 300 p.

Walser, Ray. **France's search for a battle fleet: naval policy and naval power, 1898-1914**. New York: Garland, 1992. (Modern European history. France). 306 p.

HISTORY OF NAVAL STRATEGY/WARFARE -- GENERAL -- GERMAN

Bird, Keith W. **German naval history: a guide to the literature**. New York: Garland Pub., 1985. (Military history bibliographies; vol. 7); (Garland reference library of social science; v. 215). 1121 p.

Breyer, Siegfried and Gerhard Koop. **Die deutsche Kriegsmarine, 1935-1945: Schiffe, Bewaffnung, Männer, Ausrüstung, Einsätze**. Friedberg: Podzun-Pallas, c1985-<c1991>. 7 v.
Incomplete Contents: Bd. 3. Die Ubootwaffe, Marine-Kleinkampfverbände, Seefliegerkräfte, Häfen un Bauwerften, die Angehörigen der Kriegsmarine mit den höchsten Tapferkeitsauszeichnungen, Versenkungserfolge gegen Kriegsschiffe, Uniformen, Dienstgrad- und Laufbahnabzeichen -- Bd. 4. Die Verluste der deutschen Überwasserseestreitkräfte.

_____ and _____. **The German navy at war, 1935-1945 [translated by Edward Force]**. West Chester, Pa.: Schiffer Pub., c1989-. 2 v.
Partial contents: V. 2: The U-Boat
English translation of: Die deutsche Kriegsmarine, 1935-1945.

Epkenhans, Michael. **Die wilhelminische Flottenrüstung, 1908-1914: Weltmachtstreben, industrieller Fortschritt, soziale Integration**. München: R. Oldenbourg, 1991. (Beiträge zur Militärgeschichte; Bd. 32). 488 p.
A slight revision of the author's thesis (doctoral)--Westfälische Wilhelms-Universität Münster, Sommersemester 1989.

Hobson, Rolf. **Imperialism at sea: naval strategic thought, the ideology of sea power, and the Tirpitz Plan, 1875-1914**. Boston: Humanities Press, 2001. (Studies in Central European histories). 358 p.

Koop, Gerhard und Siegfried Breyer. **Die Schiffe, Fahrzeuge und Flugzeuge der deutschen Marine von 1956 bis heute**. Bonn: Bernard & Graefe Verlag, c1996. 560 p.

Lapp, Peter Joachim and Siegfried Breyer. **Die Volksmarine der DDR: Entwicklung, Aufgaben, Ausrüstung**. Koblenz: Bernard & Graefe, 1985. 319 p.

Majoli, Sergio. **Dal baltico al pacifico per il passaggio à nord-est: storia di una nave corsara tedesca**. Roma: Rivista maritima, 1999. 62 p.

Pertek, Jerzy. **Od Reichsmarine do Bundesmarine, 1918-1965**. [Wyd. 1]. [Pozna´n] Wydawn. Pozna´nskie, 1966. 420 p.

Showalter, Dennis E. **German military history, 1648-1982: a critical bibliography**. New York: Garland, 1984. (Military history bibliographies; vol. 3); (Garland reference library of social science; v. 113). 331 p.
NPS/DKL Location: GENERAL DD101 .S46 1984

Stang, Knut. **Das zerbrechende Schiff: Seekriegsstrategien- und Rüstungsplanung der deutschen Reichs- und Kriegsmarine 1918-1939**. Frankfurt am Main; New York: P. Lang, c1995. (Europäische Hochschulschriften. Reihe III, Geschichte und ihre Hilfswissenschaften; Bd. 630). 424 p.
Originally presented as the author's thesis (Ph.D.) -- Universität Göttingen, 1994.

Tirpitz, Alfred von. **Der Aufban der deutschen Weltmacht**. Stuttgart: J.B. Cotta'sche Buchhandlung Nachfolger, 1924. 460 p.

_____. **Der Aufbau der deutschen Weltmacht**. Einmalige Ausg. Hamburg, Deutsche Hausbucherei, [c1929]. 460 p.

Witthöft, Hans Jürgen. **Lexikon zur deutschen Marinegeschichte**. 1. Aufl. Herford: Koehler, 1977-1978. 2 v.

Wolter, Gustav-Adolf and Alexander Meurer. **Die See, Schicksal der Völker. Eine Geschichte d. Seefahrt u. d. Seemacht als Einf. in d. Weltgeschichte**. 6., überarb. u. erg. Aufl. Herford: Koehler, [1973]. 309 p.
1st-5th editions published under title: Seekriegsgeschichte in Umrissen by A. Meurer.

HISTORY OF NAVAL STRATEGY/WARFARE -- GENERAL -- INDIAN

Jasjit Singh, ed.. **Maritime security**. New Delhi: Institute for Defence Studies and Analyses, 1993. 182 p.

Kohli, Sourendra Nath. **The Indian Ocean & India's maritime security**. New Delhi: United Service Institution of India, 1981. (USI national security lectures; 5). 99 p.

_____. **Sea power and the Indian Ocean: with special reference to India**. New Delhi: Tata McGraw-Hill, c1978. 168 p.

Naidu, G. V. C. **The Indian navy and Southeast Asia**. New Delhi: Knowledge World, in Association with Institute for Defence Studies and Analyses, 2000. 208 p.

Roy-Chaudhury, Rahul. **India's maritime security**. New Delhi: Knowledge World in association with Institute for Defence Studies and Analyses, 2000. 208 p.

_____. **Sea power and Indian security**. 1st English ed. London; Washington: Brassey's, 1995. 222 p.

Sakhuja, Vijay. **Confidence building from the sea: an Indian initiative**. New Delhi: Knowledge World, c2001. 128 p.
"A United Service Institution of India project."

HISTORY OF NAVAL STRATEGY/WARFARE -- GENERAL -- IRANIAN

De Guttry, Andrea and Natalino Ronzitti, eds. **The Iran-Iraq War (1980-1988) and the law of naval warfare**. Cambridge, England: Grotius Publications, 1993. 573 p.

El-Shazly, Nadia El-Sayed. **The Gulf tanker war: Iran and Iraq's maritime swordplay**. New York: St. Martin's Press, 1998. 403 p.

Navias, Martin S. and E.R. Hooton. **Tanker wars: the assault on merchant shipping during the Iran-Iraq conflict, 1980-1988**. London; New York: I.B. Tauris, 1996. 244 p.

Sreedhar, Kapil Kaul. **Tanker war: aspect of Iraq-Iran war, 1980-88**. New Delhi: ABC Pub. House, 1989. 143 p.

Walker, George K. **The tanker war, 1980-88: law and policy**. Newport, RI: Naval War College, 2000. (International law studies v. 74). 640 p.

HISTORY OF NAVAL STRATEGY/WARFARE -- GENERAL -- ITALIAN

Bagnasco, Erminio. **Le costruzioni navali della regia marina italiana (1861-1945)**. Roma: Rivista maritima, [1996]. 63 p.

Gabriele, Mariano and Giuliano Friz. **La politica navale italiana dal 1885 al 1915**. Roma: Ufficio storico della Marina militare, 1982. 333 p.

Giordano, Virgilio. **La vocazione marinara dell'Italia**. Palermo: ILA palma, 1996. 111 p.

Mallett, Robert. **The Italian Navy and Fascist expansionism, 1935-40**. London; Portland, OR: Frank Cass, 1998. (Cass series--naval policy and history; 7). 240 p.
NPS/DKL Location: GENERAL DG571 .M255 1998

Minardi, Salvatore. **Il disarmo navale italiano (1919-1936): un confronto politico-diplomatico per il potere marittimo**. Roma: Ufficio storico della marina militare, 1999. 421 p.

HISTORY OF NAVAL STRATEGY/WARFARE -- GENERAL -- JAPANESE

Hori, Motoyoshi. **Kaiyo boetgaku nyumon**. 1978. 274 p.

Jensen, Gustav. **Japans seemacht; der schnelle aufstieg im kampf um selbstbehauptung und gleichberechtigung in den jahren 1853-1937**. Berlin, K. Siegismud, 1938. 379 p.

_____. **Seemacht Japan, der schnelle Aufstieg im Kampf um Selbstbehauptung und Gleichberechtigung in den Jahren 1853-1943**. 2. erweiterte Aufl. Berlin, K. Siegismund, 1943. 421 p.

Jentschura, Hansgeorg, Dieter Jung, and Peter Mickel. **Die japanischen Kriegsschiffe: 1869 - 1945**. München: J. F. Lehmann, 1970. 515 p.

_____. **Warships of the Imperial Japanese Navy, 1869-1945** [translated by Antony Preston and J. D. Brown]. Annapolis, Md: Naval Institute Press, c1977. 284 p. English translation with revisions of Die japanischen Kriegsschiffe, 1869-1945. NPS/DKL Location: GENERAL VA653 .J52

_____. **Warships of the Imperial Japanese Navy, 1869-1945** [translated by Antony Preston and J. D. Brown]. London: Arms & Armour, 1999. 284 p.

Sekine, Gunpei. **Kokoku no kiki 1936-nen ni sonaeyo**. Tokyo: Heisho Shuppansha, Showa 8 [1933]. 244 p.

Takagi, Sokichi. **Jidenteki Nihon Kaigun shimatsuki: Teikoku Kaigun no uchi ni himeraretaru eiko to higeki no jijo**. Tokyo: Kojinsha, 1995. 474 p.

_____. **Jidenteki Nihon Kaigun shimatsuki: Teikoku Kaigun no uchi ni himeraretaru eiko to higeki no jijo**. Tokyo: Kojinsha, 1995. 474 p.

Ueda, Tetsu. **Shi ren: Nihon kiki kaiiki no giso**. Tokyo: Kosaido Shuppan, [Showa 58 i.e. 1983]. 285 p.

HISTORY OF NAVAL STRATEGY/WARFARE -- GENERAL -- KOREAN

Kang, Yong-o. **Haeyang chollyak uro p`urobon Han-Il kasang Tokto haejon**. Ch`op`an. Soul-si: Yon'gyong Munhwasa, 2000. 299 p.

Kim, Tal-chung and Doug-Woon Cho, eds. **Korean sea power and the Pacific era**. [Seoul]: Institute of East and West Studies, Yonsei University, c1990. (East and West studies series; 15). 313 p.

HISTORY OF NAVAL STRATEGY/WARFARE -- GENERAL -- NEW ZEALAND

New Zealand. Navy. **Maritime doctrine for the Royal New Zealand Navy**. [Wellington]: The Navy, c1997. 101 p.

Wright, Matthew. **Blue water kiwis: New Zealand's naval story**. Auckland, NZ: Reed Books, c2001. 237 p.

HISTORY OF NAVAL STRATEGY/WARFARE -- GENERAL -- NORWEGIAN

Børresen, Jacob. **Kystmakt: skisse av en maritim strategi for Norge**. [Oslo]: J.W. Cappelen: Europa-programmet, c1993. 349 p.

HISTORY OF NAVAL STRATEGY/WARFARE -- GENERAL -- POLISH

Drozdowski, Marian Marek. **Powrót Polski nad Baltyk 1920-1945: antologia tekstów historycznych**. Warszawa: Oficyna Wydawnicza "Typografika": Komisja Bada´n Dziejów Warszawy Instytutu Historii PAN, 1997. 267 p.

Ordon, Stanislaw. **Polska Marynarka Wojenna w latach 1918-1939; problemy prawne i ekonomiczne**. Gdynia: Wydawnictwo Morskie, 1966. 312 p.

Peszke, Michael Alfred. **Poland's navy: 1918-1945**. New York: Hippocrene Books Inc., c1999. 222 p.

HISTORY OF NAVAL STRATEGY/WARFARE -- GENERAL -- RUSSIAN/SOVIET

Breyer, Siegfried. **Enzyklopädie des sowjetischen Kriegsschiffbaus**. Herford: Koehler, c1987-<1991>. 3 v.

_____. **Guide to the Soviet Navy** [translated by M. W. Henley]. Annapolis, MD, United States Naval Institute, 1970. 353 p.
Revised and expanded English translation of Die Seerüstung der Sowjetunion.

_____. **Die Seerüstung der Sowjetunion**. München: J. F. Lehmann, 1964. 269 p.

_____. **Soviet warship development**. 1st English language ed. London: Conway Maritime Press, 1992-. v.
English translation of: Enzyklopädie des sowjetischen Kriegsschiffbaus.

_____ and Norman Polmar. **Guide to the Soviet Navy**. 2d ed. Annapolis, MD: Naval Institute Press, c1977. 610 p.

Originally published as a revised. and expanded English translation of Die Seerüstung der Sowjetunion.
NPS/DKL Location: GENERAL VA573 .B82

Daniel, Richard W. **A history of Russian and Soviet naval developments**.
Monterey, CA: Naval Postgraduate School; Springfield, VA: Available from National Technical Information Service, 1988. (ADA194343). 59 p.
Thesis (M.S. in Systems Tech.), Naval Postgraduate School, March 1988.
Abstract: *This thesis seeks to provide an historical understanding of Russian and Soviet naval developments. This historical basis is provided to complement technological analysis of Soviet naval concepts and systems. The origins of Soviet naval traditions are examined, beginning with the establishment of the ancient Russian state of Kiev, the birth of the Tsarist Navy (under Peter I), the origins of the Communist State and Navy, and concluding with the Soviet naval developments during the Second World War. In examining these developments significant naval victories (Sweden, 1721; and Tchesme, 1770) and defeats (Crimean, 1853; and Tsushima, 1905) are noted, along with non-combat administrative reforms. The employment of the Russian Navy in World War One and the Soviet Navy in World War Two are also examined. The conclusion is drawn that the primary mission of the Soviet Navy is to support the Soviet Army in a continental theater. This conclusion is based on the historical failure of the Russian and Soviet Navies in conducting blue-water operations (inferring a notion of perceived futility in attempting these operations), the historical success in conducting coastal operations in support of the army (inferring the utility of these types of operations), and the historical land combat bias of the Russian and Soviet Militaries.*
NPS/DKL Location: THESIS D14776

Gebhardt, James. F., trans. **The Soviet war at sea**. Minneapolis, MN: East View Publications, 1994-. v.
Contents: v. 1. Submarine warfare

Polmar, Norman. **Guide to the Soviet navy**. 3rd ed. Annapolis, MD: Naval Institute Press, c1983. 465 p.
Second edition. by S. Breyer and N. Polmar.
NPS/DKL Location: GENERAL/REFERENCE VA573 .P598 1983

_____. **Guide to the Soviet navy**. 4th ed. Annapolis, MD: Naval Institute Press, c1986. 536 p.
NPS/DKL Location: GENERAL/REFERENCE VA573 .P598 1986

_____. **The Naval Institute guide to the Soviet Navy**. 5th ed. Annapolis, MD: Naval Institute Press, c1991. 492 p.
Revised edition of: Guide to the Soviet Navy, 1986.
NPS/DKL Location: REFERENCE VA573 .P598 1991

_____, ed. **Soviet naval developments; prepared at the direction of the Chief of Naval Operations by the Director of Naval Intelligence and the Chief of Information, Department of the Navy, Washington, D.C**. Annapolis, MD: Nautical and Aviation Pub. Co. of America, c1979. 119 p.
Originally published under the title Understanding Soviet naval developments by the Office of Naval Intelligence, 1974.
NPS/DKL Location: GENERAL VA573 .S64

Schulz-Torge, Ulrich. **Die sowjetische Kriegsmarine**. Bonn: Wehr und Wissen, 1976-1981. 2 v.
NPS/DKL Location: GENERAL VA573 .S3

HISTORY OF NAVAL STRATEGY/WARFARE -- GENERAL -- SWEDISH

Jervas, Gunnar. **USA:s maritima strategi: hot eller möjlighet för Sverige?** Stockholm: Centralförbundet Folk och försvar, [1987]. (Försvar i nutid; 1987:3). 57 p.

Salicath, Carl P. **Svenske oppfatninger av maktendringene i norskehavet**. Oslo: Norsk utenrikspolitisk institutt, [1987]. (NUPI rapport, 0800-000X; nr. 109). 145 p.

HISTORY OF NAVAL STRATEGY/WARFARE -- GENERAL -- TURKISH

Gürdeniz, Cem and Erdogan Yücelis, eds. **Cumhuriyet donanmasi, 1923-2000 = The fleet of the Republic, 1923-2000**. 1. baski. Istanbul: Seyir Hidrografi ve Osinografi Daire Baskanligi, 2000. 201 p.

HISTORY OF NAVAL STRATEGY/WARFARE -- WWI – GENERAL

Gill, Charles and William Stevens. **The War on the Sea: Battles, Sea Raids, and Submarine Campaigns**, Vol. 4, *in* Hart, Albert Bushnell, ed. Harper's pictorial library of the world war. New York, London, Harper, [c1920]. 12 v.

Groos, Otto, ed. **Der Krieg in der Nordsee**. Berlin, E.S. Mittler, 1922-. (Der Krieg zur See, 1914-1918). 5 v.

_____. **La guerra nel mare del Nord [Walter Gladisch, Raffaele de Courten, and Waldimiro Pini, translators]**. [Livorno, Topografia della R. Accademia navale, 192?]. 3 v.
Italian translation of Der Krieg in der Nordsee.

_____. **Hokkai kaisenshi**. [Tokyo]: Kaigun Gunreibu, [1926-1932?]. 5 v.
Japanese translation of Der Krieg in der Nordsee

_____. **War in the North sea [translated by E.C. Magdeburger]**. [Newport, R.I., Naval war college, Intelligence dept., 1931]. 3 v. in 2 v.
English translation of Der krieg in der Nordsee.

_____. **The War in the North Sea. Volume I. From the Beginning of the War to the First of September 1914. Part 1. Preface, Contents, Chapters 1-4, Appendices 1-13; Charts 1-17**. Newport, RI: Naval War College, 1937. (ADA003031). 236 p.

Abstract: *Translated from the official German naval history of the First World War, this account is based upon the war diaries of the commanding officers involved and makes extensive use of the texts of naval staff memoranda andinstructions, operation and war orders, and German wireless messages. Concentrating on the North Sea area, the volume begins with an examination of the background of the conflict and the preliminary German preparations for the warvis-a-vis those of the British Navy, and then proceeds to the early operations of the two fleets during August 1914. The detailed treatment reflects theimportance of this initial stage of the contest, in which are exhibited the maritime planning and determinations that set the course of the entire naval conduct of the war: the blockade by Britain and the German employment of the submarine and mine warfare. Relevant appendices, tables, and charts accompany each part.*
English translation of Der krieg in der Nordsee

_____. **The War in the North Sea. Volume I. From the Beginning of the War to the First of September, 1914. Part 2. Chapters 5,6; Appendices 14-21; Charts 18-35**. Newport, RI: Naval War College, 1937. (ADA003032). 220 p.
English translation of Der krieg in der Nordsee.

Halpern, Paul G. **A naval history of World War I**. Annapolis, MD: Naval Institute Press, c1994. 591 p.
NPS/DKL Location: RESERVE D580 .H34 1994

Hoehling, Adolph A. **The Great War at sea: a history of naval action, 1914-1918**. Crowell, [1965]. 336 p.
NPS/DKL Location: GENERAL D580 .H6

_____. **The Great War at sea: a history of naval action 1914-18**. New York: Galahad Books, [1975] c1965. 336 p.
Originally published: New York, Crowell, 1965.

_____. **The Great War at sea: a history of naval action, 1914-18**. Westport, CT: Greenwood Press, 1978, c1965. 336 p.
Reprint of the edition published by Crowell, New York, 1965.

Newbolt, Henry John, Sir. **A naval history of the war**. London, Hodder and Stoughton limited, [1920]. 350 p.

Plaschka, Richard Georg. **Matrosen, Offiziere, Rebellen: Krisenkonfrontationen zur See, 1900-1918: Taku, Tsushima, Coronel/Falkland, "Potemkin," Wilhelmshaven, Cattaro**. Wien: H. Bohlau, 1984. (Veroffentlichungen des Osterreichischen Ost- und Sudosteuropa-Instituts; Bd. 12-13). 2 v.

Reynolds, Francis Joseph et al, ed. **The story of the Great War**. New York, P.F. Collier and son, [c1916-1920]. 16 v.
NPS/DKL Location: BUCKLEY D521 .R4

HISTORY OF NAVAL STRATEGY/WARFARE -- WWI -- AMERICAN

Coletta, Paolo Enrico. **American naval history: a guide.** 2nd ed. Lanham, MD: Scarecrow Press, 2000. 933 p.
Originally published: Lanham, MD: University Press of America, c1988 as A selected and annotated bibliography of American naval history, c1988.
NPS/DKL Location: GENERAL E182 .C63 2000

_____. **A survey of U.S. naval affairs, 1865-1917**. Lanham, MD: University Press of America, c1987. 265 p.
NPS/DKL Location: GENERAL VA57 .C65 1987

Feuer, A. B. **The U.S. Navy in World War I: combat at sea and in the air**. Westport, CT: Praeger, 1999. 197 p.

Frothingham, Thomas Goddard. **The naval history of the World War**. Cambridge, Harvard University Press, 1924-26. 3 v.
"This work has been compiled from data provided by the historical section, United States Navy." Contents: [v. 1] Offensive operations, 1914-1915. -- [v. 2] The stress of sea power, 1915-1916. -- [v. 3] The United States in the War, 1917-1918.
NPS/DKL Location: BUCKLEY D580 .F9

Gleichauf, Justin F. **Unsung sailors: the Naval Armed Guard in WWII**. Annapolis, MD: Naval Institute Press, c1990. 432 p.
NPS/DKL Location: GENERAL D769.45 .G57 1990

Sprout, Harold Hance & Margaret Sprout. **The rise of American naval power, 1776-1918**. Princeton, Princeton University Press, 1939. 398 p.

_____. **Toward a new order of sea power: American Naval policy and the world scene, 1918-1922**. [2nd ed.]. Princeton, NJ: Princeton Univ. Pr., 1943. 336 p.
NPS/DKL Location: GENERAL E182 .S75

Whitaker, Herman. **Hunting the German shark; the American navy in the underseas war**. New York, The Century co., 1918. 310 p.

HISTORY OF NAVAL STRATEGY/WARFARE -- WWI -- AUSTRALIAN

Jose, Arthur W. **The Royal Australian Navy, 1914-1918**. St. Lucia, Qld.: University of Queensland Press in association with the Australian War Memorial, 1987. (The Official history of Australia in the war of 1914-1918; v. 9). 649 p.
Originally published: Sydney: Angus and Robertson, 1928.

HISTORY OF NAVAL STRATEGY/WARFARE -- WWI -- AUSTRIAN

Gesellschaft vom Silbernen Kreuz. **Viribus unitis, Österreich-Ungarn und der Weltkrieg.** Wien, 1919. 1106 p.

Martiny, Nikolaus von, ed. **Bilddokumente aus Österreich-Ungarns Seekrieg, 1914-1918, mit Schilderungen der wichtigsten Kampfhandlungen zur See unter Benützung in- und ausländischer amtlicher Quellen, Tagebuchaufzeichnungen und Berichte.** Graz, Leykam-Verlag, 1939. 2 v.

_____. **Bilddokumente aus Österreich-Ungarns Seekrieg 1914-1918, mit Schilderungen der wichtigsten Kampfhandlungen zur See unter Benützung in- und ausländischer amtlicher Quellen, Tagebuchaufzeichnungen und Berichte.** 2. Aufl. Graz, Akademische Druck- u. Verlagsanstalt, 1973. 2 v.

Mayer, Horst F. and Dieter Winkler. **Als die Adria österreichisch war: Österreich-Ungarns Seemacht.** 4. Aufl. [Wien]: Verl. der Österr. Staatsdr., 1989. 223 p.

Prasky, Friedrich. **Die Tegetthoff-Klasse: Modellbau, Technik, Geschichte.** Hamburg: Mittler, 2000. 271 p.

Schaumann, Walther. **Ende einer Seemacht: Österreich-Ungarn 1900 - 1918; [Fotodokumente zu Österreich-Ungarns Marinegeschichte 1900 - 1918].** Klosterneuburg; Wien: Mayer, [1995]. 216 p.

Sokol, Anton E. **Seemacht Österreich die kaiserl. u. königl. Kriegsmarine 1382 - 1918.** 1. Aufl. Wien, München, Zürich: Molden, 1972 224 p.
German translation of The imperial and royal Austro-Hungarian navy.

Sokol, Hans Hugo. **Des Kaisers Seemacht: d. k.k. österr. Kriegsmarine 1848 bis 1914.** Wien; München: Amalthea, c1980. (Die k.k. österreichische Kriegsmarine in dem Zeitraum von 1848 bis 1914; Teil 3); (Geschichte der k.u.k. Kriegsmarine; Teil 3,3). 333 p.

Sondhaus, Lawrence. **The naval policy of Austria-Hungary, 1867-1918: navalism, industrial development, and the politics of dualism.** West Lafayette, IN: Purdue Univ. Press, c1994. 441 p.
NPS/DKL Location: GENERAL VA473 .S663 1994

HISTORY OF NAVAL STRATEGY/WARFARE -- WWI -- BRITISH

Beesly, Patrick. **Room 40: British naval intelligence 1914-18.** 1st American ed. San Diego: Harcourt Brace Jovanovich, c1982. 338 p.
Originally published: London: Hamilton, 1982.
NPS/DKL Location: GENERAL D639.C75 B43 1982

Bennett, Geoffrey Martin. **Naval battles of the First World War**. New York, Scribner, [1969, c1968]. (British battles series). 319 p.
Originally published: London: Batsford, 1968.
NPS/DKL Location: GENERAL D580 .B4

Bourne, J. M. **Britain and the Great War, 1914-1918**. Reprinted with corrections. London; New York: E. Arnold; New York, NY: Distributed in the USA by Routledge, Chapman and Hall, 1994, c1989. 257 p.
NPS/DKL Location: GENERAL D546 .B68 1989

Chatterton, E. Keble. **Dardanelles dilemma; the story of the naval operations**. London, Rich & Cowan, ltd., [1935]. 320 p.

Edwards, Kenneth, Commander. **The grey diplomatists**. London, Rich & Cowan, ltd. [1938]. 328 p.
NPS/DKL Location: BUCKLEY DA89 .E2

Ewing, Alfred Washington. **The man of room 40; the life of Sir Alfred Ewing**. London and Melbourne, Hutchinson & co. ltd., [1940]. 295 p.

Groos, Otto. Die **Grenzen der britischen Seemacht**. Berlin: Mittler, 1940. 32 p.

Hough, Richard Alexander. **The Great War at sea, 1914-1918**. Oxford [Oxfordshire]; New York: Oxford University Press, 1986, c1983. 353 p.
NPS/DKL Location: GENERAL D581 .H56 1986

Keegan, John. **The price of admiralty: the evolution of naval warfare**. New York, NY, USA: Penguin, 1990, c1988. 353 p.
Originally published: London: Hutchinson, 1988.
NPS/DKL Location: GENERAL V53 .K44 1990

_____. **The price of admiralty: the evolution of naval warfare**. New York: Viking, 1989. 292 p.
NPS/DKL Location: GENERAL V53 .K44 1989

Marder, Arthur Jacob. **From the dreadnought to Scapa Flow: the Royal Navy in the Fisher era, 1904-1919**. Oxford University Press, 1961-1969. 5 v.
NPS/DKL Location: GENERAL VA454 .M2

Sumida, Jon Tetsuro. **In defence of naval supremacy: finance, technology, and British naval policy, 1889-1914**. Boston: Unwin Hyman, 1989. 377 p.
NPS/DKL Location: GENERAL VA454 .S86 1989

HISTORY OF NAVAL STRATEGY/WARFARE -- WWI -- GERMAN

Berghahn, Volker Rolf. **Der Tirpitz-Plan; Genesis und Verfall einer innenpolitischen Krisenstrategie unter Wilhelm II**. Dusseldorf, Droste Verlag, [c1971]. (Geschichtliche Studien zu Politik und Gesellschaft, Bd. 1). 640 p.

Gross, Gerhard Paul. **Die Seekriegfuhrung der Kaiserlichen Marine im Jahre 1918**. Frankfurt am Main; New York: P. Lang, c1989. (Europaische Hochschulschriften. Reihe III, Geschichte und ihre Hilfswissenschaften; Bd. 387). 574 p.

Kelly, Patrick James. **The naval policy of imperial Germany 1900-1914**. Washington, DC: Georgetown University, 1970. 538 l.
Thesis (Ph.D.), Georgetown University, 1970.

Linnenkohl, Hans. **Alternativen und Möglichkeiten deutscher Seemacht 1898 - 1918**. Viernheim, Heppenheimer Str. 70: H. Linnenkohl [Selbstverl.], 1978. 282 p.

Luntinen, Pertti. **Saksan keisarillinen laivasto Itämerellä; aikeet, suunnitelmat ja toimet**. Helsinki: SHS, 1987 [i.e. 1988]. (Historiallisia tutkimuksia, 0073-2559; 143). 262 p.
Summary in German.

Plagemann, Volker, ed. **Übersee: Seefahrt u. Seemacht im dt. Kaiserreich**. München: Beck, 1988. 404 p.

Smith, Allen. **German submarine warfare, May 1915 to January 1917: an analysis of the provocative events leading the U.S. from a strict observance of neutrality to intervention in Europe, including the effects of these events on U.S. policy prior to entrance into World War II**. s.l.: s.n., [197-?]. 22 l.

Reuter, Ludwig von. **Scapa Flow: das grab der deutschen flotte**. 2 durchgesehene auflage. Leipzig: K. F. Koehler, 1921. 155 p.

Wegener, Wolfgang. **The naval strategy of the World War [translated and with an introduction and notes by Holger H. Herwig]**. Annapolis, MD: Naval Institute Press, c1989, 1929. (Classics of sea power). 231 p.
Originally published: Berlin: E.S. Mittler, 1929.
English translation of Die Seestrategie des Weltkrieges.
NPS/DKL Location: GENERAL D581 .W413 1989

HISTORY OF NAVAL STRATEGY/WARFARE -- WWI - ITALIAN

Rainero, Romain. **Raffaele Rossetti: dall'affondamento della "Viribus Unitis" all'impegno antifascista**. Settimo Milanese: Marzorati, c1989. 286 p.

HISTORY OF NAVAL STRATEGY/WARFARE -- WWII – GENERAL

Bath, Alan Harris. **Tracking the axis enemy: the triumph of Anglo-American naval intelligence**. Lawrence, KS: University Press of Kansas, c1998. (Modern war studies). 308 p.
NPS/DKL Location: GENERAL D810.S7 B35 1998

Dunnigan, James F. and Albert A. Nofi. **The Pacific war encyclopedia**. New York: Facts On File, c1998. 2 v.
NPS/DKL Location: REFERENCE D767.9 .D86 1998

Ellis, Lionel Frederic et al. **Victory in the West**. London, H. M. Stationery Off., 1962-1968. (History of the Second World War. United Kingdom military series). 2 v.
Contents: v. 1. The Battle of Normandy.-v.2. The Defeat of Germany.

Gailey, Harry A. **The war in the Pacific: from Pearl Harbor to Tokyo Bay**. Novato, CA: Presidio, c1995. 534 p.
NPS/DKL Location: GENERAL D767 .G68 1995

Greene, Jack and Alessandro Massignani. **The naval war in the Mediterranean 1940-1943**. London: Chatham; Rockville Center, NY: Sarpedon, 1998-1999. 352 p.
NPS/DKL Location: GENERAL D766 .G745 1998

_____ and _____. **The naval war in the Mediterranean 1940-1943**.
Annapolis, MD: Naval Institute Press, 2002. 352 p.

Grove, Eric, ed. **The Defeat of the enemy attack on shipping, 1939-1945**. Aldershot, Hant; Brookfield, VT: Ashgate for the Navy Records Society, c1997. (Publications of the Navy Records Society; vol. 137). 380 p.
A revised edition of the Naval Staff history, volumes 1A (text and appendices) and 1B (plans and tables). Based on a two-volume confidential manuscript issued in 1957 by the Naval Staff.

Hughes, Terry and John Costello. **The battle of the Atlantic**. Dial Press/J. Wade, 1977. 314 p.
Also published: [London]: Collins, c1977; as The battle of the Atlantic, by John Costello and Terry Hughes.
NPS/DKL Location: GENERAL D770 .H94

Ireland, Bernard. **Collins/Jane's warships of World War II**. Glasgow: HarperCollins Publishers, 1996. 255 p.

_____. **Jane's naval history of World War II**. New York: HarperCollins Publishers, c1998. 256 p.

_____. **Jane's schwimmende Bastionen: Schiffe des II. Weltkriegs; Technik, Taktik, Bewaffnung**. Augsburg: Bechtermünz, 2000. 256 p.
German translation of Jane's naval history of World War II

_____. **The war in the Mediterranean 1940-1943**. London: Arms and Armour; New York, NY: Sterling Pub. Co., c1993. 224 p.

Leutze, James R. **Bargaining for supremacy: Anglo-American naval collaboration, 1937-1941**. Chapel Hill: University of North Carolina Press, c1977. 328 p.
NPS/DKL Location: GENERAL   D750 .L62

Levine, Alan J. **The Pacific War: Japan versus the allies**. Westport, CT: Praeger, 1995. 200 p.

Marder, Arthur Jacob. **Old friends, new enemies: the Royal Navy and the Imperial Japanese Navy**. Oxford: Clarendon Press; New York: Oxford University Press, 1981-1990. 2 v.

Philbin, Tobias R. **The lure of Neptune: German-Soviet naval collaboration and ambitions, 1919-1941**. Columbia, S.C.: University of South Carolina Press, c1994. (Studies in maritime history). 192 p.
Contents: Background and context: Soviet-German relations -- The national canvas for naval issues -- Navy to navy: coexistence and interface -- Navy to navy: competition -- Nazi-Soviet naval relations -- The naval dimension of the Hitler-Stalin pact -- Operations -- Basis nord -- Cruiser "L": from Germany with reticence -- Submarines and merchant cruisers -- Conclusion.

Poolman, Kenneth. **The winning edge: naval technology in action, 1939-1945**. Annapolis, MD: Naval Institute Press, c1997. 235 p.
NPS/DKL Location: GENERAL   VF347 .P66 1997

Puleston, W. D. **The influence of sea power in World War II**. New Haven, Yale University Press, 1947. 310 p.
NPS/DKL Location: BUCKLEY   D770 .P9

Sadkovich, James J., ed. **Reevaluating major naval combatants of World War II**. New York: Greenwood Press, 1990. (Contributions in military studies, no. 92). 203 p.
NPS/DKL Location: GENERAL   D770 .R44 1990

Smith, Peter Charles. **Hold the narrow sea: naval warfare in the English Channel, 1939-1945**. Ashbourne, Derbyshire: Moorland; Annapolis, MD: Naval Institute Press, c1984. 255 p.

HISTORY OF NAVAL STRATEGY/WARFARE -- WWII -- AMERICAN

Bischof, Gunter, and Robert L. Dupont, eds. **The Pacific War Revisited**. Baton Rouge: Louisiana State Univ. Press, 1997. 220 p.

Creswell, John. **Sea warfare, 1939-1945**. Rev. and augm. ed. Berkeley: University of California Press, 1967. 343 p.
NPS/DKL Location: GENERAL D770 .C9

_____. **Sea warfare, 1939-1945; a short history**. London, New York: Longmans, Green, [1950]. 344 p.

Galantin, I. J. **Submarine admiral: from battlewagons to ballistic missiles**. Urbana: University of Illinois Press, c1995. 345 p.
NPS/DKL Location: BUCKLEY CT7.A42 G34 1995

Hoopes, Townsend and Douglas Brinkley. **Driven patriot: the life and times of James Forrestal**. New York: Knopf, 1992. 587 p.
NPS/DKL Location: GENERAL CT6 .O7 H66 1992

Lacy, James L. **Within bounds: the Navy in postwar American security policy**. Alexandria, VA: Center for Naval Analyses, Naval Strategy Group, 1983. (CNA; 83-1178). 616 p.
NPS/DKL Location: GENERAL VA58.4 .L32 1983

HISTORY OF NAVAL STRATEGY/WARFARE -- WWII -- AUSTRALIAN

Stevens, David, ed. **The Royal Australian Navy in World War II**. St. Leonards, Australia: Allen & Unwin, c1996. 212 p.

HISTORY OF NAVAL STRATEGY/WARFARE -- WWII -- BRITISH

Barnett, Correlli. **Engage the enemy more closely: the Royal Navy in the Second World War**. 1st American ed. New York: Norton, 1991. 1052 p.
Originally published: London: Hodder & Stoughton, 1991.
NPS/DKL Location: GENERAL D770 .B28 1991

Collier, Basil. **The defence of the United Kingdom**. H.M. Stationery Off., 1957. (History of the Second World War. United Kingdom military series). 557 p.
NPS/DKL Location: GENERAL D759 .C6

Gray, Edwyn. **Operation Pacific: the Royal Navy's war against Japan, 1941-1945**. Annapolis, MD: Naval Institute Press, [1990?]. 267 p.

Grenfell, Russell. **Sea power**, by T124 [pseud.]. London, J. Cape, [1941]. 252 p.
NPS/DKL Location: BUCKLEY VA454 .G7

_____. **Sea power**. Garden City, NY, Doubleday, Doran & company, inc., 1941. 244 p.

_____. **Sea power**. [New and rev. ed.] London, J. Cape [1941]. 261 p.

_____. **Sea power in the next war**. London, G. Bles, [1938]. (The next war, a series ed. by Captain Liddell Hart). 183 p.

_____. **Die Seemacht im nächsten Krieg** [Dt. Übers. v. Rudolf Stoff]. Berlin: A. Nauck & Co., 1939. 221 p.
German translation of Sea Power in the next war.

_____. **Die Seemacht im nächsten Krieg** [Dt. Übers. v. Rudolf Stoff]. Zürich: Scientia, 1939. 221 p.
German translation of Sea Power in the next war.

Jackson, Robert. **The Royal Navy in World War II**. Annapolis, MD: Naval Institute Press, c1997. 176 p.
NPS/DKL Location: GENERAL D771 .J24 1997

Kemp, Peter Kemp. **Key to victory; the triumph of British sea power in World War II**. [1st American ed.]. Boston, Little Brown, [c1957]. 382 p.
Originally published: London: F. Muller, 1957, as Victory at sea.
NPS/DKL Location: BUCKLEY D771 .K3

_____. **Victory at sea, 1939-1945**. London: White Lion Publishers, 1976. 383 p.
American edition published in 1957 as Key to victory.

Kohnen, David. **Commanders Winn and Knowles: Winning the U-boat War with Intelligence, 1939-1943**. Krakow, Poland: Enigma Press, 1999. 168 p.

Lenton, H. T. **British & empire warships of the Second World War**. London: Greenhill Books; Annapolis, MD: Naval Institute Press, 1998. 766 p.
NPS/DKL Location: REFERENCE VA456 .L3997 1998

_____ and J. J. Colledge. **Warship losses of World War II; British and Dominion fleets**. London, I. Allan [1965]. 64 p.

_____ and _____. **Warships of World War II**. 2nd ed. Shepperton: Allan, 1973. 653 p.
American ed. originally published: Garden City, NY, Doubleday [1968, c1964] as British and Dominion warships of World War II.

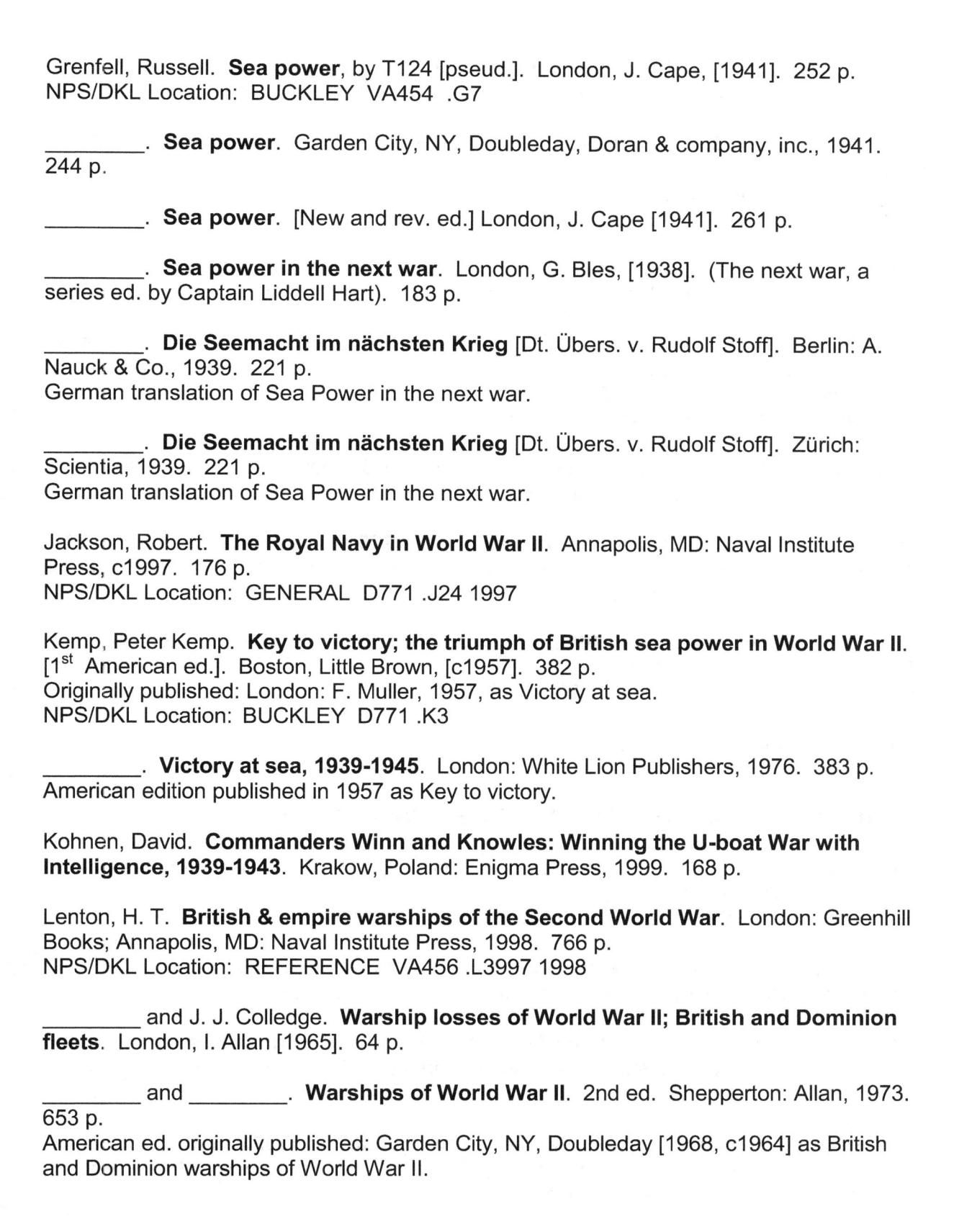

Lund, Paul and Harry Ludlam. **Atlantic jeopardy**. London: W. Foulsham. 1990. 1 v
Contents: PQ17 - convoy to hell; Trawlers go to war; Night of the U-Boats

Macintyre, Donald G. F. W. **The naval war against Hitler.** New York, Scribner [1971]. 376 p.
Originally published: London: Batsford, 1971.
NPS/DKL Location: GENERAL D771 .M15

Madsen, Chris. **The Royal Navy and German naval disarmament, 1942-1947**. London; Portland, OR: F. Cass, 1998. (Naval policy and history; 4). 277 p.

McLachlan, Donald. **Room 39; a study in Naval Intelligence**. [1st American ed.] New York, Atheneum, 1968. 438 p.
Originally published: London, Weidenfeld & Nicolson, 1968.
NPS/DKL Location: GENERAL D771 .M18

Roberts, John Arthur. **British warships of the Second World War**. Annapolis, MD: Naval Institute Press, c2000. (Blueprint series). 160 p.

Roskill, Stephen Wentworth. **The war at sea, 1939-1945**. London: H. M. Stationery Off., 1954-1961. (History of the Second World War: United Kingdom military series). 3 v. in 4.
NPS/DKL Location: GENERAL D771 .R7 1954

_____. **White ensign; the British Navy at war, 1939-1945**. Annapolis, MD: U.S. Naval Institute, [1960]. 480 p.
NPS/DKL Location: GENERAL D771 .R73

T124 [pseudo.]. see Grenfell, Russell.

HISTORY OF NAVAL STRATEGY/WARFARE -- WWII -- GERMAN

Ambrosius, Hans Heinrich. **Die schlacht im Atlantik** [tr. M. Th. Hillen]. Amsterdam, Uitgeverij Westland, 1942. 64 p.
German translation of De slag op den Atlantischen oceaan.

Assmann, Kurt. **Deutsche Seestrategie in zwei Weltkriegen**. Heidelberg, K. Vowinckel, 1957 [c1956]. (Die Wehrmacht im Kampf Bd. 12). 215 p.

Bathe, Rolf. **Der kampf um die Nordsee; chronik des luft-und seekrieges im winter 1939/40 und des norwegischen feldzuges**. Oldenburg i.O./Berlin, G. Stalling, 1941. 310 p.

Bekker, Cajus. **Das Bildbuch der deutschen Kriegsmarine: 1939 - 1945**. Genehmigte, ungekürzte Taschenbuchausg., 9. Aufl. München: Heyne, 1994. 239 p.
Originally published: München: Heyne, 1979. 239 p.

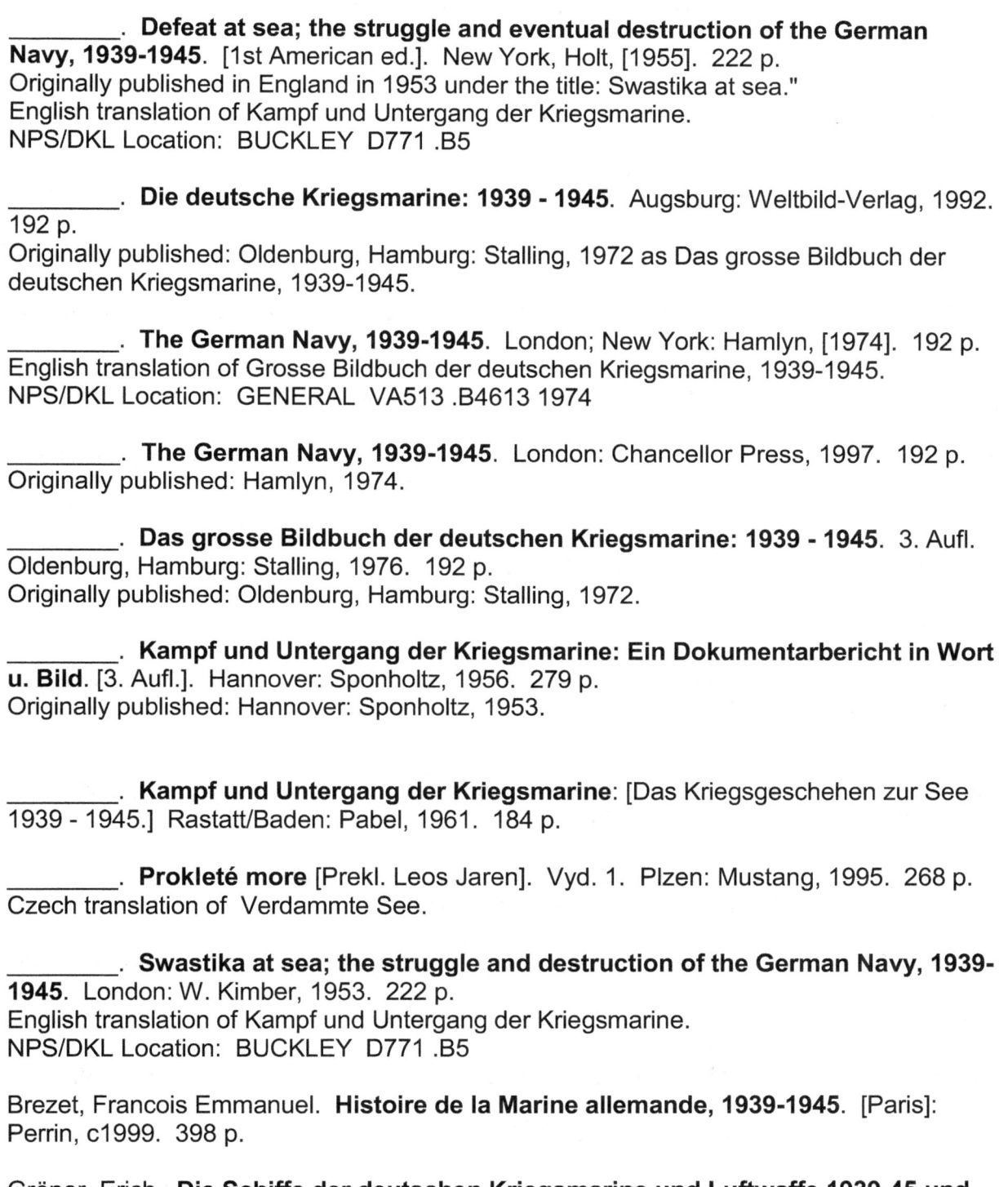

_____. **Defeat at sea; the struggle and eventual destruction of the German Navy, 1939-1945**. [1st American ed.]. New York, Holt, [1955]. 222 p.
Originally published in England in 1953 under the title: Swastika at sea."
English translation of Kampf und Untergang der Kriegsmarine.
NPS/DKL Location: BUCKLEY D771 .B5

_____. **Die deutsche Kriegsmarine: 1939 - 1945**. Augsburg: Weltbild-Verlag, 1992. 192 p.
Originally published: Oldenburg, Hamburg: Stalling, 1972 as Das grosse Bildbuch der deutschen Kriegsmarine, 1939-1945.

_____. **The German Navy, 1939-1945**. London; New York: Hamlyn, [1974]. 192 p.
English translation of Grosse Bildbuch der deutschen Kriegsmarine, 1939-1945.
NPS/DKL Location: GENERAL VA513 .B4613 1974

_____. **The German Navy, 1939-1945**. London: Chancellor Press, 1997. 192 p.
Originally published: Hamlyn, 1974.

_____. **Das grosse Bildbuch der deutschen Kriegsmarine: 1939 - 1945**. 3. Aufl. Oldenburg, Hamburg: Stalling, 1976. 192 p.
Originally published: Oldenburg, Hamburg: Stalling, 1972.

_____. **Kampf und Untergang der Kriegsmarine: Ein Dokumentarbericht in Wort u. Bild**. [3. Aufl.]. Hannover: Sponholtz, 1956. 279 p.
Originally published: Hannover: Sponholtz, 1953.

_____. **Kampf und Untergang der Kriegsmarine**: [Das Kriegsgeschehen zur See 1939 - 1945.] Rastatt/Baden: Pabel, 1961. 184 p.

_____. **Prokleté more** [Prekl. Leos Jaren]. Vyd. 1. Plzen: Mustang, 1995. 268 p.
Czech translation of Verdammte See.

_____. **Swastika at sea; the struggle and destruction of the German Navy, 1939-1945**. London: W. Kimber, 1953. 222 p.
English translation of Kampf und Untergang der Kriegsmarine.
NPS/DKL Location: BUCKLEY D771 .B5

Brezet, Francois Emmanuel. **Histoire de la Marine allemande, 1939-1945**. [Paris]: Perrin, c1999. 398 p.

Gröner, Erich. **Die Schiffe der deutschen Kriegsmarine und Luftwaffe 1939-45 und ihr Verbleib**. 8., erw. Aufl. München: Lehmann, 1976. 126 p.
Originally published: München, J. F. Lehmann, 1954.

Hinsley, Francis Harry. **Hitler's strategy**. Cambridge [Eng.] University Press, 1951. 254 p.
NPS/DKL Location: GENERAL D770 .H6

Koburger, Charles W. **Steel ships, iron crosses, and refugees: the German Navy in the Baltic, 1939-1945**. New York: Praeger, 1989. 133 p.

Lenton, H. T. **German warships of the Second World War**. New York: Arco Pub. Co., 1976, c1975. 396 p.
Originally published: London: Macdonald and Jane's, 1975.

Lochner, R. K. **Als das Eis brach: der Krieg zur See um Norwegen 1940**. München: Heyne, c1983. 735 p.

Lützow, Friedrich. **Die heutige Seekriegsführung**. Berlin: Verlag "Die Wehrmacht", 1940-1943. 11 v.
Contents: 1. Mit U-Boot und Minen gegen englische Hunger-Blockade -- 2. Seerübertum. Schlachtschiffkampf. Die Landung in Norwegen -- 3. Krieg im Kanal, Mittelmeer und Ozean -- 4. Deutschland im Angriff, England in Not -- 1. [i.e. 5] Seekrieg auf allen Meeren -- 2. [i.e. 6] S-S-S aus London -- 3. [i.e. 7] Schlacht im Atlantik. Aufmarsch am Mittelmeer -- 8. USA-Einmischung. Kampf gegen Sowjet-Seemacht -- 9. Von Binnenwassertragen und neuen Seekriegsgebieten -- 10. Japan schlägt los -- 11. Kampf um Frachtraum und Seeherrschaft.

Salewski, Michael. **Die deutsche Seekriegsleitung 1935-1945**. Frankfurt am Main, Bernard & Graefe, 1970-1975. 2 v.
Contents: Bd. 1. 1935-1941.--Bd. 2. 1942-1945.

Showell, Jak P. Mallmann. **German Navy handbook, 1939-1945**. Thrupp, Stroud, Gloucestershire: Sutton, 1999. 274 p.

_____. **The German Navy in World War Two: a reference guide to the Kriegsmarine, 1935-1945**. London: Arms and Armour Press, 1979. 224 p.

Thomas, Charles S. **The German Navy in the Nazi era**. Annapolis, MD: Naval Institute Press, 1990. 284 p.
Originally published: London: Unwin Hyman, Ltd., 1990.

Tuleja, Thaddeus V. **Twilight of the sea gods**. [1st ed.]. New York, Norton, [1958]. 284 p.
NPS/DKL Location: BUCKLEY D771 .T88

_____. **Twilight of the sea gods**. Westport, CT: Greenwood Press, 1975, c1958. 284 p.
Reprint of the 1st ed. published by Norton, New York, 1958.

Von der Porten, Edward P. **The German Navy in World War II**. New York, T. Y. Crowell, [1969]. 274 p.
Revised edition published as Pictorial history of the German Navy in World War II, 1976.
NPS/DKL Location: GENERAL D771 .V9

_____. **The German Navy in World War II**. New York: Galahad Books, [1975] c1969. 274 p.

_____. **Pictorial history of the German Navy in World War II**. Rev. ed. New York Crowell, c1976. 367 p.
Published in 1969 as The Germany Navy in World War II.

HISTORY OF NAVAL STRATEGY/WARFARE -- WWII -- ITALIAN

Bagnasco, Erminio. **Le armi delle navi italiane nella seconda guerra mondiale**. [Parma]: Albertelli, [1978]. 198 p.

Fraccaroli, Aldo. **Italian warships of World War II**. London, Allan, 1968. 204 p.

_____. **Marina militare italiana 1946**. Milano: Hoepli, [1946?]. 201 p.

Sadkovich, James J. **The Italian Navy in World War II**. Westport, CT: Greenwood Press, 1994. (Contributions in military studies no. 149). 379 p.

HISTORY OF NAVAL STRATEGY/WARFARE -- WWII -- JAPANESE

Dull, Paul S. **A battle history of the Imperial Japanese Navy, 1941-1945**. Annapolis MD: Naval Institute Press, c1978. 402 p.
NPS/DKL Location: GENERAL D777 .D83

Edwards, Bernard. **Blood and bushido: Japanese atrocities at sea, 1941-1945**. Bernard Edwards. Upton-upon-Severn, Worcs: Self Publishing Association, 1991. 253 p.

Fukui, Shizuo. **Japanese naval vessels at the end of war**. Old Greenwich, CT: We, inc. [1970?]. [168] p.
Reprint of the 1947 ed., with an appendix "Japanese naval vessels [during] World War II" added.

_____. **Shusen to teikoku kantei: Waga kaigun no shuen to kantei no kisu = Japanese naval vessels survived: their post-war activities and final disposition**. Tokyo: Shuppan Kyodo Publishers, 1961. (Showa gunkan gaishi 3). 224 p.

Parillo, Mark P. **The Japanese merchant marine in World War II**. Annapolis, MD: Naval Institute Press, c1993. 308 p.

Potter, John Deane. **Yamamoto; the man who menaced America**. New York, Viking Press [1965]. 332 p.
"Target A, Pearl Harbor's most secret weapon" p.72-115.
Originally published: London, Heinemann, 1965, as Admiral of the Pacific: the life of Yamamoto.
NPS/DKL Location: GENERAL DS890.Y25 P6 1965

_____. **Yamamoto: the man who menaced America**. New York: Paperback Library, 1965. 351 p.
Originally published: London Heinemann, 1965 as Admiral of the Pacific, the life of Yamamoto.

Tanaka, Toshiyuki. **Hidden horrors: Japanese war crimes in World War II**. Boulder, Co: Westview Press, 1996. (Transitions--Asia and Asian America.). 267 p.
English translation of Shirarezaru senso hanzai.

_____. **Shirarezaru senso hanzai: Nihongun wa Osutorariajin ni nani o shita ka**. Tokyo: Otsuki Shoten, 1993. 263 p.

Uematsu, Sonkei. **Kokumin kaigun-dokuhon**. Shohan. Tokyo: Sankaido Shuppanbu, Showa 19 [1944]. 246 p.

United States. Joint Army-Navy Assessment Committee. **Japanese naval and merchant shipping losses during World War II by all causes**. Washington: U.S. Govt. Print. Off., 1947. (NAVEXOS P. P-468). 1 v.
NPS/DKL Location: GENERAL D777 .U5

HISTORY OF NAVAL STRATEGY/WARFARE -- WWII – POLISH

Kosiarz, Edmund. **Baltyk w ogniu**. Wyd. 1. Gda´nsk: Wydawn. Morskie, 1975. (Seria Wojny morskie; 4). 245 p.

_____. **Baltyk w ogniu**. Wyd. 2. Gda´nsk: Wydawn. Morskie, 1985. (Seria Wojny morskie; 14). 252 p.

_____. **Druga wojna ´swiatowa na Baltyku**. Wyd. 1. Gda´nsk: Wydawn. Morskie, 1988. 718 p.

_____. **Wojna na Baltyku, 1939**. Wyd. 1. Gda´nsk: Krajowa Agencja Wydawnicza, 1988. 414 p.

HISTORY OF NAVAL STRATEGY/WARFARE -- WWII – RUSSIAN/SOVIET

Aprelkov, Aleksei Vasil´evich and L.A. Popov. **Iz morskikh glubin: k istorii podvodnykh lodok "Cheliabinskii komsomolets" i "Leninskii komsomol".** Cheliabinsk: Cheliabinskii obl. sovet veteranov voiny, truda, vooruzhennykh sil i pravookhranitel´nykh organov, 1996. 50 p.

Gorshkov, Sergei Georgievich. **Red star rising at sea** [translated by Theodore A. Neely, Jr.]. [Annapolis, MD]: Naval Institute Press, c1974. 150 p.
From a series of articles originally published in Morskoi sbornik.
NPS/DKL Location: GENERAL VA573 .G63

Herrick, Robert Waring. **Soviet naval theory and policy: Gorshkov's inheritance.** Newport, RI: Naval War College Press; Washington, DC: For sale by the Supt. of Docs., U.S. G.P.O., 1988. 318 p.
NPS/DKL Location: GENERAL V55.S65 H47 1988

_____. **Soviet naval theory and policy: Gorshkov's inheritance.** Annapolis, MD: Naval Institute Press, 1989, c1988. 318 p.
NPS/DKL Location: GENERAL V55.S65 H47 1989

HISTORY OF NAVAL STRATEGY/WARFARE -- SPECIAL TOPICS

ANTISUBMARINE WARFARE (ASW) -- GENERAL

Aigner, Franz. **Unterwasserschalltechnik: Grundlagen, Ziele und Grenzen: Submarine Akustik in Theorie und Praxis.** [Wien]: [s.n.], 1922. 322 p.

_____. **Unterwasserschalltechnik: Grundlagen, Ziele und Grenzen (Submarine Akustik in Theorie und Praxis).** Berlin: Krayn, 1922. 322 p.

Cote, Owen. and Harvey Sapolsky. **Antisubmarine warfare after the Cold War.** Cambridge, MA: Security Studies Program, Massachusetts Institute of Technology, [1997]. (MIT security studies conference series). 22 p.

Cox, Albert W. **Sonar and underwater sound.** Lexington, MA: Lexington Books, [1974]. 144 p.

Daniel, Donald C. **Anti-submarine warfare and superpower strategic stability.** Urbana: University of Illinois Press, c1986. 222 p.
Originally published: London: Macmillan, 1985. (Studies in international security; [24]).
NPS/DKL Location: GENERAL V214 .D36 1986

Federation of American Scientists. **Special issue, antisubmarine warfare**. Washington, DC: F.A.S., 1986. (F.A.S. public interest report; 39, no. 6, June-July 1986). 12 p.
NPS/DKL Location: GENERAL V214 .S63 1986

Franklin, George. **Britain's anti-submarine capability, 1919-1939**. London: Portland, OR: Frank Cass, 2003. (Cass series--naval policy and history; 17). 208 p.
NPS/DKL Location: GENERAL V214 .F73 2003

Gerken, Louis. **ASW versus submarine technology battle**. Chula Vista, CA: American Scientific Corp., c1986. 753 p.
NPS/DKL Location: GENERAL V214 .G47 1986

Heppenheimer, T. A. **Anti-submarine warfare: the threat, the strategy, the solution**. Arlington, VA: Pasha Publications, c1989. 250 p.

Janzon, Bo. **The anti-submarine warfare project at the Swedish Defence Research Establishment (FOA), Sweden: an outline of activities**. Stockholm: FOA, [1988]. 16 p.

Miller, David. **An illustrated guide to modern sub hunters**. New York: Arco Pub., c1984. 160 p.
NPS/DKL Location: GENERAL V214 .M55 1984

_____. **The new illustrated guide to modern sub hunters**. New York: Smithmark, c1992. 160 p.

Moore, John Evelyn and Richard Compton-Hall. **Submarine warfare: today and tomorrow**. Bethesda, MD: Adler & Adler, 1987, c1986. 308 p.
Originally published: London: M. Joseph, 1986.
NPS/DKL Location: GENERAL V210 .M66

Price, Alfred. **Aircraft versus submarine: the evolution of the anti-submarine aircraft, 1912 to 1972**. [Annapolis, MD]: Naval Institute Press, 1973. 268 p.
NPS/DKL Location: GENERAL V214 .P9

_____. **Aircraft versus submarine: the evolution of the anti-submarine aircraft, 1912 to 1980**. Rev. ed. London; New York: Jane's Publishing, 1980. 272 p.
Previously published: London: Kimber, 1973 as Aircraft versus submarine: the evolution of the anti-submarine aircraft, 1912 to 1972.

_____. **Flugzeuge jagen Uboote** [Dt. Übers. von Hans u. Hanne Meckel]. Stuttgart: Motorbuch-Verlag, 1976. 373 p.
German translation of Aircraft versus submarine.

_____. **Patrol aircraft vs. submarine**. Shrewsbury: Airlife, 1991. (Aggressors; v.4). 63 p.

Stefanick, Tom. **Strategic antisubmarine warfare and naval strategy**. Lexington, MA: Lexington Books, c1987. 390 p.
NPS/DKL Location: GENERAL V214 .S74 1987

Suzdalev, Nikolai Ivanovich. **Podvodnye lodki protiv podvodnykh lodok**. Moskva, Voenizdat, 1968. 163 p.

Tsipis, Kosta. **Tactical and strategic antisubmarine warfare**. Cambridge, MA: MIT Press, 1974. (Stockholm International Peace Research Institute SIPRI monograph). 148 p.
NPS/DKL Location: GENERAL V214 .S8

Williams, Mark. **Captain Gilbert Roberts R. N. and the anti-u-boat school**. London: Cassell, 1979. 186 p.

ANTISUBMARINE WARFARE (ASW) – GENERAL -- WWI

Grant, Robert M. **U-boats destroyed: the effect of anti-submarine warfare 1914-1918**. Putnam, [1964]. 172 p.
NPS/DKL Location: GENERAL D581 .G7

Hackmann, Willem Dirk. **Seek & strike: sonar, anti-submarine warfare, and the Royal Navy, 1914-54**. London: H.M.S.O., 1984. 487 p.
NPS/DKL Location: GENERAL V214.H33 1984

Humphreys, Roy. **The Dover Patrol, 1914-18**. Phoenix Mill, Thrupp, Stroud, Gloucestershire: Sutton, 1998. 216 p.

Messimer, Dwight R. **Find and destroy: antisubmarine warfare in World War I**. Annapolis, MD: Naval Institute Press, c2001. 298 p.
NPS/DKL Location: GENERAL D590 .M47 2001

ANTISUBMARINE WARFARE (ASW) -- GENERAL -- WWII

Doscher, J. Henry. **Subchaser in the South Pacific: a saga of the USS SC-761 during World War II**. 1st ed. Austin, TX: Eakin Press, c1994. 110 p.

Franks, Norman L. R. **Dark sky, deep water; [first hand reflections on the anti-U-boat War in Europe in WWII]**. 1st pbk. ed. London: Grub Street, 1999. 218 p. Originally published: London: Grub Street, 1997.
NPS/DKL Location: GENERAL V214.5 .F735 1999

_____. **Search, find and kill**. Rev. and updated ed. London: Grub Street, c1995. 274 p.

_____. **Search, find and kill; Coastal Command's U-boat successes**. Bourne End: Aston, 1990. 168 p.

Job, Glenn T. **A bumpy ride: the tale of U.S.S. PC(C) 1168, a World War II subchaser and the men who manned her, 1943-1954**. Ridgefield Park, NJ: Thomsen Litho, 1996. 20 p.

Keefer, Louis E. **From Maine to Mexico: with America's private pilots in the fight against Nazi U-boats**. 1st ed. Reston, VA: COTU Pub., c1997. 535 p.

Purdon, Eric. **Black company; the story of Subchaser 1264**. Washington, R. B. Luce [1972]. 255 p.
NPS/DKL Location: GENERAL D810.N4 P9

_____. **Black company: the story of Subchaser 1264**. Annapolis, MD: Naval Institute Press, c2000. (Bluejacket books). 255 p.

Roberts, Douglas L. **Rustbucket 7: chronicle of the USS PC 617 during the Great War, 1942-1946**. Newcastle, ME: Mill Pond, c1995. 159 p.

Schoenfeld, Maxwell Philip. **Stalking the U-boat: USAAF offensive antisubmarine operations in World War II**. Washington: Smithsonian Institution Press, c1995. (Smithsonian history of aviation series). 231 p.

Stafford, Edward P. **Subchaser**. Annapolis, MD: Naval Institute; London: Greenhill, 2003. (Bluejacket books). 251 p.
Originally published Annapolis, MD: Naval Institute, c1988.
NPS/DKL Location: GENERAL D773 .S78 2003

Sternhell, Charles M. and Alan M. Thorndike. **Antisubmarine warfare in World War II**. Washington, DC: s.n., [1990?], c1946. (OEG report; no. 51). 331 p.
Prepared by Operations Evaluation Group, formerly operations Research Group, Office of the Chief of Naval Operations; Originally: "Confidential".
Contents: Pt. 1. History of anti-submarine operations by Charles M. Sternhell.-- Pt. 2. Anti-submarine measures and their effectiveness by A. M. Thorndike.
NPS/DKL Location: GENERAL D780 .S74 1990

Treadwell, Theodore R. **Splinter fleet: the wooden subchasers of World War II**. Annapolis, MD: Naval Institute Press, c2000. 274 p.

Whinney, Bob. **The U-boat peril: an anti-submarine commander's war**. Poole; New York: Blandford Press; New York, NY: distributed in the United States by Sterling Pub. Co., 1986. 160 p.

_____. **The U-boat peril: a fight for survival**. London: Cassell, 1986. (Cassell military classics). 160 p.
Originally published: Poole: Blandford, 1986 as The U-boat peril: an anti-submarine commander's war.
NPS/DKL Location: GENERAL D784.G7 W45 1986B

Williams, Mark. **Captain Gilbert Roberts R. N. and the anti-u-boat school**. London: Cassell, 1979. 186 p.

Y'Blood, William T. **Hunter-killer: U.S. escort carriers in the Battle of the Atlantic**. Annapolis, MD: Naval Institute Press, c1983. 322 p.
NPS/DKL Location: GENERAL D770 .Y34 1983

ANTISUBMARINE WARFARE (ASW) -- NPS THESES AND TECHNICAL REPORTS

Adams, Brian S.. **Analysis of the Effects of Energy Spreading Loss and Transmission Loss on Low Frequency Active Sonar Operations in Shallow Water**. Monterey, CA: Naval Postgraduate School; Springfield, VA: Available from National Technical Information Service, 1997. (ADA341298). 74 p.
Thesis (M.S. in Physical Oceanography), Naval Postgraduate School, Sept. 1997.
Abstract: *Energy Spreading Loss (ESL) is qualitatively defined as the reduction in peak power level due to energy spreading of a transmitted acoustic pulse in tune. An analysis of the impact of bathymetric geometry and sediment type on ESL and TL associated with the Low Frequency Active/Compact Low Frequency Active (LFA /CLFA) sonar operations was conducted utilizing the FEPE, FEPE SYN and EXT TD programs to model the time spreading of the acoustic pulse due to multipath propagation in shallow water. Both a Blackman windowed pulse and a Continuous Wave (CW) pulse were used in this analysis. The Blackman pulse had a center frequency of 244 Hz with a bandwidth of 24 Hz. The CW pulse had a center frequency of 244 Hz with a bandwidth of 0.0625 Hz. Model inputs were a geoacoustic description of the Tanner Bank region off the coast of San Diego and a typical late summer sound speed profile taken from the MOODS database. ESL and TL's impact on low frequency active sonar operations was determined as a function of bathymetry, sediment type, sound speed profile, and pulse length. The results showed that ESL is inversely related to pulse duration and at low frequencies is relatively uninfluenced by sediment type. When pulse lengths were reduced to less than 1 second, ESL became appreciable (> 6 dB one way) and was an important segment of the active sonar equation. TL was found to be the dominating factor in LFA/CLFA operations for pulse lengths greater than 1 second and was greatly influenced by sediment type and sound speed profile.*
NPS/DKL Location: THESIS A2246
Electronic acccess:http://handle.dtic.mil/100.2/ADA341298

Adler, Vance Erick. **Digital Processing of Acoustic Signals with Application to an ASW Signal Processor**. Monterey, CA: Naval Postgraduate School; Springfield, VA: Available from National Technical Information Service, 1973. (AD7750383). 102 p.
Naval Postgraduate School Monterey Calif. Dec 1973. 102 p.
Thesis (M.S. in A.E.), Naval Postgraduate School, 1973.

Abstract: *There is a growing need within the Navy for methods for detecting discrete narrowband signals in a non-stationary background. This paper concerns itself with the application of digital processing and spectral analysis techniques toward that goal. The use of the fast Fourier Transform in estimating the power spectrum of a signal is described. The method involves sectioning the time record, making 'raw' estimates of the spectrum from these sections, and averaging these 'raw' estimates. It is shown that more stable estimates are available if the segments are overlapped and an optimum amount of overlap for the case of the Hanning window is found. It is shown that the stability of these spectral estimates can be interpreted as processing gain in the case of a discrete narrowband signal in additive noise. And finally, a brief description of signal detection theory applied to a human observer is presented to emphasize the flexibility that a human operator can bring to a signal detection system.*
NPS/DKL Location: THESIS A254

Aktan, Zafer Mutlu. **Probabilistic Approach to ASW Deployment in Shallow Waters**. Monterey, CA: Naval Postgraduate School; Springfield, VA: Available from National Technical Information Service, 1992. (ADA257589). 87p.
Thesis (M.S. in Operations Research), Naval Postgraduate School, Sept. 1992.
Abstract: *The Advanced Air Deployable Array (AdDA), which is a modern air-dropped fiber optic ASW device, provides an opportunity for the rapid enclosure of a hostile submarine in shallow waters. This thesis explores the effect of the deployment depth, and effect of using longer or shorter AdDA array segments, on the performance of eighth proposed AdDA deployment tactics which employ single or dual aircraft. It is shown that when the AdDA sinking rate is considered, several of the proposed tactics become infeasible for certain depth and submarine speed combinations. Still, today fiber optics offer unique capabilities for solving some of the U.S. Navy's and the Turkish Navy's problems in the future.*
NPS/DKL Location: THESIS A334545

Armo, Knut Rief. **The relationship between a submarine's maximum speed and its evasive capability**. Monterey, CA: Naval Postgraduate School; Springfield, VA: Available from National Technical Information Service, 2000. (ADA382255). 55 p.
Thesis (M.S. in Operations Research), Naval Postgraduate School, 2000.
Abstract: *The experiences of submarine warfare from WWI and WWII have generally dictated maximum speed when designing conventional submarines. Technological development of submarine and antisubmarine weapons, however, requires examination of submarine warfare and tactics. This thesis focuses on a coastal conventional submarine's ability to survive, as a function of its maximum speed, when attacked by a light antisubmarine warfare (ASW) torpedo. It also evaluates the maximum speed with which the submarine should be equipped to ensure a specified probability of survival. The measure of effectiveness (MOB) is the probability that the submarine, operating up to maximum speed and launching only one set of countermeasures, is not caught by the torpedo. The investigation builds on a discrete event simulation model. The systems simulated are a submarine, a light ASW torpedo, and a countermeasure system consisting of one decoy and four jammers. The results show that maximum speed of a submarine does affect the submarine's evasive performance between 12 and 18 knots. The simulated model reached a maximum probability of survival at 18 knots. That result should be regarded as a minimum since a real life system might require a higher maximum speed to reach its greatest probability of survival.*
NPS/DKL Location: THESIS A6787
Electronic access: http://library.nps.navy.mil/uhtbin/hyperion-image/00Jun_Armo.pdf
Electronic access: http://handle.dtic.mil/100.2/ADA382255

Arnold, Ronald R. II. **Comparison of P-3C Acoustic Processing Capability with Acoustic Operator Capability**. Monterey, CA: Naval Postgraduate School; Springfield, VA: Available from National Technical Information Service, 1987. (ADA181635). 57p.
Thesis (M.S. in Information Systems) Naval Postgraduate School, March 1987.

Abstract: *This thesis sought to determine if the requirements for operation of the acoustic processing equipment now installed aboard P-3C aircraft is too complex for the acoustic operators, given their current amount of training. This was accomplished by using a test scenario designed to test for all of the skills and knowledge required by acoustic operator in the performance of his duties during the passive portion of the prosecution of a target. The results seem to suggest that the students that successfully complete the P-3C 'Antisubmarine Warfare Operator' rating training pipeline are acquiring an acceptable level of operator capability. In addition, this study seems to suggest that fleet operators who are recognized in fleet squadrons as master journeyman, are operating their ASW acoustic processing equipment to its fullest capability and without apparent operator deficiencies.*
NPS/DKL Location: THESIS A7263

Arnote, Stanley Dean. **Consideration of the Carrier-Based Tactical Support Center Installation Design**. Monterey, CA: Naval Postgraduate School; Springfield, VA: Available from National Technical Information Service, 1977. (ADA039352). 49p.
Thesis (M.S. in Systems Tech.)--Naval Postgraduate School, 1977.
Abstract: *This thesis analyzes the aircraft carrier based Tactical Support Center (CV-TSC) installation design from a human factors viewpoint. Starting with the threat, the mission of the CV-TSC is defined. A modular concept between man and machine is developed. Man's role, tasks, and functions are identified and form the basis for recommended changes to the CV-TSC aboard the USS Constellation (CV-64) and general recommendations for all CV-TSC installations. (Author)*
NPS/DKL Location: THESIS A729

Aydin, Erhan. **Screen Dispositions of Naval Task Forces Against Anti-Ship Missiles**. Monterey, CA: Naval Postgraduate School; Springfield, VA: Available from National Technical Information Service, 2000. (ADA377605) 96p.
Thesis (M.S. in Operations Research) Naval Postgraduate School, March 2000.
Abstract: *Ship defense in convoy operations against Anti-Surface Missiles (ASM) has been an important aspect of Naval Warfare for the last two decades. Countries in a state of conflict often conduct threatening operations in their own territories in order to slow or stop the enemy merchant ship traffic through the straits or littoral waters. Such littoral scenarios, the quantity and capability of ASM's in non-NATO countries pose a significant threat to the safe operation of the NATO forces in the waters off of potentially hostile shores. In these operations the goals of the tactical commander are to design an optimal reaction platform (formation) and to determine an optimal strategy that will help him in multi-threat encounters. The scope and design in most anti-air warfare studies have been limited to evaluating the effectiveness of detecting sensors and weapon systems in a regular screen formation. The proposed model's (Disposition Mission Model - DMM) characterization, however, is based on how to perform an effective, defensive disposition from a task force. In DMM we focus on usage of a graphical user interface and provide a user-friendly environment for analyzing new tactics in screen formations. The model, with its user interface, allows the user to build and run a convoy simulation, and see the results comparatively on the same interface. The analysis using this model has yielded significant insights towards the defense of a convoy by way of regression methods. It has been seen that positioning the escort ships within the threat sector reduces the damage on the HVU and also balances the defensive load of each defense ship for the incoming missiles. The model, with its graphical interface and simulation components, provides an initial approach for future analysts, not only in anti-air warfare defense of screen formations, but also in the areas of anti-surface and anti-submarine warfare.*
NPS/DKL Location: THESIS A99423
Electronic access: http://library.nps.navy.mil/uhtbin/hyperion-image/00_Aydin.pdf

Bacon, Daniel Keith. **Integration of a Submarine into NPSNET**. Monterey, CA: Naval Postgraduate School; Springfield, VA: Available from National Technical Information Service, 1995. (ADA304337). 99p.
Thesis (M.S. in Computer Science) Naval Postgraduate School, Sept. 1995.

Abstract: *In the current version of NPSNET there are two problems that prevent users or this virtual environment from achieving a realistic training experience. First, the motion of the vehicles is not built around realistic, physically based models. In particular, the motion of computer-generated sea-going vehicles is not based on the hydrodynamic models that reflect the motion of actual ships moving through water. Second, vehicles in NPSNET are currently controlled by a single individual; they lack the capability to be controlled by a team. This misrepresents the many actual military vehicles-submarines, tanks, helicopters, and others- that must be controlled by several people working together. The approach taken was to update the submersible vehicle class in NPSNET in two ways. A physically-based hydrodynamic model was used to control the vehicle's motion through the virtual world. In addition, a network communications protocol was implemented to enable several remote individuals to control the same vehicle simultaneously. The result of this work is the creation of a computer-generated submersible vehicle whose motion is determined by a real-time hydrodynamic model so it moves through the virtual world according to physically based models. This submersible is also capable of being controlled by several remote individuals-effectively the same team members who would perform the job in the actual vehicle. This ultimately results in a more realistic user experience as well as a more effective training tool for NPSNET.*
NPS/DKL Location: THESIS B1056

Beckes, Michael Edward; Nicholas Pegau Burhans, and Robert Edgar Gump. **Passive Environmental ASW Prediction System (PEAPS).** Monterey, CA: Naval Postgraduate School; Springfield, VA: Available from National Technical Information Service, 1975. (NPS-71BE75031; ADA010813). 170 p.
Thesis (M.S. in Systems Technology)--Naval Postgraduate School, 1975.
Issued as Naval Postgraduate School Technical Report (NPS-71BE75031) in conjunction with Alan B. Coppens and Don E. Harrison.
Abstract: *PEAPS (Passive Environmental ASW Prediction System) is a relatively unsophisticated model which accepts input source and receiver parameters and then predicts sound propagation characteristics in an ocean environment, the corresponding transmission loss, and the probability of detection. The program was written for a programmable desk-top calculator for immediate deployment and operational testing aboard small ASW platforms. The program is also available in a form suitable for larger computers.*
NPS/DKL Location: THESIS B3354

Bjorklund, Bruce R. **Probabilistic Observations on Antisubmarine Warfare Tactical Decision Aid (ASWTDA).** Monterey, CA: Naval Postgraduate School; Springfield, VA: Available from National Technical Information Service, 1990. (ADA227540). 54 p.
Thesis (M.S. in Operations Research) -- Naval Postgraduate School, March 1990.
Abstract: *The goal of this thesis is to examine the methodology used in the Antisubmarine Warfare Tactical Decision Aid (ASWTDA) in development by Sonalysts, Incorporated of Waterford, Connecticut under Navy contract. ASWTDA is a Computer Assisted Search (CAS) program which is designed as a tool to assist platform, unit or force commanders afloat and ashore in making tactical ASW decisions. First, a Classical Computer Assisted Search program is described as a basis of comparison for the methodology employed in ASWTDA. Then, the operations as performed in ASWTDA are described, followed by a probabilistic analysis. In the analysis sections, probabilistic support for the applied methodology is provided where applicable, and conceptual problems and possible solutions are cited where appropriate. Keywords: Target motion, Probabilistic analysis. (kr)*
NPS/DKL Location: THESIS B545445

Bliss, John Robert. **Modification of the TASDA Computer Program for Inflight Use**. Monterey, CA: Naval Postgraduate School; Springfield, VA: Available from National Technical Information Service, 1973. (AD769831). 195 p.

Thesis (M.S. in Operations Research)--Naval Postgraduate School, 1973.
Abstract: *TASDA, acronym for Tactical Airborne Sonar Decision Aid, is a computer simulation designed to select optimum sonobuoy pattern spacings given enviornmental parameters and submarine mode of operation. The program was designed to operate in a Tactical Support Center for briefing of flight crew personnel. Analytical methods and statistical models are used to investigate the TASDA program with a view towards modifying it for future aircraft inflight utilization. Some improvements are made to the TASDA model which reduce program run time and core storage requirements. A modified version of the TASDA program is developed as an initial step toward an inflight model. (Author)*
NPS/DKL Location: THESIS B564

Bobbitt, Richard B. **Escape Strategies for Turboprop Aircraft in a Microburst Windshear**. Monterey, CA: Naval Postgraduate School; Springfield, VA: Available from National Technical Information Service, 1991. (ADA243090). 247 p.
Thesis (M.S. in Aeronautical Engineering) Naval Postgraduate School, March 1991.
Abstract: *A quantitative analysis was carried out on the performance of turboprop aircraft within a microburst windshear. The objective of the analysis was to provide specific flight procedures for optimal navigation through the windshear. The microburst windshear model uses in the analysis embodied the severe characteristics of the microburst encountered by Delta Flight 191 during an approach to landing at Dallas/Ft. Worth, 2 August 1985. Different escape strategies were tested using the flight performance characteristics of the U.S. Navy's P-3 'Orion' and T-44 'Pegasus' aircraft. The three flight phases investigated were approach to landing, takeoff, and the low altitude ASW mission. Results from the analysis were coupled with the pilot's view point from which conclusions were drawn. The results of the analysis support a constant-pitch-angle escape procedure. The same procedural steps can be used for both aircraft in any configuration or situation with the difference being the degree of pitch to employ. The conclusions are in a format for integrating specific microburst escape procedures within the NATOPS programs for the P-3 and T-44.*
NPS/DKL Location: THESIS B5886

Boice, Frank B. **Probability Model for a Convoy Threatened by a Submarine Launched Missile**. Monterey, CA: Naval Postgraduate School; Springfield, VA: Available from National Technical Information Service, 1965. (AD475379). 115 p.
Thesis (M.S. in Operations Research)-- Naval Postgraduate School, 1965.
Abstract: *The advent of the missile firing submarine has added yet another dimension to the problem of defending convoys and task groups during ocean transit. The specific situation wherein the submarine must surface to fire a relatively short range missile against a convoy of ships is considered. The model developed considers several different problem parameters. It enables the calculation of probability of detection of the submarine, probability of killing the submarine before a particular missile is fired, and the expected number of missiles that the submarine will fire. Selected results from randomly selected parameter values are also presented. (Author)*
NPS/DKL Location: THESIS B643

Bond, Charles Pruitt, King; Burnett, Howard Burnett.; Grehawick, Gregory Daniel. **An Analysis of the Utilization of the In-Flight Technician in the P3C Community**. Monterey, CA: Naval Postgraduate School; Springfield, VA: Available from National Technical Information Service, 1975. (ADA014553). 76 p.
Thesis (M.S. in Management)--Naval Postgraduate School, 1975.
Abstract: *This thesis investigated the number of In-Flight-Technicians assigned to a Navy P3C squadron, their contributions to the squadron's ASW capability in their dual roles as in-flight and ground repairmen, and the adequacy of the In-Flight-Maintenance-Kit. Tradeoffs between the number of In-Flight Technicians and ground avionics workers were evaluated as were various methods of the In-Flight Technician's ground and airborne utilization. Potential benefits associated with In-Flight Technician*

assignment to Intermediate and Depot Level maintenance activities were also examined. The In-Flight Technician's contributions to the squadron's ASW capability were measured in Equivalent Aircraft Units which were a function of how many repairs were corrected and the impact on ASW capability of the systems repaired. (Author)
NPS/DKL Location: THESIS B6785

Breemer, Jan S. **Anti-Submarine Warfare: A Strategy Primer**. Monterey, CA: Naval Postgraduate School; Springfield, VA: Available from National Technical Information Service, 1988. (NPS-56-88-014; ADA199554). 20 p.
Abstract: *This report reviews the naval planner's basic menu of operational anti submarine warfare (ASW) strategical choices. Basic ASW strategies, discussed from a historical perspective, are: (1) destruction of the submarine (2) containment of the submarine, and (3) limiting the submarine's efficiency. The report has been prepared for inclusion in the International Military and Defense Encyclopedia (IMADE), scheduled for publication by Pergamon-Brassey's in 1991-92.*
NPS/DKL Location: FEDDOCS D 208.14/2:NPS-56-88-01

Brennan, Peter J. **Analysis of the Manpower Costs Associated with the Helicopter Air Wing Commander Concept**. Monterey, CA: Naval Postgraduate School; Springfield, VA: Available from National Technical Information Service, 1998. (ADA343363). 120 p.
Thesis (M.S. in Management) Naval Postgraduate School, March 1998.
Abstract: *This thesis presents an analysis and comparison of manpower costs of three options for the United States Navy Helicopter force structure through the year 2020. The first option, the basic plan, leaves the force structure as it is today. The second option assumes the mission to support the Military Sealift Command (MSC) is outsourced and combines the Helicopter Combat Support (HC) and Helicopter Antisubmarine Warfare (HS) communities into a community referred to as HSC. The third option realigns the force along missions performed by the SH-60Bs and CH-60 under a Helicopter Air Wing Commander (HAWC). All three options support the requirements set forth in the Helo Master Plan (HMP) and are based on the acquisition of the CH-60 helicopter along with the upgrade of all SH-60Bs and SH-60Fs to SH-60Rs. The analysis involved developing manning levels, by pay grade, for the three options and deterniining the differences in those manning levels. Manpower costs were allocated to the total personnel requirements, and differences in costs among the options were calculated.*
NPS/DKL Location: THESIS B803415
Electronic access: http://handle.dtic.mil/100.2/**ADA343363**

Brownsberger, Nicholas Mason. **Estimation of Sonobuoy Position Relative to an Aircraft Using Extended Kalman Filters**. Monterey, CA: Naval Postgraduate School; Springfield, VA: Available from National Technical Information Service, 1979. (ADA078280). 151 p.
Thesis (M.S. in A.E. and Degree of A.E.)--Naval Postgraduate School, 1979.
Abstract: *In airborne antisubmarine warfare there is a need to more accurately determine the positions of sonobuoys on the surface of the water. This report develops two algorithms which employ extended Kalman filters to determine estimated position. The bearing from the aircraft to the sonobuoy is the primary measurement. Range information is not available. The first algorithm is a six-state filter which was reduced from the 13-state system developed by the Orincon Corporation. Its states include relative position, relative velocity, and inertial misalignments. The second algorithm includes two cascaded Kalman filters. The primary two-state filters etimates sonobuoy position. A secondary filter estimates drift from information obtained from the primary filter. Both algorithms successfully estimated sonobuoy position for simulated aircraft data. The effect of aircraft-to-sonobuoy range, the frequency of measurement, and changes in altitude are also analyzed. (Author)*
NPS/DKL Location: THESIS B8243

Brownsweiger, Jeffrey Scott. **S-3 Viking Weapon System Improvement Program: Financial Management Implications**. Monterey, CA: Naval Postgraduate School; Springfield, VA: Available from National Technical Information Service, 1990. (ADA247150) 55 p.
Thesis (M.S. in Management) Naval Postgraduate School, Dec. 1990.
Abstract: *In response to recent significant improvements in soviet Submarine Technologies the Navy developed the Weapon System Improvement Program for the S-3A Viking. This program is an example of the dynamic nature of the environment within which the program manager operates. It provides the program manager with little control over certain events and the effects they have on their programs. An effective program manager will realize these limitations exist and attempt to strategically and flexibly manage the resources available to him as effectively and efficiently as his/her political environment will allow. However, this sometime happens at the expense of contractor inefficiencies and at a higher cost to the Government. In the DOD/DON world of scarce resources a thorough analysis of the competitive environment may provide useful insight into the S-3 Program Office and their efforts to complete the S-3 WSIP.*
NPS/DKL Location: THESIS B82435

Bush, John Richard. **Aviation Squadron Organization Development of the Navy's Light Airborne Multi-Purpose System (LAMPS) Mk III**. Monterey, CA: Naval Postgraduate School; Springfield, VA: Available from National Technical Information Service, 1981. (NPS-54-81-017, ADA113863). 183 p.
Thesis (M.S. in Management)--Naval Postgraduate School, 1981. Also issued as Naval Postgraduate School Technical Report (NPS-54-81-017).
Abstract: *This paper examines the evolution of the U.S. Navy's SH-6OB, LAMPS Mk III aircraft and squadron methodology. It analyzes current HSL organization design and introduces alternative organization structures to support this new helicopter community when it is introduced in the fleet in 1983-84. It begins with a statement of the issue which includes a concise historical overview of the LAMPS program and discusses its tactical and support missions. It next examines the conventional naval air squadron organization methodology from which LAMPS squadrons are designed and manned. A statistical analysis of operational fleet HSL squadrons is presented which concludes that conventional squadron design methodology does not support the unique LAMPS community. Four general alternative organization models are proposed followed by a discussion of the possible utilization of the Naval Flight Officer in the LAMPS System. The paper concludes with a summary of the proposals from which organization redesign may result and offers recommendations to that process. (Author)*
NPS/DKL Location: THESIS B9218

Buterbaugh, Thomas A. **Multivariate Analysis of the Effects of Academic Performance and Graduate Education on the Promotion of Senior U.S. Navy Officers.** Monterey, CA: Naval Postgraduate School; Springfield, VA: Available from National Technical Information Service, 1995. (ADA300188). 97 p.
Thesis (M.S. in Management) Naval Postgraduate School, June 1995
Abstract: *This thesis develops multivariate models to estimate the effects of undergraduate academic performance and fully-funded graduate education on promotion to the ranks of Commander (0-5) and Captain (0-6) in the U. S. Navy. Using data extracted from the Officer Promotion History Files, two sample populations were selected for analysis: officers who appeared before the Commander promotion boards between fiscal years 1981 and 1994, and those who appeared before the Captain promotion boards during this same period. These data sets were further categorized into five warfare communities and two separate time periods; the period between 1981-1989 (the pre-drawdown), and the period between 1990-1994 (the drawdown). Ordinary least squares (OLS) and maximum likelihood log it regression models were employed to estimate the probability of being promoted to these two ranks. The findings reveal that*

graduate education and academic performance have positive effects on promotion probability for some, but not all, of the communities over the various time periods. Recommendations for further study are included.
NPS/DKL Location: THESIS B932

Cagle, Clifford Monroe. **An Application of Multidimensional Scaling to the Prioritization of Decision Aids in the S-3A.** Monterey, CA: Naval Postgraduate School; Springfield, VA: Available from National Technical Information Service, 980. (ADA092405). 134 p.
Thesis (M.S. in Operations Research)--Naval Postgraduate School, 1980.
Abstract: *This thesis presents an application of Multidimensional Scaling (MDS) used in the prioritization of ASW decision functions in the S-3A. The ASW decision space was divided into 14 discrete decision functions for the purposes of this analysis. The problem of developing a prioritization methodology was approached from two independent directions. First, an unconstrained sorting task was preformed to provide input to Multidimensional Scaling algorithm. The result of this analysis provided a three dimensional representation of the decision space with dimensional interpretation. Second, a series of ranking tasks were preformed to provide input to an Unfolding Analysis algorithm. The Generalized Distance Model was selected as the model most representative of the ranking data. The decision function coordinates for the MDS algorithm and the decision function coefficients for the Unfolding Analysis algorithm were combined in a regression-like equation to provide a prioritization methodology for the 14 decision functions of the S-3A ASW decision space.*
NPS/DKL Location: THESIS C169

Cahill, David Blake. **Tactical Display of CZ Propagation Area.** Monterey, CA: Naval Postgraduate School; Springfield, VA: Available from National Technical Information Service, 1977. (ADA042173). 36 p.
Thesis (M.S. in Systems Technology)--Naval Postgraduate School, 1977.
Abstract: *The need for an enhanced display of the geographic distribution convergence-zone propagation in a region of shallow and variable bottom topography is demonstrated. A preliminary display is generated utilizing point-by-point computer processing of the convergence-zone minimum depth requirements with a large bathymetric data base for a region north of the Azores. The trade-offs between the accuracy of the preliminary output and cost (computer time and space) are discussed.*
NPS/DKL Location: THESIS C175

Caldwell, James F. Jr. **Investigation and Implementation of an Algorithm for Computing Optimal Search Paths.** Monterey, CA: Naval Postgraduate School; Springfield, VA: Available from National Technical Information Service, 1987. (ADA185927). 88 p.
Thesis (M.S. in Operatons Research) Naval Postgraduate School, Sep. 1987.
Abstract: *A moving target is detected at long range with an initial position given by a probability distribution on a grid of N cells. Also located on the grid is a searcher, constrained by speed, who must find an optimal search path in order to minimize the probability of target survival by time T. A branch-and-bound algorithm designed by Professors Eagle and Yee of the Naval Postgraduate School in Monterey, California, is successfully implemented in order to solve this problem. Within the algorithm, the problem is set up as a nonlinear optimization of a convex objective function subject to the flow constraints of an acyclic N x T network. Lower bounds are obtained via the Frank-Wolfe method of solution specialized for acyclic networks. This technique relies on linearization of the objective function to yield a shortest path problem that is solvable by dynamic programming. For each iteration, the lower bound can be found by use of a Taylor first order approximation. Implementation of this algorithm is accomplished by the use of a Fortran program which is run for several test cases. The characteristics of the solution procedure as well as program results are discussed in detail. Finally, some real world applications along with several*

questions requiring further research are proposed. Keywords: Thesis; Integer programming; Antisubmarine warfare.
NPS/DKL Location: THESIS C186275

Calvano, Charles N., Robert C. Harney, David Wickersham, Ioannis Farsaris, Philip Malone, David Ruley and Nathan York. **Surface Warfare Test Ship Design**. Monterey, CA: Naval Postgraduate School; Springfield, VA: Available from National Technical Information Service, 2000. (NPS-ME-00-001; ADA374332). 260 p.
Abstract: *A systems engineering approach to the design of a ship conversion to satisfy the requirements for a Surface Warfare Test Ship (SWTS) to be employed by the Port Hueneme Division of the Naval Surface Warfare Center is presented. The ship described would meet test needs for future weapons and sensor systems and provide limited test capability for future hull, mechanical and electrical systems. The current Self Defense Test Ship is over 45 years old, approaching the end of its useful life. A conversion of a decommissioned SPRUANCE (DD 963) class ship is the basis for the replacement Surface Warfare Test Ship. The study proceeds from mission needs and operational requirements through a functional analysis and study of threat weapons to be employed against the SWTS. After summarizing the characteristics of a SPRUANCE Class ship, the study reports an analysis of four alternative conversion schemes. The alternatives are described, with the rationale for choosing that considered best. The chosen alternative is then described and analyzed in several important areas of concern including combat systems functionality, signature characteristics, engineering plant and habitability for test personnel. The fitness of the proposed design for several special evolutions is also described, and alternatives for further enhancing performance are presented.*
NPS/DKL Location: FEDDOCS D 208.14/2:NPS-ME-00-001
Electronic access: http://handle.dtic.mil/100.2/ADA374332

Cary, Steven H. **Use of the Radio Shack TRS-80 Model 100 Computer for Combat Modeling**. Monterey, CA: Naval Postgraduate School; Springfield, VA: Available from National Technical Information Service, 1986. (ADA175327). 117 p.
Thesis (M.S. in Operations Research) Naval Postgraduate School, 1986.
Abstract: *The primary purpose of this thesis is to demonstrate some principles of combat modeling using programs for the Radio Shack TRS-80 Model 100 computer. In addition to the combat modeling, the thesis includes several utility programs for the M100 of interest to students of operations analysis. The combat modeling programs include an antisubmarine warfare (ASW) detection simulation, a Kalman filter, and a Lanchester differential equation simulation. The utility programs include a matrix algebra program, a numerical double integration program for zero degree of difficulty problems. The integration program is also written as a subroutine that can be included in other programs. The matrix algebra includes a simultaneous linear equation solving subroutine which can be used in other programs. All programs are written in M100 BASIC. Documentation includes an explanation of the input required, the output produced, and the components of each program, and sample problems. The chapter on geometric programming includes a tutorial on the mathematical basis for that technique. (Author)*
NPS/DKL Location: THESIS C27436

Chang, Peng-tso. **Evaluation and Improvement of the ASW System Evaluation Tool**. Monterey, CA: Naval Postgraduate School; Springfield, VA: Available from National Technical Information Service, 1992. (ADA248081). 144 p.
Thesis (M.S. in Computer Science) Naval Postgraduate School, March 1992.
Abstract: *The Antisubmarine Warfare System Evaluation Tool (ASSET) is a generic high-level antisubmarine warfare (ASW) modeling tool, designed to aid ASW personnel in the development and refinement of ASW top-level warfare requirements and the ASW Master Plan. The primary objective of this thesis is to analyze and implement the improvements suggested in previous evaluations of various sub-areas of ASSET. The glimpse rate model for submarine detection used in ASSET has been*

substituted with compound Lambda-Sigma jump model. There is a different target radiated frequency in each environmental region. Each target will have its own detection rate to reflect the differences in its operating characteristics. Multiple engagements between platforms are used to eliminate the limitations of interaction between opponent platforms. The glimpse rate model is used to determine detection opportunities of maritime patrol aircraft (MPA) and to approximate a continuous-looking sensor pattern. A different criterion of selecting search probability area (SPA) and MPA pairs using the ratio of MPA's time on-station over the SPA size was implemented. The feasibility of converting current ASSET code to CLOS was investigated. In addition, part of the code was converted to CLOS.
NPS/DKL Location: THESIS C37156

Channell, Ralph Norman. **Naval Model Priorities for RAND Strategy Assessment System**. Monterey, CA: Naval Postgraduate School; Springfield, VA: Available from National Technical Information Service, 1988. (NPS-56-88-023, ADA204851). 24 p.
Abstract: *This paper discusses specific accomplishments and problems regarding naval models in the RAND Strategy Assessment System (RSAS), and makes recommendations for priorities desired by the Naval Postgraduate School in the RAND program for FY 89. Emphasis is on the improvements needed to conduct research on the inter-relationship between warfare at sea and the war ashore. Partial Contents: Naval Warfare Priorities- Carrier Battle Group Improvements; Nuclear Forces; Antisubmarine Warfare; Strategic Lift - Sea; Ocean Surveillance; Amphibious Warfare; Mine Warfare; Logistics. (kr)*
NPS/DKL Location: FEDDOCS D 208.14/2:NPS-56-88-023

Chuan, Edmund Cheong Kong. **A Helicopter Submarine Search Game**. Monterey, CA: Naval Postgraduate School; Springfield, VA: Available from National Technical Information Service, 1988. (ADA201212). 56 p.
Thesis (M.S. in Operations Research) Naval Postgraduate School, Sept. 1988.
Abstract: *This thesis examines a two-person zero sum game where a submarine, after revealing his position by causing a flaming datum, is hunted by a helicopter which arrives on the scene after a time delay. Various helicopter and submarine strategies are explored and simulation runs are used to determine the detection probability (payoffs) for each combination of helicopter and submarine strategy. The value of the game (detection probability) with the related optimal strategies is then obtained using linear programming. A modified random search equation is also derived using probabilities of detection obtained from different combinations of parameters used in the game. Similar and related games are also discussed with emphasis on the differences in assumptions made and approaches taken in order to solve the problem. (sdw)*
NPS/DKL Location: THESIS C4783

Ciboci, John William. **Verification of Mazeika's Method of Thermocline Depth Prediction for the Northeast Pacific Ocean**. Monterey, CA: Naval Postgraduate School; Springfield, VA: Available from National Technical Information Service, 1966. (AD803657). 65 p.
Thesis (M.S. in Oceanography)--Naval Postgraduate School, 1966.
Abstract: *Mazeika's method for forecasting mixed-layer (thermocline) depth of the upper ocean layers is discussed along with a newer version of this method developed by James. Using Mazeika's method primarily, a verification for the Northeast Pacific Ocean was completed with data from Ocean Weather Stations PAPA (50N, 145W) and NOVEMBER (30N, 140W) and a point named MIDPOINT (40N, 140W). The results indicate Mazeika's method is successful at Station PAPA more than seventy five percent of the time during the heating season followed by a rapid decline as the cooling season begins. The method should be useful in the entire Central Subarctic Domain as described by John P. Tully. The method fails at NOVEMBER and MIDPOINT producing less than thirty percent success in prediction. James' version did not improve the results obtained at Station NOVEMBER. This failure appears to be due to the controlling parameters for processes in the Subtropic or Transitional oceanographic regions (which include NOVEMBER and MIDPOINT); these differ from parameters controlling oceanic processes in the*

Pacific Subarctic region (Station PAPA), which resemble those involved in the Atlantic region for which Mazeika's method was developed. Climatology data which can be used to obtain surface and 400-foot level temperature are also tested. The results indicate these data are very useful and accurate in determining the stability index required of Mazeika's method.
NPS/DKL Location: THESIS C4788

Coleman, Richard Lewis. **Barrier Search**. Monterey, CA: Naval Postgraduate School; Springfield, VA: Available from National Technical Information Service, 1974. (AD787368). 50 p.
Thesis (M.S. in Operations Research) Naval Postgraduate School, 1974.
Abstract: *Simulation generated probabilities for an Anti-Submarine Barrier composed of submarines are examined from a statistical viewpoint. A probabilistic model which is not generally considered to be applicable to this case is demonstrated to be statistically supported. The applicability of the model is justified probabilistically. A statistical estimating relationship is then developed to estimate the sole input parameter from submarine FIGURE-OF-MERIT.*
NPS/DKL Location: THESIS C5346

Conner, George W.; Mark A. Ehlersa and Kneale T. Marshall. **Countering Short Range Ballistic Missiles**. Monterey, CA: Naval Postgraduate School; Springfield, VA: Available from National Technical Information Service, 1993. (NPSOR-93-009, ADA261056). 36 p.
Abstract: *Concepts commonly found in ASW search are used to model the flow and detect mobile launchers for short range ballistic missiles. Emphasis is on detection and destruction of the launcher before launch. The benefit of prehostility intelligence and pre-missile-launch prosecution, the backbone of successful ASW, is revealed through the analysis of a circulation model which reflects the standard operations of a third world mobile missile launcher during hostilities. A decision model is constructed and analyzed to give insight into the development of pre-hostility intelligence policies.*
NPS/DKL Location: FEDDOCS D 208.14/2:NPS-OR-93-009

Conner, Hilton L. **Application of Color-Coding in Airborne Tactical Displays**. Monterey, CA: Naval Postgraduate School; Springfield, VA: Available from National Technical Information Service, 1979. (ADA067558). 96 p.
Thesis (M.S. in Systems Tech.) Naval Postgraduate School, 1979.
Abstract: *This thesis analyzes the operational environment and task variables of the Tactical Coordinator in the S-3A for possible application of color coding in the display symbology in the multi-Purpose display. Beginning with the ASW threat to the carrier force under the CV concept, the missions of the S-3A are presented. The roles, tasks and functions of the Tactical Coordinator are identified and form the basis for an analysis of the need of color in airborne displays. Current display design requirements and discrepancies in the S-3A are discussed as a basis for areas of color application. Color research recently conducted is reviewed with the results directed toward the symbology currently used in airborne displays. (Author)*
NPS/DKL Location: THESIS C7154

Coppens, Alan B. and James V. Sanders. **An Introduction to the Sonar Equations with Applications**. Monterey, CA: Naval Postgraduate School; Springfield, VA: Available from National Technical Information Service, 1976. (NPS-61SD76071, ADA030034). 124 p.
Abstract: *This report provides an introduction to the SONAR equations for those interested in underwater sound as applied to ASW but lacking either the mathematical background or the time for a more rigorous presentation. Earlier versions of these notes were developed for Continuing Education*

courses presented at Moffett Field, California, and Naval Torpedo Station, Washington. Additionally, these notes have been in demand for certain courses at the Naval Postgraduate School. While this is the text for these courses and should be supplemented by lectures, we have attempted to design the material so that it is reasonably self-explanatory, communicating many of the essential concepts without requiring extensive verbal amplification. The unusual format has been deliberately chosen to facilitate these goals, and our experiences in presenting these materials have seemed to justify this choice. It is assumed that the reader has some familiarity with trigonometric functions and either has or will develop with the aid of the appendix the facility of handling scientific notation and logarithmic operations. (Author)
NPS/DKL Location: GENERAL VK388 .C7

_____, Harvey Arnold Dahl and James V. Sanders. **An Introduction to the Sonar Equations with Applications** (Revised). Monterey, CA: Naval Postgraduate School; Springfield, VA: Available from National Technical Information Service, 1979. (NPS-61-79-006, ADA071133). 144 p.
Revision of Rept. no. NPS-61SD-76-071, ADA030034.
Abstract: *This report provides an introduction to the Sonar equations for those interested in underwater sound as applied to ASW but lacking either the mathematical background or the time for a more rigorous presentation. While this material should be supplemented by lectures or a study guide, we have attempted to design the material so that it is reasonably self-explanatory, communicating many of the essentials concepts without requiring extensive verbal amplification. The unusual format has been deliberately chosen to facilitate these goals, and our experiences in presenting these materials have seemed to justify this choice. It is assumed that the reader has some familiarity with trigonometric functions and either has or will develop will the aid of the appendix the facility of handling scientific notation and logarithmic operations.*
NPS/DKL Location: GENERAL VK388 .C71

Cox, Stephen A. **Satellite Applications to Acoustic Prediction Systems**. Monterey, CA: Naval Postgraduate School; Springfield, VA: Available from National Technical Information Service, 1982. (NPS68-82-005, ADA125027). 151 p.
Thesis (M.S. in Systems Tech.) Naval Postgraduate School, 1982.
Also issued as Naval Postgraduate School Technical Report NPS68-82-005.
Abstract: *Predicting the thermal structure of the oceans is of importance to the Naval tactician, logistician, or search and rescue coordinator. Understanding the structure of the oceans provides valuable insights to those who must utilize the oceanic environment effectively in their day to day operations. Today, recent information about an area is limited to point observations of single bathythermographs. Few models produce an accurate picture of the ocean environment that can be used for updating tactics to conform to a changing situation. Producing a reliable prediction of conditions for a large area, while using limited resources, is the basic objective of this paper. Satellite infrared imaging of the ocean surface has been used effectively to map sea surface temperature patterns. Such sea surface temperature patterns can be used, along with climatology, to identify subsurface thermal structure in an ocean area according to results of this study. More accurate inputs can be made to range dependent acoustic prediction models, thus improving the antisubmarine warfare environmental predictions available to fleet users.*
NPS/DKL Location: THESIS C75956

Coyle, Gary Leonard. **An Antisubmarine Warfare Training War Game**. Monterey, CA: Naval Postgraduate School; Springfield, VA: Available from National Technical Information Service, 1978. (ADA055539). 89 p.
Thesis (M.S. in Systems Tech.)--Naval Postgraduate School, 1978.
Abstract: *The employment of manual tactical gaming in a training environment is discussed, outlining the advantages and disadvantages of this method of training in the context of shipboard requirements. A*

two-sided, manual tactical war game is described and rules provided for play of the game. The utility of the game in assisting Commanding Officers and Training Officers in training junior officers using the Personnel Qualification Standard (PQS) System is described, with recommendations for further use of the game as a possible tactical training tool.
NPS/DKL Location: THESIS C777

Craigie, Kyle M. **Assessment of Atmospheric Influence on Surveillance Radar Performance in Littoral Zones.** Monterey, CA: Naval Postgraduate School; Springfield, VA: Available from National Technical Information Service, 1993. (ADA273045). 87 p.
Thesis (M.S. in Systems Engineering (Electronic Warfare)) Naval Postgraduate School, Sept. 1993.
Abstract: *Acoustic sensors, traditionally thought of as the mainstay of modem ASW's means of detection and localization, are rapidly becoming secondary in the littoral zones to active sensors such as radar. The coastal region has a dynamic meteorological environment dominated by surface and near-surface ducts which influence sea clutter. Accurate, timely description of the effects this changing environment has on sensor performance is mandatory for the ASW tactitian to utilize his sensors. The Radio Physics Optics (RPO) program and the Engineer's Refractive Effects Prediction System (EPEPS) are used to evaluate influence of a measured environment. Both prediction systems are then applied to a Gulf of Oman winter environmental profile with five generic radars operating parameters. EREPS is used to evaluate factors affecting Wallops Flight Facility Space and Ranging Radar (SPANDAR) detected sea clutter in the littoral zone off the United States East Coast.*
NPS/DKL Location: THESIS C7829

Crawford, Bruce W. **Analysis of the Effectiveness Evaluation Process for VP Antisubmarine Warfare Fleet Replacement Squadron Aircrew Training.** Monterey, CA: Naval Postgraduate School; Springfield, VA: Available from National Technical Information Service, 1982. (ADA125012). 118 p.
Thesis (M.S. in Systems Tech.)--Naval Postgraduate School, 1982.
Abstract: *During January 1979 VP-31, the West Cost P-3 Fleet Replacement Squadron, implemented an Instructional System Development based training program. Due to monetary, manpower, and time constraints, the evaluation phase of the new training program was not completely developed or implemented. This thesis examines the current status of the external evaluation portion of the new training program in an attempt to determine the feasibility of its completion and implementation. The external evaluation plan is related to the Interservice Procedures for Instructional System Development Model. From this analysis, a better understanding of the plan is gained and recommendations for an improved external evaluation program and training system are presented. (Author)*
NPS/DKL Location: THESIS C813

Crawford, Frederick Roberts. **Submarine Radiated Noise Far-Field Beam Patterns for Discrete Frequencies from Near-Field Measurements.** Monterey, CA: Naval Postgraduate School; Springfield, VA: Available from National Technical Information Service, 1975. (ADA020106). 47 p.
Thesis (M.S. in Eng. Acoustics)--Naval Postgraduate School, 1975.
Abstract: *A theoretical model was developed which can predict discrete frequency far-field radiation patterns of submerged submarines from near-field measurements. The model developed uses the Helmholtz integral equation and the assumptions of the DRL method of near-field measurements. The DRL working formula is further modified by using a plane surface of integration and restricting the far-field points of interest to a horizontal plane containing the source. These assumptions and restrictions lead to a mathematical solution of the Helmholtz equation which is in the form of a Fourier transform. Near-field measurements on a horn speaker in an anechoic chamber were taken and the far-field beam pattern*

predicted by the model developed, using a simple computer program containing a Fourier transform routine. Computed beam patterns were in satisfactory agreement with measured far-field beam patterns, errors being concentrated in the outer side lobes from the acoustic axis. Problems which would be encountered in applying this model to at sea acoustic measurements are discussed.
NPS/DKL Location: THESIS C815

Criswell, Philip W. **Evaluation of the ASSET Campaign Model in a Regional Antisubmarine Warfare Context**. Monterey, CA: Naval Postgraduate School; Springfield, VA: Available from National Technical Information Service, 1992. (ADA257625). 86 p.
Thesis (M.S. in Applied Science) Naval Postgraduate School, Sept. 1992.
Abstract: *This thesis looks at the Antisubmarine Warfare Systems Evaluation Tool (ASSET), written by Metron, Incorporated for OP-71, and how it relates to a current threat environment. ASSET is a campaign level ASW Monte-Carlo simulation intended for developing ASW Master Plans, top-level war fighting requirements (TLWRs), appraisals, and assessments. ASSET, delivered in 1990, was written from a U.S.- Soviet conflict perspective, and needs some restructuring to be able to provide conflict Measures of Effectiveness using platforms that are expected in a regional war. Included as suggested improvements are: a conventional submarine addition with major emphasis on power plant abilities and limitations, improvements to the surface group-submarine interaction; and improvements and additions to the methods of detection available to the objects in simulation.*
NPS/DKL Location: THESIS C8732

Curley, Richard Charles. **A Sonar Detection Model**. Monterey, CA: Naval Postgraduate School; Springfield, VA: Available from National Technical Information Service, 1971. (AD728580). 58 p.
Thesis (M.S. in Operations Research)--Naval Postgraduate School, 1971.
Abstract: *The report modifies an existing sonar range prediction model for the AN/SQS-23 in such a manner as to attain detection range data in consonance with exercises from which the original data was extracted. It also shows personnel a method for incorporating more than one ship in the model. This model will assist users in ascertaining the number of units required to perform a given antisubmarine task. (Author)*
NPS/DKL Location: THESIS C949

Daly, Daniel Gerald. **Limited Analysis of Some Nonacoustic Antisubmarine Warfare Systems**. Monterey, CA: Naval Postgraduate School; Springfield, VA: Available from National Technical Information Service, 1994. (ADA281747). 55 p.
Thesis (M.S. in Applied Science) Naval Postgraduate School, March 1994.
Abstract: *The problem of Anti-Submarine Warfare early in the next century is examined. Nonacoustic detection methods including magnetic anomaly detection, laser radar, and hydrodynamic detection are examined. A simple analysis of their relative effectiveness is made.*
NPS/DKL Location: THESIS D14395

Dassler, Dale M. Jr. **Naval Ship Utility: The Soviet Perspective**. Monterey, CA: Naval Postgraduate School; Springfield, VA: Available from National Technical Information Service, 1986. (ADA177464). 99 p.
Thesis (M.A. in National Security Affairs) Naval Postgraduate School, Decmeber 1986.
Abstract: *This thesis critically reviews twenty-two articles from the Soviet Naval Digest, Morskoy Sbornik, dealing with a wide spectrum of measures of effectiveness such as individual time efficiency, ASW search effectiveness, command decision efficiency, effectiveness of A SW training, measures of force control, and others. These Soviet measures of effectiveness are categorized by level of combat*

action. Although there is some question about the specific Soviet meaning of the translations, this thesis uses the translator's rendering of the basic units of Soviet Naval organization; individual, subunit (podrazdeleniye), unit (chast'), and force (soyedineniye). The levels of combat action above force (generally agreed to be named front (front), and TVD (Teatr Voyennykh Deystviy), are not included in this study. The articles illustrate the Soviet tendency to organize their operations research along the same lines as the units of naval organization and indicate that the most basic measure of naval ship utility is combat effectiveness. (Author)*
NPS/DKL Location: THESIS D16177

Davis, William Oris And Paul Howard Donaldson. **Signal Processing for Antisubmarine Warfare**. Monterey, CA: Naval Postgraduate School; Springfield, VA: Available from National Technical Information Service, 1975. (ADA007866). 153 p.
Thesis (M.S. in Systems Tech.)--Naval Postgraduate School, 1975.
Abstract: *Signal Processing for Antisubmarine Warfare is a short course in electrical signal processing fundamentals and their applications in the field of antisubmarine warfare. It contains an introduction to Fourier transforms and their properties, sampling and quantization, filters and bandwidth requirements, random signals and noise, and an introduction to four types of processing equipment; the DELTIC, energy detectors, correlation detectors, and beamformers. Course objectives are given in terms of specific questions which a person completing the course should be able to answer. The course text and illustrative material is contained in the appendix to the thesis. The course is designed to be presented in the Fleet to the personnel involved with the operation and employment of detection equipment to provide them a better understanding of the operations accomplished by their equipment and to develop in them a better appreciation of the problems and limitations associated with signal detection in the antisubmarine warfare environment.*
NPS/DKL Location: THESIS D174832

Deffenbaugh, Robert M. **Investigation of the Statistical Decision Process for Anti-Submarine Warfare Tactical Decisions**. Monterey, CA: Naval Postgraduate School; Springfield, VA: Available from National Technical Information Service, 1964. (AD481269). 54 p.
Thesis (M.S. in Operations Research)--Naval Postgraduate School, 1964.
Abstract: *An application of the statistical decision process to the problem of ASW tactical decision is investigated. The Bayesian decision process is utilized. The basic ASW decision problem with emphasis on the uncertainty aspect of a possible submarine contact is analyzed. A mechanism is developed to formally connect the general problem areas.*
NPS/DKL Location: THESIS D235

DeLateur, Robert Emmett. **Job Enrichment in Antisubmarine Warfare**. Monterey, CA: Naval Postgraduate School; Springfield, VA: Available from National Technical Information Service, 1979. (ADA071079). 173 p.
Thesis (M.S. in Management)--Naval Postgraduate School, 1979.
Abstract: *Since the all volunteer force came into being, retention of military personnel beyond their first enlistment has become an increasingly important problem, especially for the U.S. Navy. Yearly retention conferences have been held for the purpose of developing plans to reduce turnover. The results of the latest conference brought the focus of attention to better leadership and management training of U.S. Navy personnel. Among the techniques that deals with the problems of absenteeism and turnover is job enrichment. The main thrust of job enrichment is to increase retention by increasing work satisfaction. Job enrichment as a management technique focuses on the basics of employee motivation and work behaviors. It aids the managers in identifying the components which comprise a job, and enables them to determine satisfying components that can be enhanced and dissatisfying components that can be diminished or eliminated.*

NPS/DKL Location: THESIS D27

Devereux, Francis A. **SEATAG Extension**. Monterey, CA: Naval Postgraduate School; Springfield, VA: Available from National Technical Information Service, 1982. (ADA117560). 73 p.
Thesis (M.S. in Operations Research)--Naval Postgraduate School, 1982.
Abstract: *The SEATAG EXTENSION will revise and suggest optional and alternative rules for the game SEATAG: A Sea Control Tactical Analysis Game. Alternative rules are proposed for damage assessment, detection, classification, targeting, weapon's effectiveness, and ASMD close in weapon systems. Air-to-air combat tables have been revised to include the latest additions to both the United States and Soviet naval aircraft inventories. Optional rules will incorporate electronic warfare, battle damage repairs, miniature ship model combat, and use of the Tomahawk cruise missile.*
NPS/DKL Location: THESIS D4588

Diamandopoulos, Fanourios P. **Optimal Deployment Angles for the Air-Dropped Undersea Warfare Cable in Shallow Water**. Monterey, CA: Naval Postgraduate School; Springfield, VA: Available from National Technical Information Service, 1994. (ADA280484). 56 p.
Thesis (M.S. in Operations Research) Naval Postgraduate School, March 1994.
Abstract: *The Advanced Air Deployable Array (AdDA) is an air-dropped undersea warfare device for detection of an enemy submarine in the shallow waters region. Previous studies have introduced six tactical deployment methods by C-130 aircraft. This thesis addresses one of the methods, called Bound the Expanding Farthest-On Circle. Changes in deployment rules are suggested, and feasibility conditions identified. A model is developed showing how the isolation area where the submarine is to be contained, and the number of needed array segments, can be reduced. Also, as the main work of this study, the effective deployment angles for successive AdDA cable are determined for C-130 pilots. Today these cables, because of their advantages and great utility, can give unique solutions in shallow water tactical operations.*
NPS/DKL Location: THESIS D4818

Dougherty, William A. **Computer Simulation for the Comparison of ASW Vehicles**. Monterey, CA: Naval Postgraduate School; Springfield, VA: Available from National Technical Information Service, 1965. (AD4752903). 57 p.
Thesis (M.S. in Operations Research)--Naval Postgraduate School, 1965.
Abstract: *A method is developed to analyze and compare the effectiveness of ASW vehicles. The measure of effectiveness is the probability that the vehicle, after detecting a submarine with passive sensors, can transit to the contact area and re-establish contact with the submarine. A computer simulation is developed and an example using three hypothetical ASW vehicles is illustrated. (Author)*
NPS/DKL Location: THESIS D69

Dunston, Mary Cottrell. **Study of the Effect of Design Parameter Variation on Predicted Tilt-Rotor Aircraft Performance**. Monterey, CA: Naval Postgraduate School; Springfield, VA: Available from National Technical Information Service, 1988. (ADA204856). 103 p.
Thesis (M.S. in Aeronautical Engineering) Naval Postgraduate School, Dec. 1988.
Abstract: *There is currently little data available for trend analyses of tilt-rotor aircraft performance. This study analyzed the sensitivity of predicted tilt-rotor performance to variations in six design parameters: disk loading, tip speed, solidity, download, wing loading, and wing thickness ratio. Two mission profiles were analyzed: A combat search-and-rescue mission and an antisubmarine warfare mission. A tilt-rotor preliminary design code was used to perform computer simulations; and data available from independent*

tests completed by NASA and the military were encoded in the input data decks. Results were presented as graphs of performance aspects plotted against the parameters varied. Because the study was a trend analysis, no specific conclusions were drawn but a summary was made of the more significant results. It is hoped that the results of this project can serve as a guide to preliminary selection of design parameters for tilt-rotor configurations that would be suitable for a broad range of military and civil applications.
NPS/DKL Location: THESIS D7935

Eagle, James N. **Partial Evaluation of the Integrated Tactical Decision Aid (ITDA) System**. Monterey, CA: Naval Postgraduate School; Springfield, VA: Available from National Technical Information Service, 1986. (NPS55-86-026, ADA176296). 22 p.
Naval Postgraduate School (U.S.). Dept. of Operations Research.
Abstract: *The purpose of this technical report is twofold: 1)to critically examine the ITDA ASW barrier and area search models for mathematic accuracy and modelling reasonableness, and 2) to report on the use of ITDA programs on USS CARL VINSON (CVN 70) during a 16 day period of high intensity, exercise operations.*
NPS/DKL Location: FEDDOCS D 208.14/2:NPS-55-86-026

Ellis, John Richard. **Design and Evaluation of a Sonobuoy Ranging System**. Monterey, CA: Naval Postgraduate School; Springfield, VA: Available from National Technical Information Service, 1970. (AD710357). 141 p.
Thesis (M.S. in E.E.)--Naval Postgraduate School, 1970.
Abstract: *Airborne anti-submarine warfare operations require a means of precise tactical navigation relative to an air-dropped sonobuoy pattern. Advantages and disadvantages of navigational techniques which could be used to solve this problem are discussed. An analysis is made of a previously proposed method to solve this problem by sonobuoy ranging concepts. The design of a prototype sonobuoy ranging system is described, and a preliminary evaluation is made of the accuracy of the prototype system. (Author)*
NPS/DKL Location: THESIS E397

Farris, Christy Lee and Neil John Gaffney. **Consideration of the Carrier-Based Tactical Support Center Design**. Monterey, CA: Naval Postgraduate School; Springfield, VA: Available from National Technical Information Service, 1976. (ADA025434). 134 p.
Thesis (M.S. in Systems Tech.)--Naval Postgraduate School, 1976.
Abstract: *This joint thesis analyzes the carrier-based Tactical Support Center (CV-TSC) design from a human factors engineering view-point. Beginning with the ASW threat to the carrier force under the CV concept, a definition of the mission of the CV/TSC is presented. System functions are identified and developed into man-machine relationships of the CV/TSC. A comprehensive, albeit general, description of TSC components is included as part of the system analysis. Man's role, functions and tasks in the CV/TSC are identified and form the basis for alternatives to the current TSC display/control console. (Author)*
NPS/DKL Location: THESIS F234

Feustel, Richard D. **A Joint Campaign Analysis Approach to Antisubmarine Warfare Using a Circulation Model Template**. Monterey, CA: Naval Postgraduate School; Springfield, VA: Available from National Technical Information Service, 1996. (ADA322913). 133 p.
Thesis (M.S. in Operations Research) Naval Postgraduate School, Sept. 1996.
Abstract: *To enhance insight into a war at sea, a general, aggregated and highly flexible model of the ASW campaign is offered. This thesis provides a simple and usable circulation model template. The*

generality and simplicity of the model allows for 'jointization' of an ASW campaign by allowing the user to utilize other resources to define the force mix. The model is designed, first and foremost, to examine the change in the marginal effectiveness of friendly ASW forces due to changes in force level, mix, effectiveness, and employment strategies. The model is keyed to the interaction of a threat submarine with friendly ASW forces and merchant or military shipping. Specific features of the model provide for four unique attack regimes. The in port and operational regimes control friendly attacks on a daily basis while the outbound and inbound regimes control barriers by events. The campaign model is a deliverable product programmed using Borland(registered) Delphi for use in Microsoft Windows.
NPS/DKL Location: THESIS F276

Fisher, Rory H. **Variability and Sensitivity of Coupled Mixed Layer-Acoustic Model Systems**. Monterey, CA: Naval Postgraduate School; Springfield, VA: Available from National Technical Information Service, 1981. (NPS68-81-002, ADA114302). 124 p.
Thesis (M.S. in Systems Tech.)--Naval Postgraduate School, 1981.
Also issued as Naval Postgraduate School Technical Report NPS-68-81-002 in conjunction with Calvin R. Dunlap and Roland W. Garwood.
Abstract: *This study is the first reported analysis of coupled mixed layer-acoustic model systems. This analysis emphasizes the performance of the combined systems rather than the acoustic or ocean models separately. Acoustic variability of the coupled model systems was studied in terms of the median detection range (MDR). Synoptic time variations of MDR as a function of figure of merit, frequency and receiver depth were analyzed during the month of May 1980 at OWS 'Papa' in order to provide a better insight into the operational capabilities of model systems to accurately represent the actual oceanic variability. The results of this limited analysis revealed that the model systems displayed more day-to-day acoustic (MDR) variability than did direct environmental input(BT). The capability to accurately model the thermal structure was reviewed with the following results. No significant correlation was observed betwen the EOTS model and the actual BT mixed layer depths while there appeared to be a strong positive correlation between the ODT model (driven by atmospheric forcing) and the BT mixed layer depths. Moreover, a possible lag of two days was observed in the EOTS model mixed layer depth relative to the observed mixed layer depth time series.*
NPS/DKL Location: THESIS F4665

Forrest, R. N. and James N. Eagle. **Empirical Analysis of a Submarine Motion Model**. Monterey, CA: Naval Postgraduate School Antisubmarine Warfare Academic Group.; Springfield, VA: Available from National Technical Information Service, 1991. (NPS-AW-91-002, ADA242342). 93 p.
Abstract: *This report describes an empirical analysis of a motion model that has been used to generate random submarine tracks for an antisubmarine warfare tactical decision aid. The model describes a submarine's motion as a series of transitions between the square cells of a grid that covers a defined operating region. A 3 x 3 transition matrix is associated with each cell of the grid which determines the submarine's transitions from a cell. The set of transition matrices define a Markov process. Despite its discrete nature, this Markov track generating process has been called a diffusion process in antisubmarine warfare tactical decision aid literature. The transition matrices are determined by tracks generated by an auxiliary stochastic process that is presumed to be of higher fidelity but more costly to implement than the Markov process.*
NPS/DKL Location: FEDDOCS D 208.14/2:NPS-AW-91-002

Fountoulakis, Radamanthis P. **Oceanographic and Acoustical Survey of the East Ionian Sea**. Monterey, CA: Naval Postgraduate School; Springfield, VA: Available from National Technical Information Service, 1990. (ADA241360). 97 p.
Thesis (M.S. in Engineering Acoustics) Naval Postgraduate School, Sept. 1990.

Abstract: *A study was conducted in an area off the Hellenic west coast to examine the spatial and time variability of various oceanic parameters, with special emphasis on those effecting ASW operations. Propagation loss runs were conducted using PE and RAYMODE models. The reactions of both models to different bottom morphology and sound speed profiles (seasons) were examined. Between the two models, the PE model was found to be closer to reality than RAYMODE. Results suggest that the application of these models can improve the understanding of sound propagation in the Hellenic seas. The bottom modeling program, BLUG, appears to need improvement.*
NPS/DKL Location: THESIS F663525

Franken, Jeroen. **Tactical Application of Coastal Acoustic Tomography**. Monterey, CA: Naval Postgraduate School; Springfield, VA: Available from National Technical Information Service, 1994. (ADA290219). 66 p.
Thesis (M.S. in Engineering Acoustics) Naval Postgraduate School, Dec. 1994.
Abstract: *The objective of this study is to investigate the utility of acoustic tomography for performance assessment of a generic low frequency active sonar system. The performance of the sonar is simulated using tomography-derived sound speed data versus a range independent ocean model. The ocean environment used in the simulation is 159 tomographic snapshots of the Barents Sea Polar Front, taken every 5 minutes in August 1992. The modeled sonar system consists of a 1000 Hz source with a source level of 205.5 dB and a towed horizontal array of hydrophones. The system is derived from unclassified parameters of ATAS (Active Towed Array Sonar), built by Thomson Sintra ASM and British Aerospace SEMA, and the experimental ALF sonar, designed by FEL-TNO (the Netherlands) and built by Thomson Sintra ASM. The tomographic images over a range of 26 km provide a realistic ocean in which system performance is assessed. This study used a broadband, coupled normal mode, propagation model and assumed a noise-limited condition. The probability of detection calculated as a function of time for 13 hours is compared with that estimated using a range- and time-independent assumption. The utility of coastal acoustic tomography for tactical applications is discussed. (AN).*
NPS/DKL Location: THESIS F782313

French, Thomas Penn. **Sonobuoy Location**. Monterey, CA: Naval Postgraduate School; Springfield, VA: Available from National Technical Information Service, 1970. (AD713077). 77 p.
Thesis (M.S. in Operations Research)--Naval Postgraduate School, 1970
Abstract: *In airborne anti-submarine warfare operations there is a critical requirement for maintaining an accurate relative plot of the sonobuoys with respect to the aircraft. This study proposed a method for locating sonobuoys in a pattern using aircraft-to-buoy slant range information. The method did not use triangulation procedures and attempted to minimize the restrictions placed on the aircraft. The study showed the feasibility of the proposed methodology and the approximate errors to be encountered. (Author)*
NPS/DKL Location: THESIS F86

Frost, Mark Douglas. **An ASW Campaign Model**. Monterey, CA: Naval Postgraduate School; Springfield, VA: Available from National Technical Information Service, 1980. (ADA094570). 122 p.
Thesis (M.S. in Operations Research)--Naval Postgraduate School, 1980
Abstract: *To enhance insight into a war at sea, a large-scale, aggregated, and highly flexible model of the ASW campaign is offered. The model was designed, first and foremost, to examine the change in the marginal effectiveness of friendly ASW forces due to changes of force level, force mix, and force employment strategies. The model is keyed to the interaction of the threat submarine force with friendly ASW forces and merchant or military shipping. Specific features of the model provide for threat deployment options, allocation of friendly forces, attrition to threat and friendly forces, aggregation of friendly ASW force performance, sensitivity to force levels, deployment of submarines within 'wolfpacks'*

and coordinated barrier stations, and parametric treatment of other warfare area effectiveness. The campaign model has been programmed in the APL/360 language for use on an IBM 360-67 computer. (Author)
NPS/DKL Location: THESIS F8974

Gangsaas, Aasgeir. **Barrier Search Model Using Active Bistatic Sonar to Protect a Channel**. Naval Postgraduate School, Monterey, CA. Mar 1993. (ADA264910). 55 p.
Thesis (M.S. in Operations Research) Naval Postgraduate School, March 1993.
Abstract: *Advances in nuclear and diesel-electric submarine technology have reduced the effectiveness of passive means of detection. The United States is faced with a multipolar threat in part due to the proliferation to Third World nations of advanced diesel-electric submarines. The use of active sonar must be explored to gain back the detection advantage the United States submarine force has enjoyed in the past. The use of bistatic sonar reduces the counter-detection threat resulting from active sonar.*
NPS/DKL Location: THESIS G1436

Garcia, Arturo M. **Model for Evaluating HS/HSL Community Consolidation**.
Monterey, CA: Naval Postgraduate School; Springfield, VA: Available from National Technical Information Service, 1995. (ADA296694). 64 p.
Thesis (M.S. in Operations Research) Naval Postgraduate School, March 1995.
Abstract: *A methodology is developed to assist in the evaluation of competing proposals for HS/HSL consolidation. Six criteria are developed to allow a quantitative measure of critical personnel, cost, and operational issues. The criteria are incorporated into a spreadsheet model that can evaluate five options simultaneously. Decision maker participation is required to derive a set of weights that represent the relative importance attributed to each criteria. Five options currently under consideration as candidates for consolidation are examined. Analysis is conducted to determine the effect different weight values have on the optimal solution. A sample run of the model is conducted to demonstrate its use. (KAR) P. 2.*
NPS/DKL Location: THESIS G18056

Garrett, Roger A. and Albert P. Herrlinger **Automatic Antisubmarine Track Reconstruction**. Monterey, CA: Naval Postgraduate School; Springfield, VA: Available from National Technical Information Service, 1966. (AD803652). 73 p.
Distribution Limitation now Removed.
Thesis (M.S. in Operations Research)--Naval Postgraduate School, 1966.
Abstract: *The relative distance between various Antisubmarine Warfare vehicles is an important parameter for the evaluation of sensor equipment. This thesis presents a computer model, utilizing a time dependent error navigation system, Omega, for the determination of this relative distance parameter. A methodology for predetermining a relative distance error distribution for a specific area by simulating various Antisubmarine Warfare vehicle tracks from known navigational errors has been established. Attempts to smooth estimated tracks of Antisubmarine Warfare vehicles which receive the same time dependent navigational information simultaneously was found not to be advisable as this disturbs the correlation of navigation error between the vehicles. (Author)*
NPS/DKL Location: THESIS G205

Gonzalez, Debra L. **Relationship between the Armed Services Vocational Aptitude Battery (ASVAB) and Performance in Fleet Readiness Squadron (FRS) Acoustic Operator Training**. Monterey, CA: Naval Postgraduate School; Springfield, VA: Available from National Technical Information Service, 1986. (ADA168516). 72 p.
Thesis (M.S. in Management) Naval Postgraduate School, 1986.
Abstract: *The purpose of this thesis is to determine whether the Armed Services Vocational Aptitude Battery (ASVAB) scores, specifically the composite of ASVAB subtests (AR + 2MK + GS) used to predict*

eligibility for formal training in the Aviation Antisubmarine Warfare Operator (AW) rating, can actually predict the success or failure of enlisted personnel attempting the P-3 fleet readiness squadron (FRS) Acoustic Operator syllabus. This was accomplished by computing a Pearson Product - Moment Correlation Coefficient, corrected for restriction in range, to determine the correlation between ASVAB subtest and composite scores and sucess or failure in the FRS syllabus. The results indicate that ASVAB scores are only slightly predictive of performance.
NPS/DKL Location: THESIS G5477

Goodman, Craig W. **An Investigation of the APAIR Acoustic Detection Model**. Monterey, CA: Naval Postgraduate School; Springfield, VA: Available from National Technical Information Service, 1989. (ADA218967). 51 p.
Thesis (M.S. in Operations Research), Naval Postgraduate School, Sept. 1989.
Abstract: *The subject of this thesis is an investigation of the effect of using the Lambda-Sigma jump process in the acoustic detection component of APAIR. A computer simulation was developed which is similiar to the sonobuoy field versus submarine engagement model found in apair, the Navy's general ASW model. This simulation was then modified to incorporate the Lambda-Sigma jump process and the effect of this modification is discussed. In order to check the structural validity of the simulation models, Results that were obtained by using them are compared to results that were obtained by using an analytical model called the random search model.*
NPS/DKL Location: THESIS G566

Goodwin, Francis R. **NAVTAG (Naval Tactical Game) System and Its Modification to Include the SH-60B Helicopter**. Monterey, CA: Naval Postgraduate School; Springfield, VA: Available from National Technical Information Service, 1984. (ADA152004). 77 p.
Thesis (M.S. in Information Systems)--Naval Postgraduate School, 1984.
Abstract: *The Naval Tactical Game (NAVTAG) Training Systems are to become the standard war gaming computers in fleet use to train Surface Warfare Officers in tactical operations. As modern weapons platforms are developed, they need to be modeled into NAVTAG in order that they might be included in applicable atsea engagements. In support of this objective, the SH-60B (SEAHAWK) Anti-Submarine Warfare Helicopter, which is currently not supported by NAVTAG, is incorporated into the NAVTAG system. The SH-60B is incorporated into the NAVTAG System with the full range of functions that are enjoyed by other aircraft modeled in NAVTAG. Using NAVTAG the SH-60B is tested in an AntiSubmarine Warfare (ASW) scenario developed to test its capabilities against a Soviet submarine. For comparison and testing purposes the SH-60B is also compared to the SH-2F helicopter previously modeled in NAVTAG. Both helicopters have comparable mission objectives and tactics. This is a research project to determine if NAVTAG can be modified in a research environment and with what degree of difficulty this may be accomplished. This in no way is meant to modify the Standard NAVTAG Systems that have been distributed to fleet units without the consent of the Program Manager. Originator supplied keywords include: Wargaming, NAVTAG, SH-60B helicopter.*
NPS/DKL Location: THESIS G5725

Gostlow, GA, **TAO (Tactical Action Officer) ASW (Anti-Submarine Warfare) Expert System Prototype.** Monterey, CA: Naval Postgraduate School; Springfield, VA: Available from National Technical Information Service, 1986. (ADA175258). 66 p.
Thesis (M.S. in Systems Tech.) Naval Postgraduate School, 1986.
Abstract: *The expertise required by Tactical Action Officers in a modern Anti-Submarine warfare environment of complex weaponry, minimal reaction time and ardous conditions at sea necessitate training and experience that is both exhaustive and progressive. For these officers to be effective in making accurate and timely decisions so as to effect the most appropriate responses, they must have ready access to current tactical doctrine and system performance statistics. In time of war there is no time to allow a junior Tactical Action Officer to progress to a level of competency: he must be a reliable,*

capable, fully functional warfare team member at the outset of his tour. This thesis presents a prototype Artificial Intelligence model of the TAO ASW decision making process using an expert system development tool run on a microcomputer, to train fledgling TAO's with an outlook to the potential development and capability of an operational expert system. (Author)
NPS/DKL Location: THESIS G585

Gowans, George Keith. **Variation in Thermal Structure and Geostrophic Current Between Alaska and Hawaii Determined from Synoptic Space Sections**. Monterey, CA: Naval Postgraduate School; Springfield, VA: Available from National Technical Information Service, 1969. (AD706077). 73 p.
Thesis (M.S. in Oceanography)--Naval Postgraduate School, 1969.
Abstract: *Five synoptic space sections along 158W longitude between Hawaii and the Aleutian Islands were developed from data collected by airborne expendable bathythermographs during experiment PARKA, a research project sponsored by the U. S. Navy in 1968. The sections are examined for spacial and temporal variation in thermal structure and geostrophic surface velocity. Two recently developed analysis techniques are employed. Denner's T-S gradient method, wherein thermal and haline contributions to total geostrophic velocity are distinguishable, expedites calculations and results in velocity fields comparable to those developed by the dynamic method. Thermocline parameters are developed using Boston's objective definition of the thermocline, a statistical curve-fitting technique which develops the notion of a Gaussain thermocline. Gross features of thermal structure remain fairly consistent during the heating season; however, thermal fronts are observed to vary in time and space. The distribution of isothermal lines with latitude suggests the possibility of a Taylor-column effect slightly north of Hawaii. (Author)*
NPS/DKL Location: THESIS G618

Grant, Gary Maxwell. **Preliminary Investigation of a Proposed Sonobuoy Ranging System**. Monterey, CA: Naval Postgraduate School; Springfield, VA: Available from National Technical Information Service, 1969. (AD705497). 68 p.
Thesis (M.S. in E.E.)--Naval Postgraduate School, 1969.
Abstract: *The ability to locate and determine the position of an ASW sonobuoy is an essential part of airborne anti-submarine operations. Present methods restrict the parent aircraft's operational capability and yield only marginal data. State-of-the-art frequency control makes it possible to range sonobuoys accurately with radio signals. Sonobuoy position can then be determined by combining this range data with other available information. A system is proposed to both free the parent aircraft from present restrictions and to increase the accuracy of the position information. (Author)*
NPS/DKL Location: THESIS G6578

Griffin, Charles Donald. **A Computer Simulation of Asw Interactions**. Monterey, CA: Naval Postgraduate School; Springfield, VA: Available from National Technical Information Service, 1971. (AD722581). 83 p.
Thesis (M.S. in Operations Research)--Naval Postgraduate School, 1971.
Abstract: *A model presented in this thesis is a computer simulation model of ASW interactions between a formation of high value group ships, protected by some screening ships, and some penetrating submarines. The model is designed for use as an aid in improving the ability of a proposed screening tactic in the detection of a penetrating submarine. A systematic procedure to improve a screen's effectiveness against a known submarine threat is demonstrated, and an example problem is worked using this procedure. (Author)*
NPS/DKL Location: THESIS G768

Griggs, William Clifton and John Arthur Thompson. **Methodology for the Conversion of Tactical Board Games to Computer Assisted War Games**. Monterey, CA: Naval

Postgraduate School; Springfield, VA: Available from National Technical Information Service, 1980. (ADA094617). 247 p.
Thesis (M.S. in Operations Research)--Naval Postgraduate School, 1981.
Abstract: *As part of the development of the capabilities of the Command, Control and Communications (C3) Laboratory at the Naval Postgraduate School, there was a need for an in-house computer assisted tactical war game. The objective of this thesis was to satisfy that need. An anti-submarine warfare board game, 'Up Scope', which was designed, developed and produced by Simulations Productions, Inc. of New York, was used as the model. This thesis provides an interactive computer assisted anti-submarine warfare war game called 'Up Scope' which is written in FORTRAN. This thesis also develops a framework for any future computer adaptation of a tactical board game, details a players manual and gives full documentation of the computer programming. Program listings, a sample game and several tactical scenarios are also included. (Author)*
NPS/DKL Location: THESIS G83225

Guillory, Theodore. Canada: **The decision to procure nuclear attack submarines and its significance for NATO**. Naval Postgraduate School; Springfield, VA: Available from National Technical Information Service, 1988. (ADA201669). 88 p.
Thesis (M.A. in National Security Affairs), Naval Postgraduate School, Sept. 1988.
Abstract: *In June 1987 the Canadian government announced plans to procure 10 to 12 nuclear attack submarines (SSNs). The evidence suggests that, for some Canadians, a primary purpose for this submarine program may not be to enhance the security of NATO, but instead to assert Canada's sovereignty, principally against the United States, in the Arctic region. The thesis discuss this decision and its possible implications for the security of North America and NATO. It is argued that the United States must continue to have unimpeded access to the Arctic region to counter the ever increasing threat posed by Soviet nuclear ballistic missile submarines (SSBNs). Finally the thesis suggests a possible solution to the current sovereignty debate and a potential strategy for employing these SNN to enhance the security of North America and NATO as a whole.*
NPS/DKL Location: THESIS G86344

Gunn, Lee Fredric and Peter Tocha. **An Algorithm for the Solution of Concave-Convex Games**. Monterey, CA: Naval Postgraduate School; Springfield, VA: Available from National Technical Information Service, 1971. (AD735309). 78 p.
Thesis (M.S. in Operations Research)--Naval Postgraduate School, 1971.
Abstract: *The thesis discusses the solution of concave-convex games. An algorithm is developed, a computer program written and applied to an anti-submarine warfare force allocation problem as an illustration. Techniques for handling concave-convex problems in high dimensions are included. (Author)*
NPS/DKL Location: THESIS G8647

Hall, Charles R. **A Game Theoretical Approach to the Contact Area Helicopter ASW Problem**. Monterey, CA: Naval Postgraduate School; Springfield, VA: Available from National Technical Information Service, 1964. (AD480740). 70 p.
Thesis (M.S. in Operations Research)--Naval Postgraduate School, 1964.
Abstract: *The helicopter contact area search problem is studied. The analysis used to derive current search plans is presented, and an error in this analysis is demonstrated. As a result of this examination a new model is constructed for investigating the problem. Game theory techniques are applied to the model andthe methodology required to derive optimal strategies is illustrated. Although no complete optimal strategies are derived, it is possible to derive such strategies by applying computer techniques to the results of this study. Such strategies could then be applied in the fleet.*
NPS/DKL Location: THESIS H147

Hammon, Colin P. **The Design and Construction of a Computer Simulation.** Monterey, CA: Naval Postgraduate School; Springfield, VA: Available from National Technical Information Service, 1964. (AD615652). 158 p.
Thesis (M.S. in Operations Research)--Naval Postgraduate School, 1964.
Abstract: *A procedural approach to the design and implementation of a computer simulation model is presented, with a simulation model description and computer code. The discussion and example are intended as reference material for the computer science and computer war gaming courses offered at the United States Naval Postgraduate School. The sample computer simulation was designed for the statistical analysis of the comparative effectiveness of different ASW helicopter search tactics in a variety of tactical and physical environments. This simulation has available a wide range of input parameters and is applicable to all ASW helicopters and any search plan employing ten helicopters or less. The accompanying FORTRAN computer code was written for the CDC 1604 computer, and is adaptable to the IBM 7090-94 by the inclusion of the appropriate control cards and random number generator. (Author)*
NPS/DKL Location: THESIS H175

Harvey, Lawrence Michael. **Application of the Sonar Equations to Bistatic Echo-Ranging.** Monterey, CA: Naval Postgraduate School; Springfield, VA: Available from National Technical Information Service, 1982. (ADA115731). 87 p.
Thesis (M.S. in Systems Tech.)--Naval Postgraduate School, 1982.
Abstract: *The thesis explores the phenomena unique to echo-ranging with a source widely separated from the receiver. In an asset-austere era of antisubmarine warfare, this technique serves as a tactical advantage, particularly in the passive tracking of a submarine. Particular emphasis is placed on the terms of the sonar equation most affected by the bistatic geometry: Reverberation level and target strength. The research is particularly applicable to ongoing NATO and Naval Laboratory work involving the bistatic concept in array design and for use with surface escorts in conjunction with friendly submarines.*
NPS/DKL Location: THESIS H296285

Harvey, Phillip Ivan. **An Analysis of Environmental Data for Use in Updating Low Frequency Propagation Loss Forecasts**. Monterey, CA: Naval Postgraduate School; Springfield, VA: Available from National Technical Information Service, 1972. (AD757682). 181 p.
Thesis (M.S. in Oceanography)--Naval Postgraduate School, 1972.
Abstract: *An acoustic model for low frequency (100-2400 HZ) propagation loss within a surface duct is examined. An analysis of the sensitivity of this model as a function of the governing environmental parameters is performed. The results of this analysis show that the frequency and mixed layer depth are influential over a wide range of environmental conditions and that the below layer thermal gradient becomes important at low frequencies when the layer depth is relatively shallow. Under certain conditions, a change in below layer thermal gradient of 2F/100 FT has the same resultant effect as a 25 FT change in the mixed layer depth. (Author Modified Abstract)*
NPS/DKL Location: THESIS H29629

Harvey, Robert Frank. **A Computer War Game for Determining the Cost-Effectiveness of Active Sonobuoys.** Monterey, CA: Naval Postgraduate School; Springfield, VA: Available from National Technical Information Service, 1967. (AD824131). 79 p.
Thesis (M.S. in Operations Research)--Naval Postgraduate School, 1967.
Abstract: *In order to evaluate the relative effectiveness of different active non-directional sonobuoys, a computer war game is developed. One submarine, employing one evasion tactic, is opposed by one helicopter, using five prosecution tactics. The tactic of the helicopter prior to the initial detection of the submarine is seen to be critical, and this simulation aids in determining an optimum tactic. A cost-*

effectiveness model to use data from this simulation is developed. An example, using hypothetical but realistic data, is presented to illustrate methods of determining the cost-effectiveness of each sonobuoy type when used with its optimum tactic. (Author)
NPS/DKL Location: THESIS H29634

Haskell, RD. **Vectored Intercept Model (VIM) - An Open Ocean Submarine Versus Submarine Search and Detection Simulation.** Monterey, CA: Naval Postgraduate School; Springfield, VA: Available from National Technical Information Service, 1972. (AD763672). 191 p.
Thesis (M.S. in Operations Research) Naval Postgraduate School, 1972.
Abstract: *A simulation model for the open ocean submarine versus submarine search and detection problem is presented. The objective of the simulation is to estimate the probability with which a nuclear powered attack submarine will achieve sonar detection of a nuclear powered transiting submarine using a search plan based on external intelligence. A detailed description of the model and its use are included along with a typical analysis. (Author)*
NPS/DKL Location: MICROFORM AD763672

Haworth, AG. **A War-Game Analysis of an Undeveloped Submarine Detection System.** Monterey, CA: Naval Postgraduate School; Springfield, VA: Available from National Technical Information Service, 1965. (AD475240). 30 p.
Thesis (M.S. in Operations Research)--Naval Postgraduate School, 1965.
Abstract: *The evaluation of a proposed submarine detection system by computer war gaming techniques is illustrated by a hypothetical example. A scenerio is chosen, tactics and policies established, and the tactical simulation conducted. From the results of the simulation, minimum specifications for the system to attain a given level of effectiveness are drawn. Finally, a scale is made for comparison of this system with similar systems in terms of cost per day per mile of barrier. (Author)*
NPS/DKL Location: THESIS H353

Hayes, James A. **SSBN Survivability: A Time for Confidence-Building Measures.** Monterey, CA: Naval Postgraduate School; Springfield, VA: Available from National Technical Information Service, 1982. (ADA126701). 134 p.
Thesis (M.A. in National Security Affairs)--Naval Postgraduate School, 1982.
Abstract: *This thesis will focus on the sea-based legs of the American and Soviet triads, examining a series of confidence-building measures (CBMs) that may be considered during the Strategic Arms Reduction Talks (START) that are underway in Geneva. Some proponents have argued that these CBMs, if implemented, would strengthen each side's belief in the invulnerability of nuclear-powered, ballistic missile launching submarines (SSBNs), thereby increasing strategic stability. These proposals seek to increase confidence in SSBN survivability by managing both the employment of anti-submarine warfare (ASW) forces and the development of technology that could be specifically directed against SSBNs. This thesis will consider the possible effects that five different CBMs could have on U.S. perceptions of SSBN survivability. These changes in perception will be measured against the costs that might be exacted in other areas (e.g., tactical anti-submarine warfare) by agreeing to the CBMs.*
NPS/DKL Location: THESIS H403

Hilzer, Ralph Conrad. **NANCEE: an Approach to Barrier Sonobuoy Pattern Optimization.** Monterey, CA: Naval Postgraduate School; Springfield, VA: Available from National Technical Information Service, 1974. (AD783809). 97 p.
Thesis (M.S. in Comp. Sci.)--Naval Postgraduate School, 1974.
Abstract: *NANCEE is a computer simulation program which uses convolution (or meeting) probabilities to determine which barrier type of sonobuoy pattern has the highest probability of detection for a*

transiting nuclear submarine. The program assumes that the optimum barrier is a straight-line one, two, or three row sonobuoy pattern containing not more than 48 sonobuoys. The barrier is centered on the submarine's expected line of transit, oriented perpendicular to the submarine's course, and placed far enough ahead of the submarine's position that all pattern sonobuoys are in the water and being monitored before the submarine enters the detection range of the sonobuoy pattern. If more than one barrier is found to have the highest probability of detection, the one with the least number of sonobuoys is selected as optimum. (Modified author abstract)
NPS/DKL Location: THESIS H534

Huff, Michael Thomas. **Comparison of ASTRAL and RAYMODE Propagation Loss Models and Their Use in Air ASW Platforms**. Monterey, CA: Naval Postgraduate School; Springfield, VA: Available from National Technical Information Service, 1993. (ADA275015). 64 p.
Thesis (M.S. in Applied Science) Naval Postgraduate School, Sept. 1993.
Abstract: *The development of new decision support systems for Antisubmarine warfare will entail the installation of propagation loss models on ASW aircraft. The decision to put either a range dependent or range independent model in the system will affect the predicted ranges, the overall probability of detection, and the computation time. Comparisons of the range dependent ASTRAL and range independent RAYMODE propagation loss models were made in the Eastern Mediterranean, the Gulf of Oman and the South China Sea for eight source/receiver/frequency combinations. Computation time differences between the two models were not significant at either of the source frequencies (50 Hz or 400 Hz). RAYMODE showed much better correlation with the split step PE model which was used as a standard. The ASTRAL model often predicted lower transmission losses than either RAYMODE or PE. For the short detection ranges normally encountered in air ASW the more complex range dependent models are not necessary. The RAYMODE model or a comparable range independent model will provide adequate propagation loss predictions. ASTRAL, RAYMODE, Propagation loss models.*
NPS/DKL Location: THESIS H85365

Hughes, David L. and Jack D. Sirrine. **Investigation into the Concepts of Flight Crew Utilization**. Monterey, CA: Naval Postgraduate School; Springfield, VA: Available from National Technical Information Service, 1965. (AD4752853). 40 p.
Research paper (M.S. in Management)--Naval Postgraduate School, 1965.
Abstract: *The contribution of a patrol squadron to the total ASW readiness is dependent in part upon the effectiveness with which allocated or assigned personnel are utilized. In the hope of increasing that effectiveness, organizationally induced problems encountered in the allocation and utilization of flight crew personnel were investigated. A possible solution is offered which involves minor organizational changes, establishment of a flight crew rate, and a functional application of manning level constraints.*
NPS/DKL Location: THESIS H855

Ingold, Barry W. Key Feature Identification from Image Profile Segments Using a High Frequency Sonar. Monterey, CA: Naval Postgraduate School; Springfield, VA: Available from National Technical Information Service, 1992. (ADA261926). 67 p.
Thesis (M.S. in Mechanical Engineering)--Naval Postgraduate School, Dec. 1992.
Abstract: *Many avenues have been explored to allow recognition of underwater objects by a sensing system on an Autonomous Underwater Vehicle (AUV). In particular, this research analyzes the precision with which a Tritech ST1000 high resolution imaging sonar system allows the extraction of linear features from its perceived environment. The linear extraction algorithm, as well as acceptance criteria for individual sonar returns are developed. Test results showing the actual sonar data and the sonar's perceived environment are presented. Additionally, position of the sonar relative to the perceived image is determined based on the identification of key points in the scene.*
NPS/DKL Location: THESIS I4525

Jaeger, Larry Ernest. **Computer Program for Solving the Parabolic Equation Using an Implicit Finite-Difference Solution Method and Incorporating Exact Interface Conditions**. Monterey, CA: Naval Postgraduate School; Springfield, VA: Available from National Technical Information Service, 1983. (ADA135325). 113 p.
Thesis (M.S. in Physics)--Naval Postgraduate School, 1983.
Abstract: *An Implicit Finite-Difference (IFD) computer program that incorporates exact interface conditions has been developed for solving the parabolic equation. The model preserves continuity of pressure and continuity of the normal component of particle velocity at the interface between media having different sound speeds and densities. Interface conditions are preserved for horizontal and sloping interfaces along a user-specified bottom profile. Test cases are included to demonstrate the use of the model.*
NPS/DKL Location: THESIS J2382

Jenkins, Alan K. **Modeling Convergence Zone Gain on MS-DOS Based Personal Computers**. Monterey, CA: Naval Postgraduate School; Springfield, VA: Available from National Technical Information Service, 1991. (ADA243252). 84 p.
Thesis (M.S. in Applied Science) Naval Postgraduate School, Dec. 1991.
Abstract: *The models for determining convergence zone Gain (G) were developed using a linearized Sound Speed Profile (SSP) and applying ray tracing theory. The SSP was divided into three cases; bilinear, bilinear with isospeed layer, and bilinear with mixed layer. Two analytical solutions were developed using Taylor series and binomial series expansions to determine G, one for the bilinear and bilinear with isospeed layer, and the other for the bilinear with a mixed layer. The solutions for G are exclusively a function of the SSP gradients. Each solution was compared to the solutions from ray tracing and the solutions from the Integrated Carrier Antisubmarine Warfare Prediction System (ICAPS) (which runs on mainframe computers and requires more data in addition to the SSP). When the SSP's were not too unusual, the solutions for G were fairly close when compared to ray tracing and ICAPS.*
NPS/DKL Location: THESIS J452

Joyce, Michael D. **Quadruplet Expansion of the Acoustic Pressure Field in a Wedge Shaped Ocean.** Monterey, CA: Naval Postgraduate School; Springfield, VA: Available from National Technical Information Service, 1993. (ADA275121). 40 p.
Thesis (M.S. in Applied Physics) Naval Postgraduate School, Sept. 1993.
Abstract: *In a wedge shaped ocean, the method of images is used to develop an analytical approximation of the acoustic pressure field. Contemporary work develops acoustic doublets from a combination of the source and surface reflection image using simple dipole theory. The method of images is then used to sum the dipole images. This thesis matches dipole pairs to achieve a quadruplet expansion. A computer program using the derived quadruplet equation is then created to verify the results by comparing them with the URTEXT program. Method of images, Quadruplet expansion.*
NPS/DKL Location: THESIS J8465

Kelley, Robert D. Verification of Mcdonnell'S Mixed-Layer Depth Forecasting Model. Monterey, CA: Naval Postgraduate School; Springfield, VA: Available from National Technical Information Service, 1966. (AD805629). 73 p.
Thesis (M.S. in Physical Oceanography)--Naval Postgraduate School, 1966.
Abstract: *A model based on Kitaigorodsky's application of similarity theory and modified by McDonnell to forecast the mixed-layer depth was studied. The model applies during the warming season and is based on the theory of similarity. The parameters involved in the model were determined from bathythermograph data recorded at Ocean Weather Stations November (latitude 30N, longitude 140W) and Bravo (latitude 56 30N, longitude 51W). Parameters were evaluated daily and grouped by months.*

Both seasonal and transitional MLD situations were treated. From these parameters, the form of the dimensionless function P(N), claimed by Kitaigorodsky to be universal, was determined by least squares fit to be best approximated by a second order polynomial. Forecasting equations involving P(N) were developed for each month and tested with data from the following years for both OWS ships. There is general agreement between the observed MLD and that found from the prediction equation based on the last year's P(N) for the same month and location. Month-to-month and spatial differences in P(N) cast considerable doubt on its universality, at least as determined by the parameters as currently defined.
NPS/DKL Location: THESIS K272

Kern, Deborah R. **Design and Implementation of the Acoustic Database and Acoustic Trainer for ARGOS**. Monterey, CA: Naval Postgraduate School; Springfield, VA: Available from National Technical Information Service, 1990. (ADA232052). 105 p.
Thesis (M.S. in Computer Science)--Naval Postgraduate School, June 1990.
Abstract: *ARGOS is a multimedia database prototype system currently being developed by the Computer Science department of the Naval Postgraduate School in Monterey. Its primary purpose is to provide a prototype system that could be used as a Battle Group Commander's assessment tool and a shipboard data management tool, in addition to providing increased efficiency and productivity to the Navy ships. This implementation demonstrates the contribution such a system would make to the efforts of the Anti-submarine warfare (ASW) community.*
NPS/DKL Location: THESIS K3872

Keys, Richard Toney. **Large Grain Data Flow Graph Construction and Restructuring Utilizing the ECOS Workstation System**. Monterey, CA: Naval Postgraduate School; Springfield, VA: Available from National Technical Information Service, 1994. (ADA289632). 90 p.
Thesis (Master of Computer Science)--Naval Postgraduate School, Sept. 1994
Abstract: *The U.S. Navy's new multiprocessor, the AN/UYS-2 Enhanced Modular Signal Processor (EMSP) utilizes a First-Come-First-Serve (FCFS) algorithm to transfer data. This algorithm is simple to implement but provides no mechanism to control execution of a specific application on the AN/UYS-2 which prevents performance predictions. A Large Grain Data Flow (LGDF) representation of a specific application is utilized to predict performance, with the introduction of trigger queues (dependency arcs) into the graphs to control execution. I utilized the EMSP Common Operational Software (ECOS) Workstation to execute graph representations of specific applications used by the U.S. Navy in the Anti-Submarine Warfare (ASW) arena. A complete description of the ECOS workstation, and the process of transforming specific applications into graph representations to be executed on the ECOS Workstation is demonstrated. Specifically, the Correlator Graph which represents a real-time ASW process is examined. To control and improve performance, the technique of implementing trigger queues using the ECOS Workstation is demonstrated. A basic graph is executed and referenced as a benchmark, with two reconstructed graphs executed demonstrating how trigger queues effect graph execution. The node execution times statistics indicate trigger queues control execution and will provide a mechanism to predict node performance.*
NPS/DKL Location: THESIS K3962212

Kiland, IN and Jerry Allen Kotchka. **A Game Theoretic Analysis of the Convoy-ASW Problem**. Monterey, CA: Naval Postgraduate School; Springfield, VA: Available from National Technical Information Service, 1967. (AD831654). 49 p.
Thesis (M.S. in Operations Research) Naval Postgraduate School, 1967.
Abstract: *The problem of allocation of ASW forces assigned to an oceanic convoy in a submarine warfare environment is postulated as a two-person game with the payoff function being based on the 'formula of random search.' The opponents in the game are a convoy system and a submarine system. A submarine is given the option of attacking the convoy system either from afar with surface-launched*

missiles or near with torpedoes. The convoy system is defended by units capable of destroying submarines exercising either of their options. The optimal allocation of forces for both sides is shown to be a set of pure strategies which are dependent on the parameters of the model. (Author)
NPS/DKL Location: THESIS K399

King, Edward Francis. **A Basic Approach to Evaluating and Predicting VP Crew Performance**. Monterey, CA: Naval Postgraduate School; Springfield, VA: Available from National Technical Information Service, 1973. (AD761385). 22 p.
Thesis (M.S. in Operations Research)--Naval Postgraduate School, 1973.
Abstract: *A basic approach to the problem of evaluating or predicting a crew' performance for the VP community is presented. The method uses an application of multiple regression analysis techniques to a model which has training parameters as its variables. The results would allow the squadron or wing commanding officer to predict a crew's performance before the actual flight and to determine how to allocate training time for the squadron. (Author)*
NPS/DKL Location: THESIS K427

King, Portia B. **Use of the Wavenumber Technique with the Lloyds Mirror for an Acoustic Doublet**. Monterey, CA: Naval Postgraduate School; Springfield, VA: Available from National Technical Information Service, 1985. (ADA156463). 113 p.
Thesis (M.S. in Systems Tech.)--Naval Postgraduate School, 1985.
Abstract: *This thesis examines a method to determine the depth of a point source in and isospeed ocean environment. Using the Fourier Transform on the acoustic pressure field in the range domain results in the attainment of the acoustic pressure spectrum in the wave-number domain and a characteristic nodal spacing unique to the source-receiver depths. Quantitative examination of a magnitude plot of the spectrum and use of simple mathematical formulae yield the source depth. The debilitative effects of narrowband noise and surface roughness on the pressure spectrum is recognizable in noise after the pressure field in the range domain has been lost in the noise field. The effect of surface gravity waves on the pressure spectrum is similar to that on the pressure field in the range domain: the characteristic nodal spacing is suppressed as the height of the surface waves increases. Keywords include: Lloyds mirror; Underwater acoustics; Fast fourier transform; Wavenumber technique; Pressure spectrum; Source depth determination; Acoustic propagation; Isospeed environment; Antisubmarine warfare.(Author)*
NPS/DKL Location: THESIS K4492

Kirsch, Walter J. **Expected Value Analysis of the Center for Naval Analyses Computer War Game - SEALIFT**. Monterey, CA: Naval Postgraduate School; Springfield, VA: Available from National Technical Information Service, 1965. (AD475310). 61 p.
Thesis (M.S. in Operations Research)--Naval Postgraduate School, 1965.
Abstract: *The Center for Naval Analyses Computer War Game, SEALIFT, is a Monte-Carlo simulation designed to help study sealift capabilities in an ASW environment. A mathematical model of the SEALIFT game is posited to obtain expected value results approximating those of the SEALIFT game. The model is cast in the Fortran terminology of SEALIFT and an effort is made to accurately reflect the SEALIFT flow chart logic in its development. Comparisons with SEALIFT results to determine the model's reliability and accuracy are not made. (Author)*
NPS/DKL Location: THESIS K519

Koepke, John Allen. **Conceptual Design of a Stand-Off Weapon for Maritime Patrol Aircraft**. Monterey, CA: Naval Postgraduate School; Springfield, VA: Available from National Technical Information Service, 1988. (ADA221572). 183 p.

Thesis (M.S. in Aeronautical Engineering)--Naval Postgraduate School, Sept. 1988.
Abstract: *A conceptual design of a stand-off weapon to be launched from maritime patrol aircraft for use against hostile surface combatants was performed at the request of the Naval Air Test Center. The purpose of this thesis was to study the feasibility of developing a low-cost, anti-ship missile for air ASW platforms. A mission threat analysis was conducted to determine the lethality of probable targets and to determine required missile performance characteristics. Current design methods and techniques were used to calculate the necessary missile geometry to meet the derived performance characteristics. An evaluation of navigation laws was conducted to determine the most appropriate flight profile for the missile. The control system was tailored to meet the specifications of the selected navigation law. An investigation of passive and active homing devices was conducted. A low cost seeker to adequately locate and track targets of interest was examined. A target engagement model was used to verify the missile's maneuverability. This model demonstrated that the missile could intercept highly maneuvering craft when launched from a desirable stand-off distance.*
NPS/DKL Location: THESIS K728

Kreitler, Walter M. **The Close Aboard Bastion: A Soviet Ballistic Missile Submarine Deployment Strategy**. Monterey, CA: Naval Postgraduate School; Springfield, VA: Available from National Technical Information Service, 1988. (ADA201696). 114 p.
Thesis (M.A. in National Security Affairs)--Naval Postgraduate School, Sept. 1988.
Abstract: *This thesis describes and analyzes a possible deployment posture for the Soviet ballistic missile submarine force. It examines the proposition that the Soviet Navy will establish a point defense, labeled Close Aboard Bastions (CABs), for it ballistic missile submarine fleet within the Soviet claimed 12 nautical mile territorial sea. This is a logical derivation of the currently widely held view that the Soviet will establish derivation of currently widely held view that the Soviets will establish a bastion defense for the strategic portion of their seagoing forces. The thesis concludes that the postulated CAB strategy is a viable option for the Soviet Union during a war that begins conventionally.*
NPS/DKL Location: THESIS K85451

Kunigami, Masaaki. **Optimizing ASW Search for HVU Protection using the FAB Algorithm**. Monterey, CA: Naval Postgraduate School; Springfield, VA: Available from National Technical Information Service, 1997. (ADA328621). 40 p.
Thesis (M.S. in Operations Research)--Naval Postgraduate School, March 1997.
Abstract: *This thesis researches the feasibility of a TDA (tactical decision aid) to defend a high value surface unit from an enemy submarine. Accordingly, this research adopts a FAB (forward and backward) algorithm to search for a moving target. It develops a prototype of a TDA: FABTDA which gives an optimal allocation for search aircraft.*
NPS/DKL Location: THESIS K8786

Lanman, George Maurice. **An adaptation of a Markov chain model for antisubmarine warfare carrier aircraft**. Monterey, CA: Naval Postgraduate School; Springfield, VA: Available from National Technical Information Service, 1988. (AD489085). 63 p.
Thesis (M.S. in Operations Research)--Naval Postgraduate School, 1966.
Abstract: *It is the purpose of this paper to develop a useful mathematical model of ASW aircraft availability. The increasing emphasis of systems studies dictates the use of accurate and representative models of the ASW systems. At present, many studies are using essentially the same models developed during World War II. This paper is an attempt to make use of advanced theory in a more powerful and flexible model and to make the use of the model practical and verifiable. The writer adapted the time homogeneous bivariate model as developed by F. C. Collins. This is a discrete time Markov process with a stochastic matrix of transition probabilities wherein the maintenance process is modeled as a pulsed input multiple server queue. The model was programmed in FORTRAN 63 on the CDC 1604 and then*

modified to allow for variability in the input parameters. Other modifications include an increase in the size of the model to accommodate a 16-aircraft squadron, the largest ASW squadron at present, and an explicit form solution to the maintenance queueing equations. (Author)
NPS/DKL Location: THESIS L265

Lassman, Abraham Joel. **Career Development for an Antisubmarine Warfare Officer Specialist**. Monterey, CA: Naval Postgraduate School; Springfield, VA: Available from National Technical Information Service, 1978. (ADA056382). 40 p. Thesis (M.S. in Systems Tech.)--Naval Postgraduate School, 1978.
Abstract: *Antisubmarine Warfare technology has made significant advances since World War II. However, this thesis is based on the assumption that training for ASW Surface Officers has not kept pace with this rapid technological growth. This thesis proposes that the career pattern for surface officers desiring in-depth ASW training be modified to improve this situation while allowing surface officers to maintain a viable career pattern in the Surface Warfare community. Such a career pattern seems to be feasible.*
NPS/DKL Location: THESIS L27494

Lee, Yuh-jeng. **Improving the ASW System Evaluation Tool**. Monterey, CA: Naval Postgraduate School; Springfield, VA: Available from National Technical Information Service, 1992. (NPSCS-92-015, ADA258798). 18 p.
Report.
Abstract: *This report summarizes the work on the evaluation, design, and reimplemenation of part of the ASW System Eval Tool (ASSET), performed in the Computer Science Department, Naval Postgraduate School, under the sponsorship of the Antisubmarine Warfare Division (OP-71), Office of Chief of Naval Operations. We analyzed and implemented the improvements suggested in previous evaluations of various sub-areas of ASSET. In addition, we have designed and implemented a prototype user interface shell for ASSET on a Sun Sparcstation running X windows.*
NPS/DKL Location: FEDDOCS D 208.14/2:NPS-CS-92-015

Lentz, Frederick Charles. **Integration of ASW Helicopter Operations and Environment into NPSNET**. Monterey, CA: Naval Postgraduate School; Springfield, VA: Available from National Technical Information Service, 1995. (ADA304302). 124 p. Thesis (M.S. in Computer Science)--Naval Postgraduate School, 1995.
Abstract: *Despite the increasing emphasis by the military on joint force operations, existing modelling and simulation programs, including NPSNET, fail to address joint operations and crew coordination. The problem is that previous work on NPSNET, the virtual environment and visual simulation platform developed by the Computer Science Department at the Naval PostGraduate School in Monterey, California, has focused primarily on individual ground force elements with little emphasis on naval forces or crew concepts. This restricts the practical use of the system to ground force training while ignoring joint force training with sea and air components and between crew members. One solution to is expand the capability of NPSNET by incorporating a variety of vehicles from different components of the military with the added capability of multiple workstation control of a single vehicle. The approach taken is to expand NPSNET to simulate helicopter Anti-Submarine Warfare. This work focuses on realistic helicopter flight control, multiple workstation control of a single vehicle, and interface design between workstations controlling one vehicle. NPSNET has become a more useful training tool for today's military forces by implementing more realistic helicopter flight controls and adding joint mission capabilities. The significance of this work is that a broad range of forces can receive valuable joint training and crew coordination training conducted in a virtual environment.*
Electronic access: http://handle.dtic.mil/100.2/ADA304302
NPS/DKL Location: THESIS L525

Llanos, Humberto M. **An Investigation of the Problem of Optimizing a Search Tactic for a Searchlight Type Sonar**. Monterey, CA: Naval Postgraduate School; Springfield, VA: Available from National Technical Information Service, 1972. (AD754352). 42 p.
Thesis (M.S. in Operations Research)--Naval Postgraduate School, 1972.
Abstract: *A searchlight type sonar is one of the systems that small navies use to counteract the danger which submarines present to their lines of supply and transport. In the paper, a standard search pattern for this type of sonar is compared with search patterns which are based on a consideration of the tactical value of detecting a submarine as a function of the relative location of the submarine. The results of the comparison suggest that is possible to increase the effectiveness of a searchlight type sonar by using a search pattern in which the sweep time allocated to a search sector is based on the sectors tactical importance. (Author)*
NPS/DKL Location: THESIS L757

Lowery, John L. **High Speed Marine Craft Threat: Buoyancy and Stability Requirements for a Sub-Launched Weapon System**. Monterey, CA: Naval Postgraduate School; Springfield, VA: Available from National Technical Information Service, 1999. (ADA374340). 59 p.
Thesis (M.S. in Applied Physics)--Naval Postgraduate School, 1999.
Abstract: *Military intelligence has considered various coastal scenarios in which the submarine is the only platform -available to engage waterborne infiltration forces. Torpedoes are meant for large ships, and cruise missiles are strategic weapons not to be wasted on small craft. Therefore, the submarine does not have a weapons capability to engage and destroy high-speed marine craft (HSMC) that would be used for coastal infiltration. The most practical scenario would utilize a torpedo stow for a weapon system that would be tube launched, thus ensuring the maximum cruise missile capability of the submarine with a minimal sacrifice to anti-surface and anti-submarine warfare capabilities. The maintaining of the submarine's stealth will be paramount, therefore, an off-hull launcher is desired. The weapon *needs to be highly discriminative due to high shipping traffic in coastal waters. In all, the major factors associated with the design and employment of a sub-launched weapon system for engaging HSMC are the threat, the missile, the launcher and the deployment method. In a hostile coastal environment, there are numerous targets ranging from surface threats to air threats. Missile design is dependent on the threat and can be varied for different scenarios. However, the launcher and deployment of a tube launched weapon system are only restricted by the dimensions of the torpedo tube and the buoyancy and stability of the designed system. These parameters can be quantified and modeled. This thesis focused on designing a weapon system, SEABAT, to meet the basic buoyancy and stability requirements. The results of the SEABAT design prove its feasibility as a torpedo tube launched weapon system.*
Electronic access: http://library.nps.navy.mil/uhtbin/hyperion-image/99Dec_Lowery.pdf
Electronic access: http://handle.dtic.mil/100.2/ADA374340
NPS/DKL Location: THESIS L86675

Martin, Claude F. **Call for New ASW Screen Geometries for Carrier Battlegroup Open Ocean Transits Under the 1990's Threat**. Monterey, CA: Naval Postgraduate School; Springfield, VA: Available from National Technical Information Service, 1992. (ADA257699). 143 p.
Thesis (M.S. in Applied Science)--Naval Postgraduate School, 1992.
Abstract: *A simulation model was specified. It examines United States Navy Antisubmarine Warfare Screen alternative dispositions for Carrier Battlegroups. The scenario posed is open ocean transit under the threat of an attack from foreign submarine hulls built in the 1990's. The investigation raises the issue of the appropriateness of current Navy practices and suggests that new tactics be developed. The author's thoughts are that in the 1990's there will be ever newer, more lethal, unpredictable threats to United State's maritime independence than current doctrine addresses. The full implementation of the*

simulation program has not been accomplished. A segment of verification output is shown for expository purposes only. A discussion is given on the adequacy of the model's abstractions along with their possible impact on potential results of experiments.
NPS/DKL Location: THESIS M35725

McClure, Donald James. **Experimental Design Considerations in the Operational Test and Evaluation of Airborne Acoustic Processing Systems**. Monterey, CA: Naval Postgraduate School; Springfield, VA: Available from National Technical Information Service, 1979. (ADA078270). 53 p.
Thesis (M.S. in Operations Research)--Naval Postgraduate School, 1979.
Abstract: *The antisubmarine warfare (ASW) capability of the air arms of western navies depends upon their ability to employ air-dropable sonobuoys for the detection and localization of enemy submarines. The systems which process the acoustic information from these sonobuoys are sophisticated spectrum analyzers. A constant effort to improve performance has led to frequent updates to existing systems and the periodic development of completely new processors. Assessing the significance of these improvements in the operational environment is the function of an operational evaluation. The operational evaluation of an acoustic processing system is complicated by the impact of the constantly changing acoustic propagation conditions encountered in ocean operating areas. This, along with the inherent variability in the other factors affecting passive sonar performance, makes it difficult to compare the ranges achieved from one trial to the next. However, it is imperative that the evaluation agencies report findings that are applicable over the wide variety of operating conditions which are likely to be encountered in the operational employment of the airborne acoustic processor.*
NPS/DKL Location: THESIS M1767

McDonnell, John R. **Application of Similarity Theory to Forecasting the Mixed-Layer Depth of the Ocean**. Monterey, CA: Naval Postgraduate School; Springfield, VA: Available from National Technical Information Service, 1944. (AD601580). 54 p.
Thesis (M.S. in Oceanography)--Naval Postgraduate School, 1964.
Abstract: *The thermal structure of the ocean, especially the uppermost mixed layer, greatly affects sonar ranges. In this paper, similarity theory is applied to the problem of forecasting the depth of the mixed layer during the warm season, assuming the controlling processes are secular, non-advective, and non divergent. The resulting forecast method consists mainly of two equations. Parameters used are wind, coriolis effect, the coefficient of thermal expansion and a measure of the excess heat within the mixed layer. The constants in the equations were determined using data from OWS Papa (50N, 145W). The forecast method treats both seasonal and transitional thermoclines. The method was tested with data from OWS Papa and OWS November (30N, 140W). The tests apparently indicate wide applicability of this forecast method and thus tend to corroborate the proposal by Kitaigorodsky that the mixed-layer depth is a function of a universal coefficient. (Author)*
NPS/DKL Location: THESIS M1835

McGinlay, Thomas Charles John. **Personnel and Equipment Design Concept for a Maritime Patrol Airship (Non-Rigid) to Conduct Search, Anti-Submarine Warfare, and Airborne Mine Countermeasures Missions**. Monterey, CA: Naval Postgraduate School; Springfield, VA: Available from National Technical Information Service, 1979. (ADA085144). 227 p.
Thesis (M.S. in Management)--Naval Postgraduate School, 1979.
Abstract: *A personnel and equipment design concept for a non-rigid, 100 hour endurance, Maritime Patrol Airship meeting Search and Rescue (SAR), Anti-Submarine Warfare (ASW), and Airborne Mine Countermeasures (AMCM) requirements was developed. The Maritime Patrol Airship could readily be equipped with off-the-shelf equipment. Minimal new design equipment requirements were identified. A baseline flight scenario and on station scenarios for: SAR, transoceanic ASW utilizing a passive towed*

array sonar, and AMCM were developed. Human factors task analyses and a time line analysis were constructed from the scenarios. Manning reductions resulted for each scenario (3 crewmembers for SAR, 10 crewmembers for transoceanic ASW, 7 crewmembers for AMCM). Further research areas are identified. (Author)
NPS/DKL Location: THESIS M188335

McMichael, D.L. **An on-Line Simulation of ASW in a Multi-Burst Nuclear Environment.** Monterey, CA: Naval Postgraduate School; Springfield, VA: Available from National Technical Information Service, 1966. (AD488327). 123 p.
Thesis (M.S. in E.E)--Naval Postgraduate School, 1966.
Abstract: *A general approach is documented as a guide to aid in the formulation and implementation of on-line, real time computer simulations. A computer program, MULNUC1, is developed as an on-line, real time computer simulation of antisubmarine warfare in a multiple burst nuclear environment. The principals of the game are a submarine armed with torpedoes, and two destroyers equipped with stand-off antisubmarine weapons. The simulation is intended as a demonstration of the on-line capabilities of the United States Naval Postgraduate School computer system and as a tool for further study of the factors involved in a representative ASW operational environment. (Author)*
NPS/DKL Location: MICROFORM AD488327

Nash, Wyatt J.. **Calculation of Barrier Search Probability of Detection for Arbitrary Search Tracks.** Monterey, CA: Naval Postgraduate School; Springfield, VA: Available from National Technical Information Service, 2000. (ADA378067). 103 p.
Thesis (M.S. in Operations Research)--Naval Postgraduate School, 2000.
Abstract: *The Surface Warfare Development Group is responsible for conducting the Ship Anti-submarine Warfare Readiness/Effectiveness Measuring program. They currently employ a standard set of measures for evaluating the performance of shipboard anti-submarine warfare sensors. This research investigates several new performance-based measures to determine if they are more suitable than the standard measures for evaluating the conduct of anti-submarine warfare barrier searches. The investigation simulates barrier searches to determine probability of detection, calculates the proposed measures, and compares the two. The results indicate that the proposed measures can be improved. A barrier search algorithm exploiting target-relative space ideas is developed which generalizes the classical search theory results for predicting probability of detection during barrier search.*
NPS/DKL Location: THESIS N24536
Electronic access: http://handle.dtic.mil/100.2/ADA378067

Naval Postgraduate School. **Summary of Research 1992**. Monterey, CA: Naval Postgraduate School; Springfield, VA: Available from National Technical Information Service, 1992. (NPS-08-92-002, ADA280616). 484 p.
Abstract: *This report contains 344 summaries of research projects which were carried out under funding of the Naval Postgraduate School Research Program. A list of recent publications is also included which consists of conference presentations, contributions to books, published papers, magazine articles, and technical reports. The research was conducted under the areas Administrative Sciences, Aeronautics and Astronautics, Computer Science, Electrical and Computer Engineering Mathematics, Mechanical Engineering, Meteorology, National Security Affairs, Oceanography, Operations Research, Physics, C3 Joint Academic Group, Electronic Warfare Academic Group, Antisubmarine Warfare Group, and Space Systems Academic Group.*
NPS/DKL Location: MICROFORM ADA280616

Naval Postgraduate School. **Compilation of Abstracts of Theses Submitted by Candidates for Degrees.** Monterey, CA: Naval Postgraduate School; Springfield, VA:

Available from National Technical Information Service, 1985. (NPS-012-85-003PR, ADA164988). 558 p. Report:
Abstract: *This publication contains the abstracts of theses submitted during the period 1 October 1983 - 30 September 1984 by candidates for Doctoral, Master's and Engineer's degrees at the Naval Postgraduate School, Monterey, CA 93943. Subject areas include: Aeronautical engineering, Electrical engineering, Mechanical engineering, Applied science, Computer science, Engineering acoustics, Hydrography, Information systems, Management, Meteorology and oceanography, Operations research, Systems technology(Antisubmarine warfare), Telecommunications systems management, and National security affairs.*
NPS/DKL Location: REFERENCE Z5055.U39 U6 1983-84

Naval Postgraduate School. **Compilation of Abstracts of Theses Submitted By Candidates for Degrees; Summary rept 1 Oct 1989-30 Sep 1990**. Monterey, CA: Naval Postgraduate School; Springfield, VA: Available from National Technical Information Service, 1990. (NPS-08-91-001, ADA242938). 355 p.
Abstract: *No abstract available.*
NPS/DKL Location: MICROFORM ADA242938

Naval Postgraduate School. **Summary of the Naval Postgraduate School Research Program and Recent Publications, October 1988 to September 1989**. Monterey, CA: Naval Postgraduate School; Springfield, VA: Available from National Technical Information Service, 1990. (NPS-012-90-002, ADA236052). 373 p.
Abstract: *This report contains 330 summaries of research projects which were carried out under funding of the Naval Postgraduate School Research Program. A list of recent publications are also included which consist of conference presentations, contributions to books, published papers, and technical reports. The research was conducted under the areas of: Computer science, Mathematics, Administrative Sciences, Operations research, National security affairs, Physics, Electrical and computer engineering, Meteorology, Aeronautics and astronautics, Oceanography, and Mechanical engineering, Antisubmarine warfare, Space systems, Electronic warfare, and in Command control communications.*
NPS/DKL Location: MICROFORM ADA236052

Naval Postgraduate School. **Summary of the Naval Postgraduate School Research Program and Recent Publications**. Monterey, CA: Naval Postgraduate School; Springfield, VA: Available from National Technical Information Service, 1990. (NPS-08-91-002, ADA242939). 554 p.
Abstract: *This report contains 357 summaries of research projects which were carried out under funding of the Naval Postgraduate School Research Program. A list of recent publications are also included which consist of conference presentations, contributions to books, published papers, magazine articles, and technical reports. The research was conducted under the areas of Computer Science; Mathematics; Administrative sciences; Operations research; National security affairs; Physics; Electrical and computer engineering; Meteorology; Aeronautics and astronautics; Oceanography; Mechanical engineering, C3 Joint Group; Electrical Warfare Group; ASW Group; and the Space System Academic Group.*
NPS/DKL Location: REFERENCE V422.M5 Z97 (FY1989)

Nicholson, Charles Louis. **Localization of Acoustic Transients in Shallow Water Environments.** Monterey, CA: Naval Postgraduate School; Springfield, VA: Available from National Technical Information Service, 1992. (ADA260615). 38 p.
Thesis (M.S. in Electrical Engineering)--Naval Postgraduate School, 1992.
Abstract: *Determination an of an underwater target's position using passive acoustic sensors is of considerable use for the Navy, both for anti-submarine warfare (ASW) and underwater surveillance. This*

thesis proposes and develops localization algorithms capable of passively determining the location of a transient source given some broad constraints. In particular, this thesis investigates the effect of the source signal uncertainty on localizer performance. The localization process consists of two parts. First, a time domain propagation modeling code determines the impulse response of the environment from all possible source locations to a single hydrophone. This program predicts the signal as it would appear at the receiver from a grid of possible source locations. Second, source localization results from finding the maximum correlation between the positionally dependent, numerically modeled signals and the actual received signal. The position of the maximum cross correlation reveals an estimate of source position. Using model to model correlation, this technique successfully localized acoustic sources in both Monterey Bay and Barents Sea scenarios.
NPS/DKL Location: THESIS N4873

Nissen, Richard J. **Feasibility for the Integration of ASW Information Databases**. Monterey, CA: Naval Postgraduate School; Springfield, VA: Available from National Technical Information Service, 1992. (ADA257581). 150 p.
Thesis (M.S. in Applied Science)--Naval Postgraduate School, 1992.
Abstract: *There are currently three databases supported by three different commands that collect and output basically the same type of information: PACER, SHAREM, and AIREM. These systems contain initial detection data, tracking data, environmental data, system performance data, and weapon performance data. This thesis investigates the commonalities and differences in structure and content of the three databases, and examines the feasibility of integrating PACER, SHAREM, and AIREM into a single database. The benefits of this database integration are a more comprehensive utilization of data, reduced data collection for fleet users, and a standardization of how the data is analyzed.*
NPS/DKL Location: THESIS N5934

Null, James Mark. **Perturbative Inversion of Geoacoustic Parameters in a Shallow Water Environment**. Monterey, CA: Naval Postgraduate School; Springfield, VA: Available from National Technical Information Service, 1995. (ADA294674). 99 p.
Thesis (M.S. in Meteorology and Physical Oceanography)--Naval Postgraduate School, 1995.
Abstract: *In many strategic shallow water areas the geoacoustic properties of the sub-bottom are largely unknown. In this thesis it is demonstrated that inverse theory and measured data from a single hydrophone can be used to accurately deduce the geoacoustic properties of the sub-bottom, even when the initial background geoacoustic model is a highly inaccurate guess. Since propagation in shallow water is very sensitive to the geoacoustic properties of the sub-bottom, the inverse technique developed in this thesis presents the Navy with a vitally important, practical, and inexpensive means to improve sonar performance prediction in a potentially hostile environment. To provide ground truth for the inverse technique, measured data collected during Project GEMINI were compared to the inverse solutions. Detailed, site-specific geoacoustic models were developed for two array locations and the Finite Element Parabolic Equation (FEPE) model was used to estimate transmission loss (TL). The model estimates from FEPE compared well with the measured data and the detailed geoacoustic models were considered as ground truth. To test the efficacy of the technique, initial background geoacoustic models were constructed assuming no a priori information of the bottom. The resultant inverse solution was used to predict the geoacoustic properties at each of the sites. The final results were in excellent agreement with the measured data and the resulting inverse technique TL estimates were as good or better than the Th estimates obtained from the detailed, site-specific geoacoustic models.*
NPS/DKL Location: THESIS N9485
Electronic access: http://library.nps.navy.mil/uhtbin/hyperion-image/95_Null.pdf
Electronic access: http://handle.dtic.mil/100.2/ADA294674

Offenberg, Jerome William. **Marginal Oscillator For A Modified Free Precession Magnetometer.** Monterey, CA: Naval Postgraduate School; Springfield, VA: Available from National Technical Information Service, 1969. (AD705083). 41 p.
Thesis (M.S.in Electrical Engineering)--Naval Postgraduate School, 1969.
Abstract: The primary interest concerns the use of a magnetometer for mine detection, anti-submarine warfare, salvaging and other related naval perations. The original concept of a modified free precession magnetometer using the Overhauser Effect was formulated by A. Abragam. The objective of this thesis was to develop an improved marginal oscillator for the magnetometer. (Author)
NPS/DKL Location: THESIS O248

Ong, Seow Meng. **A Mission Planning Expert System with Three-Dimensional Path Optimization for the NPS Model 2 Autonomous Underwater Vehicle.** Monterey, CA: Naval Postgraduate School; Springfield, VA: Available from National Technical Information Service, 1990. (ADA226395). 192 p.
Thesis (M.S. in Computer Science.)--Naval Postgraduate School, 1990.
Abstract: *Unmanned vehicle technology has matured significantly over the last two decades. This is evidenced by its widespread use in industrial and military applications ranging from deep-ocean exploration to anti-submarine warfare. Indeed, the feasibility of short range, special-purpose vehicles (whether autonomous or remotely operated) is no longer in question. The research efforts have now begun to shift their focus on development of reliable, longer range, high-endurance and fully autonomous systems. One of the major underlying technologies required to realize this goal is Artificial Intelligence (AI). The latter offers great potential to endow vehicles with the intelligence needed for fully automated and extended range capability; this involves the increased application of AI techniques to support mission planning and execution, navigation and contingency planning. This thesis addresses two issues associated with the above goal for Autonomous Underwater Vehicles (AUV's). Firstly, a new approach is proposed for path planning in underwater environments that is capable of dealing with uncharted obstacles and which requires significantly less planning time and computer memory. Secondly, it explores the use of expert system technology in the planning of AUV missions. (KR)*
NPS/DKL Location: THESIS O5825

Panoff, Timothy J. **Reference Path Generation and Tracking of Marine Vehicles.** Monterey, CA: Naval Postgraduate School; Springfield, VA: Available from National Technical Information Service, 1992. (ADA257742). 64 p.
Thesis (M.S. in Mechanical Engineering)--Naval Postgraduate School, 1992.
Abstract: *This thesis analyzes the problem of accurate path keeping for marine vehicles. The reference path is generated automatically through the use of a critically damped second order model. An appropriate shift in the time axis allows a smooth path with zero overshoot regardless of the initial conditions. Control design for the physical system is achieved through the use of optimum control and linear quadratic regulator techniques. Results are presented for general maneuvering scenarios in the horizontal plane and demonstrate the validity of the models used in the research.*
NPS/DKL Location: THESIS P145755

Peck, Robert Louis. **A Systematic Approach Toward Developing ASW Tactics Based on Plausible Soviet Resource Allocation.** Monterey, CA: Naval Postgraduate School; Springfield, VA: Available from National Technical Information Service, 1974. (AD777216). 37 p.
Thesis (M.S. in Operations Research)--Naval Postgraduate School, 1974.
Abstract: *The thesis relates the fact that, in the past, our ASW community has placed great (and justifiable) emphasis in detection and classification of submarines, while a serious lag in tactical procedures has developed. In order to alleviate this problem, it was felt that a systematic approach be*

taken which uses the principles of Operations Research. By examining submarine warfare from the viewpoint of the Soviet Union, a resource allocation problem was devised which compares the various submarine classes and the possible mission areas in which they may be assigned. Characteristics and available numbers of submarines were estimated, and the resulting allocation of forces was determined. (Modified author abstract)
NPS/DKL Location: THESIS P319

Philipp, Brian F. **Categorical Analysis of Weapon System Accuracy Trial (WSAT) Data**. Monterey, CA: Naval Postgraduate School; Springfield, VA: Available from National Technical Information Service, 1992. (ADA257573). 47 p.
Thesis (M.S. in Applied Science)--Naval Postgraduate School, 1992.
Abstract: *This thesis contains an analysis of the last five years of Antisubmarine Warfare (ASW) Weapon System Accuracy Trial (WSAT) data from both the Atlantic and Pacific Fleet. The analysis is conducted in an effort to provide recommendations to be applied toward future evolution of the ASW Test Program for surface ships. A statistical chi-square test is conducted on Fleet and Navy wide data to determine which ASW combat system material categories are most prone to degradation. Additionally, a critical examination of the existing WSAT data base is provided with an aim toward promoting future statistical analysis. Results of this thesis indicate that degradation to weapons delivery systems like torpedo tubes and ASROC launchers are statistically significant with respect to other material categories. The analysis also shows how the existing WSAT data base can easily be modified and adapted for further use to document inspections on existing ships as well as new construction ships with future material systems.*
NPS/DKL Location: THESIS P4654

Pollak, Kenneth D. **Real-Time Enhancement of a Climatology or Forecast of Ocean Thermal Structure Using Observed Ocean Temperatures**. Monterey, CA: Naval Postgraduate School; Springfield, VA: Available from National Technical Information Service, 1984. (ADA151118). 66 p.
Thesis (M.S. in Oceanography)--Naval Postgraduate School, 1984.
Abstract: *Vertical temperature profiles observed in the eastern North Pacific were used to examine the feasibility of extrapolating an observation from one location to another. The technique, referred to as simple enhancement, is a special case of the Gandin (1963) optimum interpolation methodology. Application to Navy ASW (Antisubmarine Warfare) operations is considered. The technique requires the use of a trial value and a local observation. Trial values are obtained from a climatology and a synoptic analysis /forecast system provided by the Fleet Numerical Oceanography Center. An enhanced temperature profile is calculated by adding an observed anomaly (i.e., observation minus trial value) to the trial value at the desired location. Calculations of mean and RMS errors indicate that simple enhancement can provide a closer estimate to actual conditions than unenhanced climatology. The mixed layer depth cannot be extrapolated accurately to new locations presumably due to mesoscale eddies, fronts, internal waves and small scale fluctuations at the base of the mixed layer. Experiments at different locations and seasons would be required for a complete assessment of the application to ASW operations.*
NPS/DKL Location: THESIS P686

Recalde, Cesar J. **Reactive ASW Sonar Search Tactical Decision Aid**. Monterey, CA: Naval Postgraduate School; Springfield, VA: Available from National Technical Information Service, 1996. (ADA322381). 164 p.
Thesis (M.S. in Operations Research)--Naval Postgraduate School, 1996.
Abstract: *This thesis develops, implements and tests a Tactical Decision Aid for a Reactive Target ASW Active Search. The mode uses a Bayesian Filtering Process to fuse information from a real world search conducted by several assets with information from a Monte Carlo Simulation that encompasses five*

hundred equally likely different possible initial positions and behaviors of the real target. A Reactive Target Model resembles the behavior of a target that is always aware and reacts because of the presence and activity of the searchers. An initial 'prior', or best estimate of the location of the target is updated using the movement of the simulated targets, the negative information conveyed in an unsuccessful search over a period of time and the positive information implied in a contact report. The search effort is measured using a Fixed Scan Stochastic Model that solves the Sonar Equation limited by noise and reverberation. As a result of updating the prior, a 'posterior' distribution is obtained. The Law of Total Probabilities is used to render a probability map of the location of the Target by mapping color intensities to probabilities. A recursive expression for evaluating a contact report is also developed.
Electronic access: http://handle.dtic.mil/100.2/ADA322381
NPS/DKL Location: THESIS R2576

Ridings, William V. **Analysis of Effects of Variable Factors on Weapon Performance**. Monterey, CA: Naval Postgraduate School; Springfield, VA: Available from National Technical Information Service, 1993. (ADA267166). 61 p.
Thesis (M.S. in Operations Research)--Naval Postgraduate School, 1993.
Abstract: *Statistical analysis provides a powerful tool for modern decision makers. Unfortunately, this tool can be a two-edged sword. Improper or erroneous analysis can result in incorrect and costly decisions. Many analysis errors can be traced to the misapplication of statistical methods. When examining experimental data, it is first necessary to determine the true nature of that data, specifically, the structure from which the data is drawn. This determination will then be a primary factor in the choice of statistical tests. This thesis examines an analysis performed by Surface Warfare Development Group (SWDG). The SWDG analysis is shown to be incorrect due to the misapplication of testing methods. A corrected analysis is presented and recommendations suggested for changes to the testing procedures used by SWDG. Additionally, a computer program to perform basic Analysis of Variance (ANOVA) tests is provided to be appended to the current SWDG statistical software.*
NPS/DKL Location: THESIS R4516

Rinaldi, Ronald Daniel. **The Probable Distribution of Whales as False Sonar Targets in the North Pacific Ocean by Analysis of Whaling Data**. Monterey, CA: Naval Postgraduate School; Springfield, VA: Available from National Technical Information Service, 1972. (AD743085). 181 p.
Thesis (M.S. in Oceanography)--Naval Postgraduate School, 1972.
Abstract: *False Sonar targets present a serious unpredicted problem to U. S. Navy ASW units. It is believed that planning and operations could be enhanced by a forecasting capability for whale distribution. As a possible solution to this problem, a modified form of the 'Transect Method of population estimation' is applied to whaling data to calculate probable numbers of false targets per 1000 nautical miles of steaming with a 1000 yard sonar range. Japanese and Russian whale fishery data are analyzed by the 'q' and Expected Catch methods of population dynamics to obtain two independent estimates of the populations of fin, sei and sperm whales. The mean of the two estimates is applied to the equation along with a term for assumed ideal sonar conditions. (Author)*
NPS/DKL Location: THESIS R557

Robbins, Douglas L. **Decision-Making Process of an Antisubmarine Warfare Commander**. Monterey, CA: Naval Postgraduate School; Springfield, VA: Available from National Technical Information Service, 1986. (ADA176049). 49 p.
Thesis (M.S. in Systems Tech.)--Naval Postgraduate School, 1986.
Abstract: *This thesis represents a study of the decision-making process of an Anti-submarine Warfare Commander (ASWC). Several real world operational issues are analyzed and discussed as to how they can influence his thought process when making decisions. One approach to model this individual's thought process was accomplished by ALPHATECH, INC. By utilizing an ASW scenario, it evaluates how*

an ASWC makes his tactical decisions to track submarines based upon pieces of received acoustical information. In order to improve this model's representation of a realistic operational environment, a conceptual ASWC decision-making model is provided here.
NPS/DKL Location: THESIS R5912

Roll, Raymond B. **Analysis of the Advantages and Disadvantages Associated with the Consolidation of the HS And HSL Missions and Communities**. Monterey, CA: Naval Postgraduate School; Springfield, VA: Available from National Technical Information Service, 1994. (ADA280069). 78 p.
Thesis (M.S. in Management)--Naval Postgraduate School, 1994.
Abstract: *This study examines the advantages and disadvantages associated with the consolidation of the Helicopter Antisubmarine (HS) and Helicopter Anti-submarine (Light) (HSL) communities. The primary source material is generated from personal interviews of the Commanding Officers of these communities. The helicopter and mission developments of each community are researched to determine the goals, environments and technology that shape the squadron operational structures. The operational design of the current squadrons are then examined to see how they are structured to respond to these organizational constraints. This paper analyzes both sides of the consolidation issue and proposes four combined community organizations. Additionally, the advantages and disadvantages of each new structure are reviewed to make recommendations on consolidating the HS and HSL communities. This study also recommends a Project Action Team be formed to continue analyzing the consolidation of these communities and outline the steps required to implement a consolidation plan. Anti-submarine Warfare(ASW), Fleet Replacement Squadron(FRS), Helicopter Anti-submarine(HS) Helicopter Anti-Submarines(Light)(HSL).*
NPS/DKL Location: THESIS R68947

Sanchez, Jaime Cortes. **Small Computer analysis of Regional Ocean Data.** Monterey, CA: Naval Postgraduate School; Springfield, VA: Available from National Technical Information Service, 1971. (AD734974). 84 p.
Thesis (M.S. in Oceanography)--Naval Postgraduate School, 1971.
Abstract: *The study presents the results of an experiment in objective analysis of oceanographic data for a limited area. The objective analysis is designed to provide a reliable operational system for tactical use in coastal waters. It is shown that this approach makes it possible to obtain a very detailed analysis with good vertical consistency and that only a relatively small amount of highly accurate data is required. The procedure requires only a small computer and little computer time. This method will provide a basis for short-time forecasts of oceanographic parameters using only small computer centers or even time sharing systems.*
NPS/DKL Location: THESIS S1579

Sandel, Clyde D and Mary F. Gleason. **Enlistment Standards as Related to Performance in Aviation Antisubmarine Warfare Operator and Aviation Antisubmarine Warfare Technician Ratings**. Monterey, CA: Naval Postgraduate School; Springfield, VA: Available from National Technical Information Service, 1983. (ADA137558). 75 p.
Thesis (M.S. in Management)--Naval Postgraduate School, 1983.
Abstract: *The purpose of this study is to discover if the Navy's system of assigning personnel to the Aviation antisubmarine Warfare Technician (AX) and the Aviation Antisubmarine Warfare Operator (AW) ratings can be improved. A multivariable model is developed using success and failure as criterion variables. Biographical and aptitude data available at the time of enlistment are used as predictor variables. Two independent models were created using data available on personnel entering the Navy in 1976, 1977 and 1978. The models were then validated on a new sample. These models predict the future fleet performance of AX and AW personnel as measured by length of service, paygrade achieved, and*

recommendation for reenlistment. Other results and recommendations regarding implementation and future research are discussed. (Author)
NPS/DKL Location: THESIS S15796

Saunders, Norman Thomas and Harald Ziehms. **An Investigation of the Concepts of Purposeful Searches.** Monterey, CA: Naval Postgraduate School; Springfield, VA: Available from National Technical Information Service, 1972. (AD751645). 34 p.
Thesis (M.S. in Operations Research)--Naval Postgraduate School, 1972.
Abstract: *A discrete-time discrete-space search model is considered in which an observer employing an idealized detection device is searching for a uniformly distributed stationary target. The model is formulated as a discrete-time counting process, called the search process, which under weak additional conditions is uniquely determined by a sequence of probabilities. Formulas for the time-to-detection and the detection rate of a search are derived in terms of the parameters of the search process, and are applied to two special types of searches, the systematic search and the random search. Using these search types as boundary cases a purposeful search is defined, and sufficient conditions on the sequence of probabilities are established for the purposeful search. Possible extensions of the search process to less restricted models are indicated. (Author)*
NPS/DKL Location: THESIS S214

Sawrey, William Michael. **Comparison of Propagation Loss Output from the Tactical Environmental Support System (TESS v. 2.2A) and ASW Tactical Decision Aid (v. 2.1.2.1).** Monterey, CA: Naval Postgraduate School; Springfield, VA: Available from National Technical Information Service, 1993. (ADA264005). 98 p.
Thesis (M.S. in Applied Science) Naval Postgraduate School, 1993.
Abstract: *The United States Navy uses a number of different systems to predict underwater acoustic transmission loss for operational forces. Historically, these systems have used different acoustic models and supporting databases, resulting in significantly different predictions. Major efforts to bring all acoustic models and databases under configuration control in the Oceanographic and Atmospheric Master Library (OAML) have reduced, but not eliminated, differences in acoustic predictions. Comparisons of 1600 transmission loss runs from the AntiSubmarine Warfare Tactical Decision Aid (ASWTDA) and the Tactical Environmental Support System (TESS) were made in the Mediterranean and Sea of Japan for the months of January and July. All inputs to the acoustic models were provided by the respective system databases. Significant differences between ASWTDA and TESS in the areas investigated are evident in regions of complex bathymetry, and these differences become more acute with higher frequency....
TESS, ASWTDA, RAYMODE, ASTRAL, PE, Transmisssion Loss.*
NPS/DKL Location: THESIS S2207

Schmidt, Dennis Arvin. **NPSNET: A Graphical Based Expert System to Model P-3 Aircraft Interaction with Submarines and Ships.** Monterey, CA: Naval Postgraduate School; Springfield, VA: Available from National Technical Information Service, 1993. (ADA271543). 89 p.
Thesis (Master of Computer Science) Naval Postgraduate School, 1993.
Abstract: *The Computer Science Department at Naval Postgraduate School in Monterey, California has developed a low-cost battlespace simulation system, known as NPSNET, to work on commercially available Silicon Graphics IRIS workstations. Initial work on NPSNET has concentrated primarily on ground-based forces with only limited work focusing on naval or maritime air forces. With the present movement of the military towards totally integrated joint force operations, there exists a need to expand existing modeling and simulation programs to include all aspects of military operations. This thesis takes a step in that direction by incorporating naval maritime air units into NPSNET, expanding its capability to include naval and Antisubmarine Warfare (ASW) units. This work focuses on several areas of research, including modeling of the P-3 aircraft, aircraft motion control, aircraft ordnance ballistics modeling,*

interstation networking using the Distributed Interactive Simulation (DIS) protocol and development of an expert system to autonomously control aircraft behavior. Graphics, P-3, Expert systems, Torpedo ballistics, Sonobuoy ballistics, CLIPS.
NPS/DKL Location: THESIS S33717

Schriner, Karl Leonard. A Study of Enlisted Training and Education in Applied Oceanography. Monterey, CA: Naval Postgraduate School; Springfield, VA: Available from National Technical Information Service, 1972. (AD751596). 225 p.
Thesis (M.S. in Oceanography)--Naval Postgraduate School, 1972.
Abstract: *The study concludes that the primary reason for present programs of enlisted training and education in oceanography is to support ASW. There is a significant lack of courses, schools, and self-study material available to enlisted personnel on the subject of oceanography. Through more than the surface ASW community in the ability to utilize ASW. ASW sonar technicians are inadequately trained in environmental effects on underwater sound propagation. To increase the oceanography knowledge of all enlisted personnel including STs and to provide enlisted ratings to better utilize training in environmental effects, several programs are proposed. These programs include an ASW sensor rating and an oceanographer rating. (Author)*
NPS/DKL Location: THESIS S348

Schroeder, Roger Glenn. **Search and Evasion Games**. Monterey, CA: Naval Postgraduate School; Springfield, VA: Available from National Technical Information Service, 1966. (TR/RP-68; AD639921). 39 p.
Abstract: *The author develops some two-person zero-sum formulations of search and evasion problems. By employing a game theoretic approach, he allows the hider, as well as the searcher, to choose a strategy. This is in contrast to most search models which assume a stationary or passive hider. Both non-sequential and sequential search games are investigated. Some interesting aspects of the non-sequential game and an example of an antisubmarine search problem are given. The sequential games consist of a sequence of moves. When the players move, they not only determine a payoff but also the probability that the game terminates before the next move. When at most a finite number of moves is allowed, he proves that a solution may be found by solving a recursive sequence of matrix games. When the number of moves is not bounded, the game is characterized by a special type of non-linear program. The solution to this program can be approximated by successive perturbations of a related linear program. He obtains the result that a pair of strategies minimaxes the expected duration of the game if and only if these strategies also maximin the probability of termination in one step.*
NPS/DKL Location: GENERAL TA7 .U62 NO.68

Shaffer, Richard M. **Evaluation of the MPA Detection and Allocation Models Utilized by the ASW Systems Evaluation Tool (ASSET).** Monterey, CA: Naval Postgraduate School; Springfield, VA: Available from National Technical Information Service, 1991. (ADA246595). 52 p.
Thesis (M.S. in Applied Science)--Naval Postgraduate School, 1991.
Abstract: *The primary objective of this thesis is to analyze and recommend improvements to the Maritime Patrol Aircraft (MPA) detection and allocation models utilized by the ASW Systems Evaluation Tool (ASSET), version 1.0. ASSET is a generic high-level ASW modeling tool, designed to aid CNO (OP-71) in the development and refinement of ASW top-level warfare requirements and the ASW Master Plan. ASSET's strengths lie in its C3I modeling of submarine, MPA, and overhead surveillance in large scale ASW campaigns. To reduce the processing time required by ASSET, the current version of the MPA detection model contains simplifications which can limit its ability to effectively simulate some MPA tactical ASW scenarios. This thesis proposes two new MPA detection models which utilize the coverage area of a user-defined sonobuoy pattern and address the limitations of the current ASSET model. Also proposed is an MPA allocation scheme which should provide a higher cumulative detection probability.*

NPS/DKL Location: THESIS S43322

Shaheen, Frederick F. **Soviet Union and Its Caribbean Allies: Strategic, Maritime, and Regional Threat to the United States**. Monterey, CA: Naval Postgraduate School; Springfield, VA: Available from National Technical Information Service, 1985. (ADA155886). 96 p.
Thesis (M.A. in National Security Affairs)--Naval Postgraduate School, 1985.
Abstract: *The Soviet Union's activity in the Carribean Basin, executed via its client-states of Cuba and Nicaragua, has created a serious threat to U.S. security in the region. This threat to U.S. security takes two forms. The first is the reality of a heavily militarized Cuba posing a significant anti-SLOC potential against Caribbean sea lanes in the event of general war. Such a scenario would tie down NATO antisubmarine warfare (ASW) assets in the Caribbean, detracting from NATO's ability to wage the ASW campaign in more critical areas such as the Central and North Atlantic. The second threat is Nicaraguan and Cuban active support of leftist insurgencies in the Basin. These efforts, at the direction of the Soviet Union, pose, not a potential, but a present-day and ongoing security concern for the United States. This thesis briefly examines the historical context of Soviet involvement in the region, and then proceeds to catalog the above mentioned threats to U.S. security, and discusses their implications.*
NPS/DKL Location: THESIS S43323

Shudde, Rex H. **Estimation of a Contact's Course, Speed and Position Based on Bearings-Only Information From Two Moving Sensors with a Program for an HP-67/97 Calculator**. Monterey, CA: Naval Postgraduate School; Springfield, VA: Available from National Technical Information Service, 1977. (NPS-55-77-43, ADA050011). 22 p.
Abstract: *This report provides a procedure for estimating a contact's course, speed and position based on bearings-only data from two moving sensors. This report also contains a program for the HP-67 /97 calculator to implement the procedure. (Author)*
NPS/DKL Location: GENERAL V214 .S56

Smith, Billy Joe and Larry Wayne Vice. **Probability and Its Applications to Antisubmarine Warfare**. Monterey, CA: Naval Postgraduate School; Springfield, VA: Available from National Technical Information Service, 1975. (ADA009913). 99 p.
Thesis (M.S. in Systems Tech.)--Naval Postgraduate School, 1975.
Abstract: *A basic course with applications of probability to ASW, the course consists of six lesson plans and a 'Study Guide.' The lesson plans are designed to give the/an instructor guidance in what to teach, the depth required and objectives the student should be able to accomplish. The 'Study Guide' provided is for the use of both the instructor and the student, and it should serve as a basic text for the course. Current ASW tactical publications were examined by the authors while developing the course, and as many of the probability applications and as much associated probability terminology from these sources as practicable (and when this could be accomplished at the 'unclassified' level) are incorporated in the course.*
NPS/DKL Location: THESIS S5737

Smith, Thomas Joseph. **Two Evaluative Models for a Family of Submarine Versus Submarine Expanding Square Search Plans**. Monterey, CA: Naval Postgraduate School; Springfield, VA: Available from National Technical Information Service, 1973. (AD7698244). 94 p.
Thesis (M.S. in Operations Research)--Naval Postgraduate School, 1973.
Abstract: *The thesis investigates the effectiveness of a search plan developed by B. O. Koopman in a submarine versus submarine search situation. Two computer simulation models allow probability of target*

detection as a function of sonar range to be used as a measure of effectiveness. The Koopman search plan is analyzed and a family of alternate search plans are developed. The choice of a particular alternate search plan is dependent on the parameters of the problem. These parameters are target speed, searcher speed, time late to the search area and total time available to conduct the search. By use of the computer programs a search plan can be chosen so as to maximize the probability of target detection at a particular sonar range for each combination of input parameters. (Author)
NPS/DKL Location: THESIS S6054

Stair, Sammy Dean. **The Application of the Kalman Filter to the Sonobuoy Reference System on the S3A**. Monterey, CA: Naval Postgraduate School; Springfield, VA: Available from National Technical Information Service, 1972. (AD753608). 33 p.
Thesis (M.S. in Operations Research)--Naval Postgraduate School, 1972.
Abstract: *In 1960 R. E. Kalman introduced a least square concept that gives optimal estimates for the state of some dynamic systems. Included is a brief historical introduction leading to his work, a summary of his work, and the application of the theory to the sonobuoy reference system used on the S3A aircraft. Also, a tutorial development of certain quantities used in the filter is presented. (Author)*
NPS/DKL Location: THESIS S6715

Story, William Ferguson. **A Short History of Operations Research in the United States Navy**. Monterey, CA: Naval Postgraduate School; Springfield, VA: Available from National Technical Information Service, 1968. (AD855532). 95 p.
Thesis (M.S. in Operations Research)--Naval Postgraduate School, 1968.
Abstract: *The thesis traces the history of the practice of and organization for operations research in the United States Navy. The author points out that operations research was being conducted in the U. S. Navy before operations research became identified as a separate science. From that point its growth, major accomplishments and organizational changes are described. The final part of the thesis outlines the organization through which the Navy conducts its operations research and systems analysis at the present. (Author)*
NPS/DKL Location: THESIS S767

Stuart, Jay C. **Optimal Evasive Trajectories of an Isotropic Acoustic Radiator.**
Monterey, CA: Naval Postgraduate School; Springfield, VA: Available from National Technical Information Service, 1975. (ADA019378). 68 p.
Thesis (M.S. in Operations Research)--Naval Postgraduate School, 1975
Abstract: *The effects of detection equipment integration time on the optimal evasive trajectory of an isotropic acoustic radiator are studied. The boundary cases of infinite and zero integration time are examined. The infinite integration time case is formulated as a control problem and a maximum principle solution is obtained. The results consist of advice as to the choice of control vectors. The zero integration time problem is formulated in ordinary differential equations and the results consist of control vector advice. The relative movement plots and control vectors of the two bounding cases are compared.*
NPS/DKL Location: THESIS S85715

Sullivan, Daniel J. **A Correlation Study of Some Factors Effecting Submarine Detection by Destroyer Mounted Sonars**. Monterey, CA: Naval Postgraduate School; Springfield, VA: Available from National Technical Information Service, 1970. (AD722573). 47 p.
Thesis (M.S. in Operations Research)--Naval Postgraduate School, 1970.
Abstract: *A FORTRAN IV computer program was employed to conduct a statistical analysis of data collected during fleet antisubmarine warfare exercises. The object of the investigation was the*

identification of those variables which had greatest influence on a destroyer's ability to detect a submarine under certain conditions. The variables were treated as a random vector arising from one of two multivariate normal populations with common covariance matrix. An artificial regression relation was formulated to facilitate development of a linear discriminant function in a subset of those variables found to be of dominant importance. This latter subset was identified by examination of multiple correlation coefficients. The discriminant function was found to be seventy five per cent effective in classifying the experimental data correctly. (Author)
NPS/DKL Location: THESIS S8584

The'berge, Marc W. **Three Case Studies of Management Information Systems**. Monterey, CA: Naval Postgraduate School; Springfield, VA: Available from National Technical Information Service, 1990. (ADA238310). 55 p.
Thesis (M.S. in Information Systems)--Naval Postgraduate School, 1990.
Abstract: *The Naval Postgraduate School must, by default, make use of teaching cases in information technology case studies oriented or based upon corporations. It has been difficult for the school to obtain such studies oriented to the military, much less the United States Navy. This thesis provides the Naval Postgraduate School with three teaching cases concerning automated information systems serving the administrative and operational needs of unit-level command organizations.*
NPS/DKL Location: THESIS T3634

Timmerman, Michael Jay. **Genetic Algorithm Based Anti-Submarine Warfare Simulator**. Monterey, CA: Naval Postgraduate School; Springfield, VA: Available from National Technical Information Service, 1993. (ADA274956). 110 p.
Thesis (Master of Computer Science)--Naval Postgraduate School, 1993.
Abstract: *This research was aimed at improving the genetic algorithm used in an earlier anti-submarine warfare simulator. The problem with the earlier work was that it focused on the development of the environmental model, and did not optimize the genetic algorithm which drives the submarine. The improvements to the algorithm centered on finding the optimal combination of mutation rate, inversion rate, crossover rate, number of generations per turn, population size, and grading criteria. The earlier simulator, which was written in FORTRAN-77, was recoded in Ada. The genetic algorithm was tested by the execution of several thousand runs of the simulation, varying the parameters to determine the optimal solution. Once the best combination was found, it was further tested by having officers with anti-submarine warfare experience run the simulation in various scenarios to test its performance. The optimum parameters were found to be: population size of eight, five generations per turn, mutation rate of 0.001, inversion rate of 0.25, crossover rate of 0.65, grading criteria of sum of the fitness values of all alleles while building the strings, and checking the performance against the last five environments for the final string selection. The use of these parameters provided for the best overall performance of the submarine in a variety of tactical situations. The submarine was able to close the target and execute an attack in 73.1% of the two hundred tests of the final configuration of the genetic algorithm.*
NPS/DKL Location: THESIS T496

Tisdale, Vance S. **Investigation of Initial Detection Models in the Search and Localization Tactical Decision Aid (SALT)**. Monterey, CA: Naval Postgraduate School; Springfield, VA: Available from National Technical Information Service, 1990. (ADA238698). 52 p.
Thesis (M.S. in Operations Research)--Naval Postgraduate School, 1990.
Abstract: *The goal of this thesis is to investigate the initial search planning phase of the Search and Localization Tactical Decision Aid (SALT) developed by METRON, Incorporated of McLean, VA. SALT is a Computer Assisted Search (CAS) program intended for use by P3 UPDATE IV crews to assist them in optimal deployment of a sonobuoy field to prosecute a submarine threat. The initial search planning phase of SALT takes as user inputs environmental data, an initial elliptical Search Probability Area, an assumed target motion model, and the duration of the search. Outputs include a recommended sonobuoy*

pattern and the probability of detection of this pattern. The investigation of this phase of the algorithm is conducted in two parts. First, a series of simulation routines is used to ensure that the probability of detection of the sonobuoy patterns generated by SALT is mathematically correct. Second, these same computerized simulation routines are used to determine if there are alternate sonobuoy patterns that result in higher probabilities of detection.
NPS/DKL Location: THESIS T555

Tobin, Albert A. **Analyst's and User's Guide to the Passive Sonar Model in the Naval Warfare Interactive Simulation System**. Monterey, CA: Naval Postgraduate School; Springfield, VA: Available from National Technical Information Service, 1984. (ADA147055). 118 p.
Thesis (M.S. in System Tech.)--Naval Postgraduate School, 1984.
Abstract: *This user's guide examines the passive sonar model used by the Naval Warfare Interactive Simulation System (NWISS). The processes by which passive sonar detections are made are discussed. The thesis includes an explanation of how to affect those processes in order to control the interaction and results of an NWISS ASW scenario. A method for determining a sonar system's figure of merit and estimating ranges of detection is presented for the benefit of the operator who prepares the scenario, as well as for the user. This method is primarily intended for use with NWISS in its tactical training role.*
NPS/DKL Location: THESIS T577

Traganza, Eugene Dewees. **The Use of Temperature and Color in Satellite Detection of Ocean Fronts and Eddies for ASW Applications. A Summary of Selected Literature Condensed and Edited**. Monterey, CA: Naval Postgraduate School; Springfield, VA: Available from National Technical Information Service, 1979. (NPS-68-79-008, ADA069955). 60 p.
Technical report, period July to September 1978.
Abstract: *The purpose of this report is to briefly update the state of the art of detection of ocean fronts and eddies by satellite sensed sea surface temperature and to consider oceanographic color characteristics which may be used to detect their presence when the sea surface temperature pattern is absent.*
NPS/DKL Location: GENERAL G70.4 .T76

Trelles Sánchez, Jorge. **Prediction of the Far-Field Beam Pattern of a Random Noise Source from Measurements Made in the Near-Field**. Monterey, CA: Naval Postgraduate School; Springfield, VA: Available from National Technical Information Service, 1976. (ADA035850). 54 p.
Thesis (M.S. in Engineering Acoustics) Naval Postgraduate School, 1976.
Abstract: *A theory is presented for computing the far field beam patterns from distributed random noise sources. The theoretical model uses the Green's Function for the wave equation and the space-time autocorrelation function for determining the radiation from a randomly vibrating area. The actual far field beam pattern of a horn speaker in an anechoic chamber was obtained, and also near field measurements were taken to obtain the correlation distance and the mean square of the particle velocity using the autocorrelation function. Finally a computer program was written to evaluate the integral wave equation by numerical methods. It was found that the critical parameters in the mathematical model were the correlation distance and the frequency limits of integration. Small variations in the correlation distance modified greatly the width of the predicted beam pattern, while changes in the limits of integration had a moderate effect. The Frequency Spectrum was obtained in the anechoic chamber and it was used to determine the limits of integration of the integral solution for the intensity field.*
NPS/DKL Location: THESIS T794

Tritten, James John. **Naval Arms Control: An Idea Whose Time Has Yet to Come**. Monterey, CA: Naval Postgraduate School; Springfield, VA: Available from National Technical Information Service, 1989. (NPS-56-89-015, ADA212743). 29 p.
Abstract: *Analysis of three major areas for naval arms control proposals: restrictions on strategic antisubmarine warfare, naval operations, and strategic antisubmarine warfare technology. Author reviews the goals of arms control and finds none of these three areas in need of regulation. Author concludes with a number of innovative areas for naval arms control in areas of doctrine, strategy, operations, and exercises with concrete recommendations and acceptable (to USN) fallback positions. (SDW)*
NPS/DKL Location: FEDDOCS D 208.14/2:NPS-56-89-015

_____. **Naval Arms Control: A Poor Choice of Words and an Idea Whose Time Has Yet to Come**. Monterey, CA: Naval Postgraduate School; Springfield, VA: Available from National Technical Information Service, 1990. (NPS-56-90-012, ADA226710). 77 p.
Abstract: *Author makes case that due to recent events, initiatives in areas of naval arms control are extremely poorly timed. These events include political changes in USSR, the changing international security environment, the new Soviet military doctrine and strategy, ongoing arms control negotiations, unarticulated U.S. and NATO goals, and changes in Soviet and U.S. planning assumptions and scenarios. Author then analyses three major areas for naval arms control proposals: (1) restrictions on strategic antisubmarine warfare, (2) naval operations, and (3) strategic antisubmarine warfare technology and fanks them on technical grounds. Author reviews the goals of arms control and finds none of these three areas in need of formal regulation. Author concludes with a number of innovative areas for naval arms control in areas of doctrine, strategy, operations, and exercises with concrete recommendations and acceptable (to USN) fallback positions. (Author) (kr)*
NPS/DKL Location: FEDDOCS D 208.14/2:NPS-56-90-012

_____. **Scenarios of Nuclear Escalation Dominance and Vulnerability**. Monterey, CA: Naval Postgraduate School; Springfield, VA: Available from National Technical Information Service, 1988. (NPS-56-88-013, ADA195668). 43 p.
Abstract: *This document examines the role of strategic missile - carrying submarines in deterrence and mission of attacking these forces during the conventional phase of a war. Strategies considered include wars originating in varying regions. Included is a discussion of varying locations for submarine deployments impacting on potential Antisubmarine Warfare campaign. Escalation considerations include vertical, horizontal, and time. It concludes with analysis of possible arms control regulations. (KR)*
NPS/DKL Location: FEDDOCS D 208.14/2:NPS-56-88-013

_____. **Withholding and Attacking SSBNs**. Monterey, CA: Naval Postgraduate School; Springfield, VA: Available from National Technical Information Service, 1988. (NPS-56-88-004, ADA189439). 23 p.
Abstract: *This report examines the role of strategic missile carrying submarines in deterrence and the mission of attacking these forces during the conventional phase of a war. It includes discussion of varying locations for submarine deployments impacting on potential antisubmarine warfare campaigns, and also analyzes possible arms control regulation of ASW.*
NPS/DKL Location: FEDDOCS D 208.14/2:NPS-56-88-004

Van Train, WA. **Submarine Barrier Analysis by a Computer War Game Method**. Monterey, CA: Naval Postgraduate School; Springfield, VA: Available from National Technical Information Service, 1961. (AD480912). 69 p.
Thesis (M.S. in ??) Naval Postgraduate School, 1961.

Abstract: *The computer war game is emerging as a vital tool for finding near-optimal solutions to current military problems. A computer war game designed to permit parametric analysis of a submarine barrier is developed. Simulation techniques both mathematical and computer, are discussed. The effects of assumptions inherent in the computer war game are described. Illustrative analyses conducted through use of this computer war game are exhibited. Potential uses and methods for improvement of the developed war game are discussed. (Author)*
NPS/DKL Location: THESIS V32

Vebber, Paul W. **Examination of Target Tracking in the Antisubmarine Warfare Systems Evaluation Tool (ASSET).** Monterey, CA: Naval Postgraduate School; Springfield, VA: Available from National Technical Information Service, 1991. (ADA245802). 129 p.
Thesis (M.S. in Applied Science)--Naval Postgraduate School, Sept. 1991.
Abstract: *The role of the Maneuvering Target Statistical Tracker (MTST), a Kalman filter tracking algorithm based on the Integrated Ornstein-Uhlenbeck (IOU) motion process, in the Antisubmarine Warfare System Evaluation Tool (ASSET) is examined and its operation described. ASSET is a campaign simulation which models open-ocean ASW scenarios featuring prosecution of hostile submarines by friendly submarines and aircraft based on cues provided by data fusion centers. The heart of each data fusion center is an MTST which integrates new contact information into tracks. Comparing the level of sophistication of the tracking algorithm to that of the contact data provided to it, a number of simplifications are proposed. These include using reduced complexity IOU prediction and Kalman filter equations; the use of preprocessed variance data together with the true position of targets to estimate, rather than explicity calculate, updated track states; and limiting contact processing based on information content. Results indicate a good simulation of tracker output is produced using a greatly simplified algorithm. This technique can be generalized to other types of simulations involving target tracking.*
NPS/DKL Location: THESIS V3615

Vermillion, George M. **Development of a Decision Aid for Passive Acoustic Localization Using Knowledge of the Environment.** Monterey, CA: Naval Postgraduate School; Springfield, VA: Available from National Technical Information Service, 1983. (ADA128604). 87 p.
Thesis (M.S. in Systems Technology)--Naval Postgraduate School, 1983.
Abstract: *No abstract available.*
NPS/DKL Location: THESIS V417

Walsh, Raymond Michael. **The Effect of False Contacts on the Probability of Detecting a Submarine.** Monterey, CA: Naval Postgraduate School; Springfield, VA: Available from National Technical Information Service, 1970. (AD713394). 65 p.
Thesis (M.S. in Operations Research)--Naval Postgraduate School, 1970.
Abstract: *A computer war game is developed to measure the effect of false contacts on the probability of detecting a submarine. The variables are the probability of correctly classifying a non-submarine contact, the probability of correctly classifying a submarine contact, and the false contact density. A scenario is developed to focus on the false contact problem while holding other ASW variables constant. It is concluded from the output of the game that the effect of false contacts is deeply embedded in the interrelationships between units. (Author)*
NPS/DKL Location: THESIS W22284

Washburn, Alan. **TBMs and the Flaming Datum Problem.** Monterey, CA: Naval Postgraduate School; Springfield, VA: Available from National Technical Information Service, 1993. (NPS-OR-93-015, ADA273215). 13 p.

Abstract: *Theater Ballistic Missile launching systems are vulnerable just after a missile is launched because the missile's track can be extrapolated backwards to the location of the launcher. The situation is similar to one where a submarine torpedoes a ship, thus creating a flaming datum near which ASW forces may concentrate a search for the submarine. This report describes how some simple analytic methods adapted from ASW can be applied to the task of locating the TBM launcher. TBM, Scud, Search.*
NPS/DKL Location: FEDDOCS D 208.14/2:NPS-OR-93-015

Waterman, Larry Wayne. **Officer Education and Training in Oceanography for ASW and Other Naval Applications**. Monterey, CA: Naval Postgraduate School; Springfield, VA: Available from National Technical Information Service, 1972. (AD741135). 218 p.
Thesis (M.S. in Oceanography)--Naval Postgraduate School, 1972.
Abstract: *The study into the knowledge and experience required for optimum performance by officers assigned to operational, R+D, and managerial duties in Anti-submarine Warfare concludes that oceanography should receive the major emphasis in an interdisciplinary graduate level program of the contributing disciplines in ASW. In planning education and training for officers in ASW and other oceanography-related duties the total Service experience should be considered. Oceanography graduate curricula are recommended which will provide knowledge for developing careers of three categories of officers who respectively will: 'specialize' in ASW; become special duty 'environmentalists'; and serve in technical management assignments. Billets are identified for each of these categories. (Author)*
NPS/DKL Location: THESIS W22997

Whalen, Harold R. III. **Preliminary Software Design for a Personal Computer-Based Antisubmarine Warfare Tactical Flight Simulator**. Monterey, CA: Naval Postgraduate School; Springfield, VA: Available from National Technical Information Service, 1985. (ADA162255). 75 p.
Thesis (M.S. in Information Systems)--Naval Postgraduate School, 1985.
Abstract: *This thesis provides a preliminary software design for an Antisubmarine Warfare Tactical Flight Simulator. The simulation uses AN/ASN-123 Tactical Navigation Set (TACNAV) display symbology and selectable graphic functions to track and localize a single fully-evasive submarine. The primary design objectives are flexibility, utility, and understandability. A composite design methodology including levels of abstraction, information hiding, coupling, and cohesion as modularization criteria is used to effect a top-down modular decomposition of the simulation. A hierarchical structure is developed and modular packaging is discussed. Some aspects of physical implementation are also discussed and appropriate recommendations made. (Author)*
NPS/DKL Location: THESIS W48429

Wichert, Terry S. **Feature Based Neural Network Acoustic Transient Signal Classification**. Monterey, CA: Naval Postgraduate School; Springfield, VA: Available from National Technical Information Service, 1993. (ADA263437). 111 p.
Thesis (M.S. in Applied Physics)--Naval Postgraduate School, 1993.
Abstract: *Utilization of neural network techniques to recognize and classify acoustic signals has long been pursued and shows great promise as a robust application of neural network technology. Traditional techniques have proven effective but in some cases are quite computationally intensive, as the sampling rates necessary to capture the transient result in large input vectors and thus large neural networks. This thesis presents an alternative transient classification scheme which considerably reduces neural network size and thus computation time. Parameterization of the acoustic transient to a set of distinct characteristics (e.g. frequency, power spectral density) which capture the structure of the input signal is the key to this new approach. Testing methods and results are presented on networks for which computation time is a fraction of the necessary with traditional methods, yet classification reliability is maintained. Neural network acoustic classification systems utilizing the above techniques are compared*

to classic time domain classification networks. Last, a case study is presented which looks at these techniques applied to the acoustic intercept problem.
NPS/DKL Location: THESIS W5725

Wilde, Robert L. **Comparative Analysis of a CV Helicopter and a JVX (Joint Services Advanced Vertical Lift) Tilt-Rotor Aircraft in an Aircraft Carrier Based ASW (Anti-Submarine Warfare).** Monterey, CA: Naval Postgraduate School; Springfield, VA: Available from National Technical Information Service, 1985. (ADA156871). 134 p.
Thesis (M.S. in Operations Research)--Naval Postgraduate School, 1985.
Abstract: *This thesis analyzes the environmental compatibility and the potential performance capabilities of two proposed types of vertical flight capable aircraft in an aircraft carrier Anti-Submarine role. The aircraft compared are the CV Helicopter(SH-60F) and an ASW variant of the Joint Services Advanced Vertical Lift (JVX) tilt-rotor aircraft. This thesis compares their adaptability and relative expected mission effectiveness by analyzing their physical dimensions and characteristics and their projected flight performance parameters. Their expected performance in a specific scenario, an ASW pouncer mission employing active dipping sonar, is analyzed using a simulation model. Keywords: Time On-Station; Search and rescue; Flight deck; Detection probability; Helicopter in-flight refueling; Tilt-rotor; and Dipping sonar.*
NPS/DKL Location: THESIS W58526

Wile, Ted Shannon. **Sealane Defense: An Emerging Role for the JMSDF.** Monterey, CA: Naval Postgraduate School; Springfield, VA: Available from National Technical Information Service, 1981. (ADA114547). 176 p.
Thesis (M.A. in National Security Affairs)--Naval Postgraduate School, 1981
Abstract: *Japan's economy, the third largest in the world, is totally dependent on the sea lines of communication for the importation of 90 percent of its energy requirements and strategic metals and for over 70 percent of its food. Despite the importance of the sealanes to Japanese security, the Japanese Maritime Self Defense Force (JMSDF) remains incapable of protecting those sealanes against interdiction. Although the JMSDF is currently the seventh largest navy in the world, future expansion has been stymied by Japan's steadfast refusal to increase defense spending above one percent of the GNP. Long-range procurement plans focus on qualitative improvements with a primary emphasis on anti-submarine warfare, a strategy which could foreshadow a building program to enable the JMSDF to control the vital sea lanes. On the other hand, political and domestic constraints on a strong military indicate a continuing reliance on the United States for Japan's security. This study examines the factors affecting military decision-making in Japan, looks into the problems and realities of sealane defense and analyzes the future prospects for the JMSDF. (Author)*
NPS/DKL Location: THESIS W5877

Wirtz, JJ. **Allies and Theater Missile Defense: The Benefits of an ASW Approach to Counterforce.** Monterey, CA: Naval Postgraduate School; Springfield, VA: Available from National Technical Information Service, 1994. (NPS-OR-94-007, ADA282801). 40 p.
Abstract: *This report describes how the philosophy that influenced Anti-Submarine Warfare operations can be used to guide counterforce attacks against mobile missiles. It explains why an ASW approach to counterforce is superior to just attacking an opponent's missile infrastructure. It also explains why this type of counterforce strategy can be based on preemption not preventive war. The impact of ASW counterforce operations are also evaluated in terms of the stability-instability paradox, crisis stability, alliance relations and deterrence. TBM, Scud, Search, Crisis-stability, Counterforce, Alliances.*
NPS/DKL Location: FEDDOCS D 208.14/2:NPS-OR-94-007

Wright, Sherman E. **A Proposed Tactical Antisubmarine Warfare Service Utilizing Resonant** Bubbles. Monterey, CA: Naval Postgraduate School; Springfield, VA: Available from National Technical Information Service, 1973. (AD772789). 43 p. Thesis (M.S. in Engineering Acoustics)--Naval Postgraduate School, 1973.

Abstract: *An experimental investigation was made of the scattering properties of a bubble cloud in a sound field in a fresh water medium. The size of the bubbles was on the order of 0.125 cm radius, and was far above resonant size for the ensonifying sound field. It was determined that the bubbles scattered coherently in the forescatter direction, and incoherently in the backscatter direction. Based upon the scattering properties of the bubble cloud, it appears feasible to develop a device that could have tactical applications in the prosecution of long-range submarine contacts held by active sonars. Such a device would utilize the principle of resonant bubbles, and would require approximately 2.5 cubic feet of air (corrected to STP) to maintain a +20dB target strength for five minutes of continuous operation. (Author)*

NPS/DKL Location: THESIS W914

ANTISUBMARINE WARFARE (ASW) -- OTHER THESES & TECHNICAL REPORTS

Acker, David D. **Evaluation of the Effectiveness of the Defense Systems Acquisition Review Council (DSARC). Volume I. Technical Report with Appendices A and B**. Arlington VA: Information Spectrum Inc.; Springfield, VA: Available from National Technical Information Service, 1983. (ADA129795; ISI-V-3824-03-VOL-1). 149 p.
Final report 1969-1982.
Abstract: *The objective of this study is to evaluate the Defense System Acquisition Review Council (DSARC) process since its inception and to assess, in a qualitative sense, the degree to which the process has proved to be effective and efficient. In contrast to earlier studies, this study focuses on both the process and the supporting procedures from the standpoint of the program by examining impacts on programs reviewed. The study focuses on two specific areas, the actual process; and the procedures. The process in defined as the basic concept of decentralized management with centralized control of key decisions. The procedures are defined as those activities required to support the process.*
NPS/DKL Location: MICROFORM ADA129795

_____. **Evaluation of the Effectiveness of the Defense Systems Acquisition Review Council (DSARC). Volume II. Part 2. Appendices J through R**. Arlington VA: Information Spectrum Inc.; Springfield, VA: Available from National Technical Information Service, 1983. (ADA129797, ISI-V-3824-03-VOL-2-PT-2). 279 p.
Abstract: *The objective of this study was to evaluate the Defense System Acquisition Review Council (DSARC) process since its inception and to assess, in a qualitative sense, the degree to which the process has proved to be effective and efficient. The study focused on both the process and the supporting procedures from the standpoint of specific programs. Study Reports contained in this volume include: ROLAND Program Study Report; Copperhead Program Study Report; SOTAS Program Study Report; AV-8B Program Study Report; LAMPS Program Study Report; TRIDENT Program Study Report; FFG Program Study Report; HARPOON Program Study Report; and TACTAS Program Study Report.*
NPS/DKL Location: MICROFORM ADA129797

Antipov, I. And J. Baldwinson. **Detection Performance Prediction Model for a Generic Anti-Submarine Warfare Radar Mode**. Salisbury (Australia):. Defence Science and Technology Organisation, Electronics and Surveillance Research, 2001. (DSTO-TR-1101); (ADA389153). 94 p.
Abstract: *This report describes the design and implementation of a detection performance prediction model that has been developed for a generic Anti-Submarine Warfare mode of a maritime surveillance radar system. The model provides a map of the predicted average probabilities of detection as a function of range and look direction compared to the wind/swell direction for a user defined target in user defined operational and environmental conditions.*
Electronic access: http://handle.dtic.mil/100.2/ADA389153

Bermingham, William and Robert Wolfe. **MIL-STD-1553B Protocol Specification for P-3 Modernization**. Warminster,PA: Naval Air Development Center Software And Computer Directorate; Springfield, VA: Available from National Technical Information Service, 1982. (ADA111679; NADC-81089-50). 84 p.
Abstract: *This report describes a protocol developed for the P-3C Modernization Program utilizing the MIL-STD-1553B Data Bus. The protocol described in this report is intended as a menu of tools to be utilized, as the interfacing requirements govern, by the I/O and applications programmers. The work formats specified are a layer below that of the application software except where specifically mentioned.*

This protocol is intended to satisfy all currently envisioned P-3C interfacing requirements which fall into the realm of the restrictions and capabilities imposed by MIL-STD-1553B.
NPS/DKL Location: MICROFORM ADA111679

Berni, A. and L.Mozzone. **Wireless Tactical Networks in Support of Undersea Research**. La Spezia (Italy): SACLANT Undersea Research Centre, 2001. 10 p. Presented at RTO information Systems Technology Panel (IST), Istanbul, Turkey 9-11 Oct 2000. This article is from ADA391919, New Information Processing Techniques for Military Systems (les Nouvelles techniques de traitement de l'information pour les systemes militaires) p16-1/16-9.
Abstract: *Emerging concepts for Anti-Submarine Warfare (ASW) and Rapid Environmental Assessment (REA) increasingly rely on communication technology in order to implement distributed information networks and to exchange information between naval units and military commands ashore. The necessary communication links could be accomplished using a variety of solutions: our main focus is on radio frequency (RF) links which offer easy deployment and flexible operations. This document illustrates how spread-spectrum techniques can be adopted to substitute and enhance existing communications systems to permit the deployment of distributed scalable networks of ships and sensors characterized by reliable performance (resistance to hostile jamming and environmental interference) and low probability of interception. An overview of real applications in ASW and REA is presented.*
Electronic access: http://handle.dtic.mil/100.2/ADA391919

Bischoff, D.E. and R.E. Palmer. **Development of Lateral-Directional Equivalent System Models for Selected U. S. Navy Tactical Aircraft**. Warminster, PA: Naval Air Development Center Aircraft and Crew Systems Technology Directorate, 1983. (ADA141672; NADC-83116-60). 100 p.
Abstract: *The high order transfer functions representing the lateral directional responses to pilot control inputs were matched with low order equivalent forms in the frequency domain for five Navy tactical aircraft: the A-6, A-7, S-3, F-14, and F-18. The candidate low order equivalent forms investigated were: 1) the complete three degree of freedom representation of roll and sideslip angle responses, and 2) the single degree of freedom roll mode and Dutch roll approximations. Acceptable models were generally obtained for both forms. Simultaneous matching of sideslip and roll angle responses and/or apriori information for the roots was required to match the full three degree of freedom forms. The equivalent system models are discussed in terms of their match statistics. The equivalent system modal parameters, when compared against the requirements of the military flying qualities specification, demonstrate level 1 flying qualities for the conditions analyzed with the exception of roll angle time delay for the A-7 and F-18 airplanes. The A-7's lateral command augmentation structure results in Level 2-3 equivalent time delays, while the F-18's control force inputs produce Level 2 equivalent time delays. (Author)*
NPS/DKL Location: MICROFORM ADA141672

Breyer, S. **KYNDA Class 20 Years Old**. Washington DC: Naval Intelligence Support Center; Springfield, VA; Available from National Technical Information Service, 1982. (ADA119812; NISC-TRANS-6875). 9 p.
English translation of Marine Rundschau (Germany, F.R.) n5 p. 252-255 May 82.

Brook, Linton F. **Pricing Ourselves out of the Market: The Attack Submarine Program**. Newport, RI: Naval War College, 1979. (ADA079599). 47 p.
Abstract: *Present high costs of nuclear attack submarines have led to reduced procurement rates and will lead to significantly reduced force levels in the 1990's. The paper examines the impact of these reduced levels in order to suggest possible steps to mitigate their severity. An analysis is first made of possible roles for submarines under a variety of wartime scenarios; submarine employment is determined to depend more on invariant Soviet naval missions than on the precise nature of a future war. The*

interaction of U. S. submarine capabilities and Soviet Navy missions suggests the most important use of submarines is in anti-submarine warfare, both for sea control and for protection of carrier power projection forces. Dealing with the projected decrease in submarine force levels by reducing missions, improving effectiveness of existing forces and building more submarines are each examined; the examination suggests that no totally satisfactory solution exists given the probability of continued austere shipbuilding budgets. The analysis concludes that a mixed approach including procurement of less expensive (and less capable) nuclear submarines after 1985, extension of service life of some existing submarines, and various other steps is required to maintain submarine warfare capabilities. (Author)
NPS/DKL Location: MICROFORM ADA079599

CALS Test Network. **Technical Engineering Drawing Transfer Using: Loral Defense Systems' Data. Supporting: Naval Undersea Warfare Center, Vertical Launch ASROC Program (Code 416). MIL-STD-1840A, MIL-D-28000A (IGES). Quick Short Test Report**. Wright-Patterson AFB, Oh: CALS Test Network; Springfield, VA: Available from National Technical Information Service, 1994. (ADA 312310; AFCTN-TR-94-097; AFCTB-ID-94-055). 57 p.
Abstract: *The purpose of the informal test, reported in this QSTR, was to analyze moral Defense System's interpretation and use of the CALS standards in transferring technical engineering data. Loral used its CALS Technical Data Interchange System to produce data, in accordance with the standards, and delivered it to the AFCTN technical staff using an electronic ftp transfer. The stated purpose of this test was to evaluate the data and not the CALS format.*
Electronic access: http://handle.dtic.mil/100.2/ADA312310

Cooper, Keith and Ted Jones. **Project Execution Plan for the Installation of the SOCAL Acoustic Range (SOAR)**. Washington, DC: Chesapeake Division Naval Facilities Engineeering Command, 1984. (CHES/NAVFAC-FPO-1-84(4); ADA168632). 72 p.
Abstract: *The Southern California Acoustic Range (SOAR) is designed to provide a 100 square mile Anti Submarine Warfare training range in 4000 feet of sea water west of San Clemente Island, California. SAR will provide accurate tracking of air, surface and submerged targets. This plan is a working document that details the mobilization, execution and demobilization of the underwater portion of the SOAR project. The overall scenario of the project is to accomplish the following; (a) Prefabrication and assemble project materials at NOSC, San Diego; (b) Conduct training near Coronado beach; (c) Mobilize the OCP SEACON and UCT-2 personnel and equipment at West Cove, San Clemente Island; (d) Land the SSL cable and deploy the SSL system at sea; (e) Land the WQC cable and deploy the WQC transducer at sea; (f) Conduct a complete as-built survey (g) Demobilize SEACON and return all equipments and (h) Prepare a detail completion report.*
NPS/DKL Location: MICROFORM ADA168632

Cote, Owen R. **The third battle: innovation in the U.S. Navy's silent Cold War struggle with Soviet submarines.** Newport, RI: Naval War College, Center for Naval Warfare Studies, 2003. (Newport paper; no. 16). 104 p.
Contents: From Holland to the second battle: the first fifty years -- Phase I of the third battle: the German Type XXI and the early Cold War, 1945-1990 -- Phase II of the third battle: ASW and the two nuclear revolutions, 1950-1960 -- Phase III of the third battle: ASW and the happy times, 1960-1980 -- Phase IV of the third battle: ASW and acoustic parity, 1980-1990 -- The fourth battle? Submarines and ASW after the Cold War.
Electronic access: http://www.nwc.navy.mil/press/npapers/np16/NewportPaper16.pdf

DiOrio, David R. **Forward .... From Under the Sea Historical Perspective and Future Vision of Submarine Littoral Warfare**. Newport, RI: Naval War College Department of Operations, 1995. (ADA293706). 21 p.

Abstract: *This paper addresses the role of submarine warfare in today's national strategy. Analysis of submarine coastal operations during the Pacific War, specifically during the final campaign to invade mainland Japan, provides insight into submarine littoral warfare today. Following the decline of the Soviet Union, U.S. forces have focused on the application of maneuver warfare against emerging regional threats. Undoubtedly, this means control of the littoral regions of the world, where joint forces, including submarines, can influence events ashore. Included within the text is a historical perspective and future vision of submarine littoral warfare as it relates to operational maneuver from the sea.*

Electronic access: http://handle.dtic.mil/100.2/ADA293706

Dobeck, Gerald Joseph. **System identification and application to undersea vehicles**. Tampa, FL: USF, 1976. 164 l.
Thesis (Ph.D.)--University of South Florida, 1976.
NPS/DKL Location: GENERAL QA402 .D63

Domselaar, Gijsbertus E. Van. **Some Models Relating to Encounters between Ships in Transit and Submarines on Patrol**. La Spezia, Italy: SACLANT ASW Research Centre, 1980. (ADA095111; SACLANTCEN-SM-145). 31 p.

Abstract: *Recently in force employment studies several high level campaign models have been developed. These models mainly describe the interactions between Soviet submarines and NATO reinforcement and resupply shipping. Values of input parameters are required, which could either be determined by submodelling or simply be given. The choice highly depends on the sensitivity of the results to these input parameters and to the available information for developing submodels. This memorandum describes submodels for two input parameters to higher models: the encounter rate between transitting NATO ships and a submarine on patrol, and the average time a submarine will spend inside the convoy defences when attempting to attack a convoy. (Author)*

NPS/DKL Location: MICROFORM ADA095111

Dutton, L.M. **Flight Test Techniques Employed in the Nimrod MR Mk 2 Weapon System Performance Trials**. Boscombe Down, England: Aeroplane And Armament Experimental Establishment; Springfield, VA: Available from National Technical Information Service, 1984. (ADA147625). 13 p.
This article is from the Proceedings of the Flight Mechanics Panel Symposium Held in Lisbon, Portugal on 2-5 Apr 84, ADA147625, p. 14-1 - 14-13.

Abstract: *The Nimrod MR Mk 2 represents a large step forward in antisubmarine warfare (ASW) technology. Therefore aircraft operators need to know not only how accurately the new systems and sensors work, but also the best ways in which to use the overall system as a fighting machine. A&AEE Boscombe Down and the RAF Central Tactics and Trials Organisation joined forces to assess the ASW performance of the Nimrod in as near an operational environment as possible. The paper shows how the potential accuracy of the Nimrod's ASW system had to be matched by the precision of trials data collection, in the air and both on and below the sea surface. To gain such precision, the aircraft were extensively instrumented and the majority of the trials were conducted at the AUTEC Range in the Bahamas. The paper continues by explaining how the trials analysis technique had to match the variety of combinations of the data which were needed to make a statement on overall system performance.*

NPS/DKL Location: MICROFORM ADA147625

Egeberg, Lansing E., Lyle D Johnson and Neil H. Farlow. **Operation DOMINIC. Shot SWORDFISH. Project Officers Report - Project 2.1. Radiological Effects from an**

Underwater Nuclear Explosion. Santa Barbara, CA: Kaman Tempo, 1981. ADA995152; DNA-POR-2004(EX) DOE-WT-2004(EX)). 84 p. Extracted version of report dated 14 Jun 63, AD345025L.
NPS/DKL Location: MICROFORM ADA995152

Ford, Robert D. **The Excedrin Headache of ASW: From U-boats to the New Boats**. Newport, RI: Naval War College, 1997. (ADA328139). 21 p.
Abstract: *US forces today are under-trained in antisubmarine warfare, at a time when the world conventionally powered submarine base is at an all-time high. The conventional submarine poses a unique and potent threat to US forces, particularly in the littoral regions where ASW is the most difficult. The lessons of World War II, in which German U-boats inflicted great damage and caused a disproportionate diversion of Allied ASW assets, and the inability of British forces to detect the single Argentine submarine San Luis in the Falklands War underscore the relevance of proper planning to deal with the submarine threat in today's joint littoral warfare arena. The approach taken by JTF and Maritime Component staffs in countering the conventional submarine is critical in the success of the maritime forces achieving dominance as an enabling force in the joint littorals.*
Electronic access: http://handle.dtic.mil/100.2/ADA328139

Friedman, M.J., L.J. Cowles and R.C. Carson, Jr. **Application of Flight Performance Advisory Systems to U.S. Navy Aircraf**t. Warminster PA: Naval Air Development Center; Springfield, VA: Available from National Technical Information Service, 1986. (ADA182150). 14 p.
This article is from 'Efficient Conduct of Individual Flights and Air Traffic or Optimum Utilization of Modern Technology for the Overall Benefit of Civil and Military Airspace Users; Conference Proceedings of the Symposium of the Guidance and Control Panel (42nd) Held in Brussels, Belgium on 10-13 June 1986,' ADA182150, p. 62-1-62-14.
Abstract: *The U.S. Navy, is currently investigating methods for improving the fuel efficiency of Navy aircraft. Fuel saving concepts include an aircraft integrated flight performance advisory system, a pre-flight mission planning program using a desk type computer and an aircraft performance advisory system using an HP-41 DV hand-held calculator. The integrated flight performance advisory system for the F/A-18, the A-7E, and the S-3 are described in detail by reviewing the displayed outputs to the pilots and describing the required inputs and their sources. Features of each aircraft system are described in accordance with the development status of the program. The preflight mission planning program using an HP-9845 desktop computer is described for the P-3C aircraft. The approach to weather, takeoff and cruise are described by specifying input and output data. Sample displays are also shown. The hand-held HP-41 CV calculator used for flight performance predictions is described for takeoff and cruise flight modes of this aircraft.*
NPS/DKL Location: MICROFORM ADA182150

General Accounting Office, National Security and International Affairs Div. **Navy Maintenance: The P-3 Aircraft Overhaul Program Can Be Improved**. Washington, DC: General Accounting Office National Security and International Affairs Div; Springfield, VA: Available from National Technical Information Service., 1987. (ADA182095; GAO/NSIAD87-157). 29 p.
Abstract: *The P-3 is a shore-based, long-range aircraft designed to combat submarines. The Navy has 24 active P-3 squadrons, 13 reserve squadrons, and 5 squadrons for training and special projects. The P-3 inventory totals 441 aircraft. During its 30-year life, a P-3 is expected to undergo six overhauls at one of two Naval Air Rework Facilities, also known as depots. The study's objective was to determine whether the Navy could reduce depot overhaul turnaround time for the P-3 aircraft by improving overhaul procedures. Topics examined include: Selecting aircraft for overhaul; Inspections needed to ensure*

overhauls are necessary; Overhauls can be scheduled more efficiently; Labor resources can be applied more efficiently; and Depots have excess overhaul capacity.
NPS/DKL Location: MICROFORM ADA182095

_____. **Undersea Surveillance: Navy Continues to Build Ships Designed for Soviet Threat**. Washington, DC: General Accounting Office National Security and International Affairs Div; Springfield, VA: Available from National Technical Information Service, 1992. (ADA258570; GAO/NSIAD93-53). 40 p.
Report to the Acting Secretary of the Navy.
Abstract: *The Navy's Surveillance Towed Array Sensor System (SURTASS) program, like other defense programs, has been caught in the midst of rapidly changing world events. SURTASS sensors 'listen' for acoustic signals from enemy submarines in the deep, open ocean. However, the submarine threat for which SURTASS was designed has declined dramatically with the collapse of the Soviet Union. The United States no longer faces a well-defined nuclear submarine threat in the deep water ocean areas where strategic naval conflict and antisubmarine warfare operations were expected to occur. Instead, the Navy faces an ill-defined, less predictable regional threat from diesel submarines operating in shallow water areas. Yet, the Navy continues to build SURTASS surveillance ships designed for the deep water threat. In light of the recent world changes, we examined (1) how the submarine threat environment has changed and (2) what changes the Navy has proposed regarding its SURTASS program.*
NPS/DKL Location: MICROFORM ADA258570

_____. **Weapon Systems Overview: A Summary of Recent GAO (General Accounting Office) Reports, Observations and Recommendations on Major Weapon Systems**. Washington, DC: GAO; Springfield, VA: Available from National Technical Information Service, 1983. (ADA133323; GAO/NSIAD83-7). 103 p.
Report to the Congress.
Abstract: *To aid the Congress in its deliberations on the fiscal year 1984 defense budget, The General Accounting Office issued, from August 1982 through July 1983, 17 reports on selected weapon systems. Chapter 2 summarizes the potential impact of GAO's recommendations and observations. Chapter 3 categorizes and summarizes the major issues highlighted in each report. These issues could have a direct impact on the systems' efficient acquisition and/or operational effectiveness. These issues formed the bases for GAO's recommendations and observations. Chapters 4 through 7 contain individual report summaries. The systems which were reviewed and which are included in this overview are: AH-64; Army Helicopter Improvement Program; Patriot; Sergeant York; Stinger POST; S-3A; CG-47; Rapidly Deployable Surveillance System; TOMAHAWK; F/A-18; Over-the-Horizon Backscatter radar; The antisatellite development program; The Wide Area Antiarmor Munitions; The B-1B bomber; Light Armored Vehicle; Advanced Medium Range Air-to-Air Missile; and Trainer aircraft.*
NPS/DKL Location: MICROFORM ADA133323

Grant, S.C. **Assessing Intelligent Software Agents for Training Maritime Patrol Aircraft Crews**. Downsview (Ontario): Defence and Civil Inst. of Environmental Medicine, 2001. (DCIEM-TR-2001-036);(ADA400717). 36 p.
Abstract: *Training simulators often require the participation of several people to play the role of supporting players in the simulated operation. Use of intelligent software agents to play the role of these personnel has the potential to reduce support staff and increase an instructor's control of training. This report evaluates a simulator prototype developed for the CP140 Aurora maritime patrol aircraft that incorporated intelligent software agents to play the roles of the Tactical Navigator and an Acoustic Sensor Operator. Human crews, intelligent agent crews, and mixed human-agent crews performed a simulated antisubmarine mission. Mission performance and crew communications were recorded and rated to determine whether the intelligent software agents could perform individual crewmember functions and whether they could provide the interaction necessary for crew coordination training. The results indicate*

that: (1) agents can perform individual crewmembers' functions; (2) agent interaction with humans is sufficient to allow humans to perform their own tasks; and (3) the agents did not interact in a way suitable for crew coordination training. It is concluded that the prototype is suitable for supporting individual training, but the agents' knowledge base must explicitly address team dynamics if crew coordination training is to be supported.
Electronic access: http://handle.dtic.mil/100.2/ADA400717

Greenhill, S., S. Venkatesh, A. Pearce, T.C. Ly. **Representations and Processes in Decision Modelling**. Victoria: Defence Science and Technology Organisation, Aeronautical and Maritime Research Lab, 2002. (DSTOGD0318);(DSTOAR012146);(ADA405860). 75 p.
Abstract: *This report contains a survey by Curtin University on decision modelling which covers: (1) Our current understanding of how we make decisions, and points out our qualities, our weaknesses and the types of aids that could help us. Of note is the theory that people commit to options even though alternatives exist once a situation has been recognised. (2) Techniques useful for eliciting and representing knowledge about how experts make decisions. (3) What is situation assessment, and how others have tried to capture the process and use the captured information. (4) The different technologies that could be employed to represent the process of situation assessment. This report represents the first step of a larger project to represent how submarine commanders assess situations.*
Electronic access: http://handle.dtic.mil/100.2/ADA405860

Guidry, M. S. Whitley and B. Markowich. **Boeing 767 Proximity Evaluation with F/A-18C and S-3B Aircraft**. Patuxent River, MD: Naval Air Warfare Center Aircraft Div.; Springfield, VA: Available from National Technical Information Service, 2001. (ADA389851; NAWCADPAX/RTR-2001/17). 74 p.
Abstract: *The Boeing Company participated in the Future Strategic Tanker Aircraft program which was intended to provide aerial refueling and aerial transport capability to the United Kingdom Royal Air Force under a Private Finance Initiative. Boeing contracted NAWCAD Patuxent River, Maryland, under a commercial service agreement to determine if an area of acceptable wake turbulence existed in the proximity of a 767 aircraft in order to perform the aerial refueling mission. This was accomplished by evaluating the 767 aerodynamic and wake turbulence effects on two receiver aircraft (F/A-18C and S-3B) at locations behind the 767, which approximated potential aerial refueling engagement areas. During the period of 22 and 23 June 2000, three F/A-18 and three S-3B flights were flown totaling 5.8 F/A-18 flight-hours, 6.7 S-3B flight-hours, and 12.5 767 flight-hours. A Lear 35 cinematography aircraft was used to document test results. The test program consisted of proximity evaluations only with no aerial refueling pods installed on the 767 aircraft and no receiver-to-"tanker" engagements. All flights were conducted within the Patuxent River restricted or local warning areas. At the positions evaluated, areas of acceptable wake turbulence existed for the F/A-18C and the S-3B in the proximity of the 767 aircraft in order to perform the aerial refueling mission. Recommend that testing continue to evaluate the 767 tanker aircraft.*
Electronic access: http://handle.dtic.mil/100.2/ADA389851

Haralabus, G., E. Capriulo and W.M. Zimmer. **SWAC 4: Broadband Data Analysis Using Sub-Band Processing**. LA SPEZIA (ITALY): SACLANT Undersea Research Centre, 2000. (SACLANT Undersea Research Centre,); (SACLANTCEN-SR-320, ADA378129). 41 p.
Abstract: *The frequency dependence of broadband active detection/localization is examined. The analysis is based on 1200 Hz (2300-3500 Hz) LFM signals acquired during the SWAC 4 sea trial. A sub-band matched filter scheme is devised according to which a replica of the transmitted pulse is segmented into ten 120 Hz sub-bands and processed independently through a matched filter detector. Comparison of target detection and ranging results indicate comparable performance for all sub-bands. However, ping-to-ping variability of the ten correlator outputs suggest that the detection performance may be improved*

by employing incoherent processing schemes. Signal-to-noise ratio is proved to be controlled mainly by noise (reverberation is the predominant noise source) rather than signal variations. The signal intensity remains proportional to the distance between source and receiver due to favorable propagation conditions. Doppler effects and sub-band detection synchronization problems which may lead to performance degradation in large time-bandwidth signal processing are addressed. A method to estimate range rate (relative velocity between source and receiver) based on single ping differential time delay between sub-band MF outputs is developed. This intra-ping technique is an alternative to the standard inter-ping method which requires multiping detection history.

Electronic access: http://handle.dtic.mil/100.2/ADA378129

_____, _____ and _____. **SWAC 4: Broadband Data Analysis Using Sub-Band Processing - Part 2**. La Spezia (Italy): SACLANT Undersea Research Centre, 2001. (SACLANTCEN-SR-337; ADA389996). 38 p.

Abstract: *The frequency dependence of reverberation is examined using the processing method as for the frequency analysis of target detection during the same experiment. In this experiment, reverberation is induced by abrupt changes in the bottom bathymetry (a 200 m sea mount). For the analysis of the received signal a sub-band matched filter scheme is devised, according to which, a replica of the transmitted pulse (2300 Hz-3500 Hz LFM signal) is segmented into ten 120 Hz sub-bands, each of which is processed independently through a matched filter detector. Following the necessary corrections for array gain and calibration, transmitted power spectrum and propagation loss, the matched filter data are compared to reveal the frequency dependence of reverberation. Due to insufficient in situ measurements, the propagation loss estimate is based on model calculations - a challenging task for the range dependent seafloor at the experimental site. After examining a large number of pings it is concluded that the reverberation energy calculated at the correlator output is comparable for all ten sub-bands. This leads to the conclusion that for the particular environment and experimental geometry, the frequency spectrum is not sufficiently wide to allow significant frequency variability which may indicate an optimum operational frequency.*

Electronic access: http://handle.dtic.mil/100.2/ADA389996

Ingenito, F., R. H. Ferris, W. A. Kuperman, S. N. Wolf. **Shallow Water Acoustics: Summary report on Phase 1**. Washington, DC: Naval Research Lab, 1978. (ADA057691, NRL-8179). 57 p.

Abstract: *In response to the Navy's need for a submarine warfare capability in shallow water areas of the oceans, NRL has been conducting a research program in shallow-water acoustics. The goal of the first phase of this program has been to determine if wave theory can be used to predict the acoustic field at long rangesfrom a submerged acoustic source. The approach used an iterative process involving trial models and at-sea measurements. The wave equation for the physical model is solved by numerical methods and implemented on a high-speed general-purpose computer. Since the acoustic field at long ranges is propagated in the discrete normal modes of the duct, special experimental methods were used to resolve individual modal fields so that their measured characteristics could be compared with predictions. This report presents a detailed description of the NRL normal-mode model in its current form and describes the experimental evaluation procedures and results. Salient features of the model include variable sound speed in the water, slowly variable water depth, statistically rough boundaries, sediment layering, and both shear-wave and compressional-wave propagation in the bottom. Although certain recognized problems remain to be solved, it has been demonstrated that the model can in most cases predict the characteristics of the signal field with sufficient accuracy to be a sueful tool in system design, performance prediction, and tactics.*

NPS/DKL Location: MICROFORM ADA057691

Inspector General, Dept of Defense. **Second Source Procedures for the ANSQQ-89 Combat System**. Arlington, VA: Dept of Defense; Springfield, VA: Available from National Technical Information Service, 1991. (ADA379828; IG/DOD-91-088). 35 p.

Abstract: *This audit was performed at the request of Representative John Conyers, Jr., Chairman of the House Committee on Government Operations. The Chairman requested that we perform an audit of the procedures used by the Navy in soliciting a second source for the production of the AN/SQQ-89 ASW Combat System. The request was based on information that GE's cost and pricing data may have been disclosed to WEC, and that WEC may not have been qualified to produce the combat system.*
Electronic access: http://handle.dtic.mil/100.2/ADA379828

Jorgensen, Jason T. **The United States Navy's Ability to Counter the Diesel and Nuclear Submarine Threat With Long-Range Antisubmarine Warfare Aircraft**. Fort Leavenworth, KS: Army Command And General Staff College, 2002. (ADA406874). 119 p.
Master's thesis.
ABSTRACT: *The threat of the Soviet Union and Communism to the United States diminished with the end of the Cold War in the early 1990s. Instead, the asymmetric threat of terrorism has spread throughout the world and become a grave danger to Amen can citizens at home and abroad. Throughout these changes in global landscape, the US Navy has adapted and given new emphasis to a variety of missions during these times of fiscal challenge. However, one of the most dangerous weapons of the Cold War, the submarine, still exists and is being proliferated widely today. Once the primary ASW aircraft used in the prosecution of submarines, the P-3C Orion, has added new equipment to perform its added warfare missions. Thus, the central focus of the thesis: Does the US Navy have the airborne capability to defend itself from current as well as projected submarine threats? The thesis will examine the relevancy of ASW today and determine whether current and future submarines pose a threat to US, its interests as well as its military. The final analysis involves an evaluation of P-3C Orion's capability to detect adversary submarines in the contemporary as well as future operating environment.*
NPS/DKL Location: MICROFORM ADA406874
Electronic access: http://handle.dtic.mil/100.2/ADA406874

Jung, P.A. and R.C. Shaw. **Water Entry Structural Technique (WEST): An Analytical Technique to Determine Frangible Nosecap Behavior During Water Entry**. San Diego, CA: Naval Ocean Systems Center; Springfield, VA: Available from National Technical Information Service, 1989. (ADA219641; NOSC-TR-1317). 251 p.
Abstract: *Described here is a computational method to design frangible nosecaps for air- and surface-launched undersea weapons (such as for ASROC, VLA, and Mk-50 torpedoes). WEST is a technique that can rapidly and accurately assess the state of stress and deformation of missile nosecaps intended to break up at water entry. WEST links the powerful geometry and FEM pre- and post-processor PATRAN, a potential-flow computer code that can calculate dynamic pressure-time histories of an arbitrary entry body, and the nonlinear FEA code ABAQUS. This code linkage has been validated through comparison with experimental work. WEST is a valuable analytical tool that reduces the design cycle time for frangible nosecaps.*
NPS/DKL Location: MICROFORM ADA219641

Karlsson, P.A.; K. Ohlson and L. Pers. **Modellering av Tradstyrning av Torped Fran Ubatsjaktenyheter med Aktiva Sonarer (Modelling of Wire-Guidance of Torpedoes Fired by Anti-Submarine Units with Active Sonars)**. Stockholm: Div. of Systems Technology, Swedish Defence Research Agency, 2001. (FOI-R-0212-SE). 32 p.
In Swedish.
Abstract: *Duels between torpedoes and ships with counter measures for protection can be simulated with the computer model MUMS. In earlier versions of the model the simulated torpedoes are autonomous after firing. Recently passive sonars have been included which allows simulation of wire-guided torpedoes from submarines. The development described in this report comprises modeling of*

wire-guided torpedoes from submarines. The development described in this report comprises modeling of wire-guided torpedoes launched from surface ships and/or helicopters with active sonars.

Kaufman, A.I. and E. Zdankiewicz. **Gauging the Military Value of Naval Infrastructure**. Alexandria, VA: Istitute for Defense Analyses, 2001. (IDA-P-3605); (IDA/HQ01-000666); (ADA394149). 47 p.
Abstract: *This paper proposes a methodology for relating investments in naval infrastructure programs to investment programs in naval structure and illustrates the utility of such a methodology in trading infrastructure for structure by applying the methodology to organic mine countermeasure and shallow water antisubmarine operations.*
Electronic access: http://handle.dtic.mil/100.2/ADA394149

Kaye, G. T., Roger Nies and Michael Lovern. **Uplink Laser Propagation Measurements Through the Sea Surface, Haze and Clouds**. San Diego, CA Naval Command Control and Ocean Surveillance Center RDT&E Div.; Springfield, VA: Available from National Technical Information Service, 1993. (ADA264687). 14 p.
Abstract: *An Airborne Optical Receiver (AOR) was developed and tested to investigate the propagation and reception of optical communications uplinks from a submerged laser source to an overflying fleet aircraft. The AOR was flown in a P-3C Orion aircraft for an at-sea test off the southern California coast in August, 1990. A green laser transmitter was suspended from the Research Platform FLIP at depths of 15 to 45 m. During six nights of operations, the AOR received the laser light at various test geometries and through clear and cloudy conditions. This represents the first optical uplink cloud experiment at visible wavelengths. Results show that optical pulses in clouds are significantly more forward-scattered than modeled. The results can be explained by Mie scattering theory. Measured cloud attenuation and pulse stretching agreed with an existing optical propagation model. Significant attenuation and signal spreading due to haze and fog was measured and compared with theory.*
NPS/DKL Location: MICROFORM ADA264687

Kotchka, Jerry Allen. **On a Bayesian methodology to the solution of the Naval ASW screen placement problem**. Columbus, OK: OSU, 1970. 182 l.
Thesis (Ph. D.)--Ohio State University, 1970.
NPS/DKL Location: THESIS K8345

Kuhta, Steven F., Carol L. Kolarik, Samuel N. Cox, Daniel C. Hoagland and James F. Dinwiddie. **TRIDENT II: Reduction to MK-6 Guidance System Inventory Objectives May Be Possible**. Washington, DC: General Accounting Office National Security And International Affairs Div; Springfield, VA: Available from National Technical Information Service., 1994. (ADA283196; GAO/NSIAD94-192). 29 p.
Abstract: *The Navy plans to have 10 Trident II submarines by the end of fiscal year 1997. Currently, it has six operational Trident II submarines and four others are under construction. Each Trident II submarine carries 24 D-5 missiles. Each D-5 missile is equipped with the MK-6 guidance system, which is comprised of an inertial measurement unit and an electronics assembly. The inertial measurement unit senses velocity and direction and relays this data to the electronics assembly, which issues flight control commands to the missile.*
NPS/DKL Location: MICROFORM ADA283196

Lahey, George F. and Dewey Slough. **Relationships between Communication Variables and Scores in Team Training Exercises**. San Diego CA: Navy Personnel Research And Development Center; Springfield, VA: Available from National Technical Information Service, 1982. (ADA110117; NPRDC-TR-82-25). 25 p.

Abstract: *As investigation was made of the practicality of assessing anti-submarine warfare (ASW) team performance by means of measures of the volume of communications. A system for classifying communications was developed based on an analysis of published data on communication rates (e.g., number of evaluative messages sent per minute) of ASW helicopter crews. Next, communication rates were recorded for ship's teams during two exercises in the 14A2 ASW team trainer. Communication rates were computed for various types of messages over the ship-to-ship and ship-to-air circuits. Rates were compared against instructor grades for individuals, subteams, and teams. Communication rates on the intership circuit tended to be negatively correlated with grades, primarily because instructors gave lower grades to teams doing excessive talking. Rates on the ship to air circuit were positively correlated with performance on the later exercise where two aircraft were used rather than one and where a much greater volume of information needed to be transmitted. On the internal circuits, few significant relationships were found between communication rates and performance. Implications of the findings for development of an objective performance measurement system for team training are discussed. (Author)*
NPS/DKL Location: MICROFORM ADA110117

Landry, Normand. **The Canadian Air Force Experience: Selecting Aircraft Life Extension as the Most Economical Solution.** Ottawa: Department Of National Defence; Springfield, VA: Available from National Technical Information Service, 2000. (ADA381871). 11 p.
Presented at RTO SCI Symposium on Aircraft Update Programmes, The Economical Alternative?, Ankara, Turkey, 26-28 Apr 1999. This article is from ADA381871 Advances in Vehicle Systems Concepts and Integration. (les Avancees en concepts systemes pour vehicules et en integration).
Abstract: *Canada like several other countries has limited resources to trade-in its outdated and ageing fleets for state-of-the-art weapon systems. With the CF188 and the CP140, the Canadian Forces (CF) have chosen, as with the CF116 before, to perform a structural and systems upgrade. These upgrades will allow the aircraft to meet their operational requirements until the first quarter of the next century. The choice for this course of action is based on option analysis studies. In the end, fleet modernisation has proven to be The most economical solution. This paper will present the approach taken and the assumptions made for the various scenarios studied to reach That conclusion. Avionics packages are readily available off-the-shelf and in most cases the decision is based mostly on structural limitations. Hence in-service failures and results of full scale fatigue tests obtained through collaborative agreements can be a cost effective way to determine The cost of ownership of each fleet. The paper will briefly talk about The concept taken for the CP140 but will use the CF188 as the demonstration test case.*
Electronic access: http://handle.dtic.mil/100.2/ADA381871

Lohrenz, M., M. Trenchard and S. Edwards. **On-Line Evaluation of Cockpit Moving-Map Displays to Enhance Situation Awareness in Anti-Submarine Warfare and Mine Countermeasures Operations.** Stennis Space Center, MS: Naval Research Lab Marine Geosciences Div., 2001. (NRL/PP/7440--01-1008, ADA393119). 4 p.
Abstract: *Cockpit moving-map systems have provided heightened situation awareness to the fighter pilot for more than ten years, but these systems have yet to be integrated into military helicopters. The Navy now plans to install a moving-map system into its new, multi-functional MH-60S helicopter, which will perform mine countermeasures (MCM), combat search and rescue, special operations, and logistics. Other H-60 variants (e.g., SH-60B) perform anti- submarine warfare (ASW), surface warfare, surface surveillance, and other missions. Naval Research Laboratory scientists were tasked to demonstrate and evaluate the potential of a cockpit moving-map for enhanced situation awareness during multi-functional helicopter missions (particularly MCM and ASW). This project consisted of three main tasks: (1) conduct a web-based survey of pilots and aircrew experienced in MCM and ASW for their preferences with respect to various environmental data that could be displayed in a moving-map; (2) demonstrate and evaluate pilot-preferred data on existing moving-map displays; and (3) recommend potential data types to be collected and displayed in a multi- mission helicopter.*

Electronic access: http://handle.dtic.mil/100.2/ADA393119

Lorch, Dan amd John Quartuccio. **S-3A Ballast Block Final Design and Engineering Tests; Final report**. Warminster, PA: Naval Air Development Center Aircraft And Crew Systems Technology Directorate; Springfield, VA: Available from National Technical Information Service, 1984. (ADA147685; NADC-84015-60). 50 p.

Abstract: *The S-3A aircraft has (4) ejection seats. Both the pilot and copilot have Comand Eject Selector levers which allow them the option to eject all crewmembers or 'Self Eject.' If one of the aft seats is unoccupied, and 'Command Eject' is selected, the unoccupied seat will accelerate ahead of the occupied seat next to it. Two hazards exist; first, the crewmember next to the unoccupied seat could be burned by the rocket plume from the empty seat which has a higher acceleration; second, the empty seat could tumble into one of the other seats because the center of gravity and the center of rocket thrust are too far apart. To eliminate these hazards it is necessary to ballast the unoccupied seat. This is presently being done with anthropomorphic test dummies, if they can be obtained. Unfortunately these dummies have various weights and are usually damaged (i.e. arms, legs, or head missing). There is no guarantee that the center of gravity is in the proper location to prevent tumbling. To correct this potentially dangerous situation the Naval Air Systems Command tasked the Naval Air Development Center to design a ballast block. After the initial prototype was developed and tested, references (1) and (2) recommended changes to be incorporated into the final design. All of these recommendations have been incorporated into the final design. The S-3A Ballast Block is a 169 pound (77 Kg) assembly of four (4) interlocking aluminum blocks. It is used to control the trajectory of an unoccupied 1E-1 ejection seat. Tests indicate that it meets all functional and structural requirements for use in the S-3A aircraft.*

NPS/DKL Location: MICROFORM ADA147685

_____ and _____. **S-3A Ballast Block Final Design and Engineering Tests; Final report**. Warminster, PA: Naval Air Development Center Aircraft And Crew Systems Technology Directorate; Springfield, VA: Available from National Technical Information Service, 1984. (ADA327372; NADC-84015-60) 58 p.

Abstract: *The third prototype S-3A Ballast Block weighs 169 pounds (77 Kg). It is an assembly of four interlocking aluminum blocks. One crewman can carry two blocks at a time into the aircraft where he can quickly assemble the unit either on the 1E-1 ejection seat or on the avionics aisleway step Restraint on the ejection seat is obtained by connecting the four quick disconnect adjuster fittings on the ejection seat to fittings on the Ballast Block. When the Assembly is placed on the avionics aisle steps it is restrained with two aluminum locking plates which are bolted to the top block. These plates extend beyond the edges of the block and fit into keyways on either side of the main bulkhead forgings directly behind the aft ejection seats. When the Block is secured on the 1E1-1 seat the overall center of gravity falls 0.72 inches below the centerline of rocket thrust. The Ballast Block meets all operational and structural requirements for safe function in the aircraft. It can be maintained at the Operational level; the only parts that may need replacement are straps which are readily available. The S-3A Ballast Block provides a simple and cost effective replacement for anthropomorphic dummies presently being used to ballast unoccupied 1E1 ejection seats.*

Electronic access: http://handle.dtic.mil/100.2/ADA327372

Madey, S.L. and J.C. Petz. **SH-60B/DDG-994 Dynamic Interface Tests; Interim report 1**. Patuxent River, MD: Naval Air Test Center; Springfield, VA: Available from National Technical Information Service, 1984. (ADA383636; NATC-RW-49R-84). 15 p.

Abstract: *NAVAIRTESTCEN was tasked by reference (a) to conduct dynamic interface (DI) testing of the SH-60B helicopter on the DDG-993 class ships. Testing was conducted on board the USS CALLAGHAN, DDG-994, from 14 through 18 May 1984. Lack of ambient winds precluded completion of day/night launch/recovery envelopes. Further testing is possible 11 through 15 June 1984. Data were collected using the DI Pilot Rating Scale (PRS) presented in enclosure (1).*

Electronic access: http://handle.dtic.mil/100.2/ADA383636

Maskell, D.M. **Navy's Best-Kept Secret: Is IUSS Becoming a Lost Art?** Newport, RI: Naval War College, Joint Military Operations Dept., 2001. (ADA401150). 73 p. Master's thesis.

Abstract: *There were a series of events that led to the consolidation and downsizing of the Integrated Undersea Surveillance System (IUSS). These events occurred simultaneously during a period when the United States was, defining its National Strategy toward the Soviet Union. What caused the IUSS Downsizing and consolidation. Budget cuts. End of the Cold War (Change in National Strategy). The Dissolution of the Soviet Union as a threat. Base realignment and closure. Were the decisions made valid now that it is 2001, eight years after the consolidation.*

Electronic access: http://handle.dtic.mil/100.2/ADA401150

Matzelevich, William W. **Real-Time Condition Based Maintenance for High Value Systems**. Arete Associates, Arlington VA,; Springfield, VA: Available from National Technical Information Service 2001. (ADA412395). 11 p.

Presented at the Meeting of the Society for Machinery Failure Prevention Technology (55th) held in Virginia Beach, VA on 2-5 Apr 2001, p. 57-67. This article is from ADA412395, **New Frontiers in Integrated Diagnostics and Prognostics**, Proceedings of the 55th Meeting of the Society for Machinery Failure Prevention Technology. Virginia Beach, Virginia, April 2-5, 2001.

Abstract: *Many industries operate high value equipment often remotely -- that requires reliable performance in severe environments. Similarly, the U.S. Navy's submarine p (TASs) stress conventional approaches to operating and maintaining this system level capability comprised of integrated hydraulic, mechanical, electronic and acoustic sub-systems. The Navy invested in a Condition Based Maintenance (CBM) proof of concept for an individual ship TAS by developing the Thinline Health Monitoring System (THMS). THMS collects real-time discrete reliability data and synchronizes this data with other historical information and the TAS's current condition assessment. As a predictive "intelligent code" it uses Bayesian Belief Networks (BBNs) to extract the full value of real-time data and provide a complete range of system performance evaluations -- from diagnosis to prediction. Drawing upon THMS' success, the U.S. Navy supported expanding this capability fleet- wide to encompass health assessments of the entire submarine TASs population. Plans have been developed to build a relational database that is accessible to a geographically separated towed systems community via the Internet for interactive analysis and diagnostics. These system level analyses and first principal processes are directly translatable to other government and commercial critical systems that cannot afford unscheduled -- or unnecessary -- maintenance.*

NPS/DKL Location: MICROFORM ADA412395

McConnell, James M. **A possible change in Soviet views on the prospects for anti-submarine warfare**. Alexandria, VA: Center for Naval Analyses, Naval Planning, Manpower, and Logistics Division, [1985]. (ADA153610); (Professional paper; 431); Soviet Union special studies, 1982-1985; 9); (Special studies series (University Publications of America, Inc.). 19 p.

Abstract: *In the summer of 1982 there was an apparent shift in Soviet views on the future potential for combating submarines. The following points trace the perceived evolution of this shift. (1) From the early 1970s, Soviet emphasis had been on the submarine's great capacity for concealment and the decreasing cost effectiveness of anti-submarine warfare (ASW) as a 'law-governed' trend extending into the foreseeable future; (2) The first sign of a new perspective came in 1979-80; here, the Soviets implied that no significant breakthrough in ASW was expected during the next five-year plan (1981-85), but they did not rule out an effective innovation after that; (3) In 1982, however, the Soviets apparently saw an operational capability arising ahead of this schedule. Using alleged U.S. views as an almost certain surrogate for their own, they indicated that a 'technological break-through' in ASW (possibly nonacoustic*

and space- based) was imminent, perhaps (this is the best interpretation) before the end of the current planning period in 1985. A new 'law-governed' trend in naval affairs was set out: the growing susceptibility of submarines to detection and the increasing cost effectiveness of ASW; (4) If Moscow is on the verge of a long- range detection capability, then one might want to speculate on the means they would develop for submarine kill. It is conceivable thay they might revive the concept, abandoned in the 1970s, of using a submarine-launched ballistic-missile (SLBM) system for hitting mobile targets as sea.
NPS/DKL Location: MICROFORM ADA153610

McElhannon, Timothy S. **Operational Maneuver and Anti-Submarine Warfare.** Newport, RI: Naval War College Dept of Operations; Springfield, VA: Available from National Technical Information Service, 1995. (ADA298144). 21 p.
Abstract: *Operational maneuver, one of the principles of operational art, is key to the Navy's doctrine in From the Sea and Forward From the Sea. The objective of operational maneuver is to strike quickly and violently to isolate and frustrate the enemy and destroy their forces and will to fight. The application of operational maneuver can enable U.S. forces to overcome the shallow-water diesel submarine threat by using speed and concentrated fires to avoid the enemy's strengths and attack their weaknesses, thus isolating, neutralizing and destroying the threat.*
Electronic access: http://handle.dtic.mil/100.2/ADA298144

McMillan, T., W.P. de La Houssaye and C.T. Johnson. **OPERATION DOMINIC, SHOT SWORD FISH. Project Officer's Report - Project 1.3b; Effects of an Underwater Nuclear Explosion on Hydroacoustic Systems**. San Diego, CA: Naval Electronics Lab; Springfield, VA: Available from National Technical Information Service, 1985. (ADA995394; DNA-POR-2003(EX); DOE-WT-2003(EX)). 62 p.
Extracted version of Report no. POR-2003(WT-2003).
Abstract: *The objectives of Project 1.2 were to determine and evaluate the effects of an underwater nuclear explosion on the operational capabilities of shipboard sonar and other types of hydroacoustic systems. Project 1.3b included all measurements at ranges greater than 10 nautical miles and the results of these measurements constitute the subject of this report. This report concerns the effects of the underwater nuclear explosion, Sword Fish, on: (a) Long-range active detection systems at the first convergence zone (25 to 30 miles); (b) Passive shipboard or submarine sonars at a few hundred miles; and (c) Long-range passive detection and surveillance at Sound Surveillance System (SOSUS) and Missile Impact Locating System (MILS) stations at several hundred to several thousand miles. A submarine station at the first convergence zone and five shipboard stations at ranges from 200 miles to 5,000 miles recorded signals from hydrophones suspended at various depths to approximately 2,000 feet. Submarines on other assignments recorded signals on standard submarine sonar equipment on a not-to interfere basis. SOSUS and MILS stations operated normally during the period and also made special magnetic-tape and strip-chart recordings of signals from single hydrophones from before burst time to several hours after burst.*
NPS/DKL Location: MICROFORM ADA995394

Montana ,P.S. **Use of Helicopters to Develop Operational Concepts for V/STOL (Vertical and Short Takeoff and Landing) Aircraft in Naval Missions**. Bethesda, MD: David W Taylor Naval Ship Research and Development Center Aviation And Surface Effects Dept.; Springfield, VA: Available from National Technical Information Service, 1983. (ADA139354; DTNSRDC/ASED-83/07). 20 p.
Abstract: *Vertical and short takeoff and landing (V/STOL) aircraft promise new operational capabilities for the Navy. In the past, new vehicle types have been slow in gaining acceptance because of the difficulty in visualizing how these new vehicles should be employed. Once built, experience gained with the vehicle evolved into an operational concept exploiting its best qualities. Now, competition for fiscal resources has reached a level from which it may be difficult to justify the development of any new vehicle*

without having a well-defined operational concept in hand. This report discusses the use of existing large helicopters to develop operational concepts for V/STOL in naval applications. (Author)
NPS/DKL Location: MICROFORM ADA139354

Murray, W.W. **OPERATION DOMINIC, SHOT SWORD FISH. Scientific Director's Summary Report**. Washington, DC: David Taylor Model Basi; Springfield, VA: Available from National Technical Information Service n, 1985. (ADA995502; POR-2007(EX); WT-2007(EX)). 233 p.
Extracted version of report dated 21 Jan 63.
No abstract available.
NPS/DKL Location: MICROFORM ADA995502

Naval Underwater Systems Center (U.S.) **Naval Underwater Systems Center Brief**. Newport, RI: NUSC, 1978. (ADA103242, NUSC-TD-5740). 17 p.
Abstract: *The Naval Underwater Systems Center was formed in 1970 by the merger of two independent laboratories of the Naval Material Command: the Naval Underwater Weapons Research and Engineering Station (NUWS), Newport, Rhode Island, and the Naval Underwater Sound Laboratory (NUSL), New London, Connecticut. These two complexes are now the principal laboratories of NUSC. In July 1971, the Atlantic Undersea Test and Evaluation Center (AUTEC) in the Bahamas was made a detachment of NUSC. A basic and applied research program supporting systems development is a major thrust at the Center, and activities at NUSC cover all phases of the Center's primary mission responsibilities as the Navy's principal research, development, test, and evaluation center for submarine warfare and submarine weapons systems. These activities include responsibilities in programs in surface ship and submarine sonars, ASW weapons, combat control, and in undersea ranges--including the management of the AUTEC range complex.*
NPS/DKL Location: MICROFORM ADA103242

Nedresky, Donald L. **Aircraft Avionics Nonsupportability and Microcircuit Obsolescence**. Patuxent River, MD: Naval Air Warfare Center Aircraft Div.; Springfield, VA: Available from National Technical Information Service, 1996. (ADA309766). 11 p.
Abstract: *The declining military budget has resulted in service life extensions for many weapons systems. Conversely, mission essential systems, such as the avionics suite, on naval aircraft extend, must contend with the scheduled phase out of subcomponents and microcircuits over the next few years. This unplanned obsolescence will have a costly impact on the ability of naval aviation to maintain weapons systems in a high state of operational readiness. Identifying the size of this problem is made more complex because provisioning data (for older systems) is often incomplete or inaccurate, making it difficult to cross obsolete part numbers to specific system applications. This paper describes a proactive process for analyzing avionics system supportability issues involving microcircuit obsolescence and other factors, such as mission criticality, reliability, supply and demand, and aircraft allowance. Based on this analysis, a comprehensive, life cycle model is developed to predict time critical mission degraders and offers solutions for solving supportability issues.*
Electronic access: http://handle.dtic.mil/100.2/ADA309766

Office Of The Assistant Inspector General For Audit (DOD) Systems And Logistics. **Acquisition: Acquisition of the Advanced Deployable System (ADS)**. Arlington, VA: OAIG-AUD, 2002. (OAIG-AUD-D-2003-004O; ADA406702). 38 p.
Abstract: *This report should be read by all who are interested in the acquisition of the Navy's Advanced Deployable System (the System). The report addresses acquisition issues that require higher management attention before the System program should be allowed to progress further through the acquisition process. The System, a Navy Acquisition Category II program, is a next-generation, ship-deployable, undersea surveillance system that is designed to operate in littoral waters, The System is*

linked to a land facility for data processing, evaluation, and reporting. The System will be used to conduct missions, such as threat port surveillance, friendly port protection, area defense, area sanitization, and strategic indications and warnings. The System will have the ability to be installed overtly or covertly, depending on the needs of the Joint Task Force Commander. The program office's estimate includes $793.7 million for research, development, test, and evaluation for all four blocks of the evolutionary acquisition strategy and $785 million for procurement for the first two blocks.
NPS/DKL Location: MICROFORM ADA406702
Electronic access: http://handle.dtic.mil/100.2/**ADA406702**

Operational Test and Evaluation Force. **Follow-On Operational Test and Evaluation of the CPU-152/A Standard Central Air Data Computer (SCADC).** Norfolk, VA: Operational Test and Evaluation Force; Springfield, VA: Available from National Technical Information Service, 1990. (ADA226310). 5 p.
Abstract: *The purpose of the evaluation was to verify that all the deficiencies from OT-IIC OPEVAL has been corrected prior to approval for full fleet introduction. The evaluation was based on the results of non-scenario operational tests conducted under Project M756, supplemented by the results of OPEVAL, developmental testing, and operational experience. Based on this evaluation, the CPU-152/A SCADC as installed in the S-3A/B aircraft is determined to be operationally effective and operationally suitable. Approval for full fleet introduction of the CPU-152/A is recommended. The SCADC uses air pressure from the pitot static system and temperature signals from the temperature probe to provide air data outputs for navigation, cockpit display, sonobuoy and weapon delivery systems, Automatic Flight Control System (AFCS), and altitude reporting. While the digital SCADC is a form, fit, and function replacement for the existing S-3 Airspeed Altitude Computer Set (AACS), it has, in addition, a Built-in-test (BIT) function allowing maintenance personnel to determine system status without removing the unit. Keywords: Antisubmarine aircraft. (KR)*
NPS/DKL Location: MICROFORM ADA226310

Optical Communications Research Lab. **The Mk 15 Destroyer-Launched Torpedo: End of an Era**. Stanford, CA: Optical Communications Research Lab; Springfield, VA: Available from National Technical Information Service, 1993. (ADA274999; NUWC-NPT-TD-10132). 19 p.
Abstract: *The Mk 15 torpedo, designed and developed by the former Naval Torpedo Station in Newport, Rhode Island, in the 1930s, was the last destroyer-launched antisurface ship weapon to see wide service use. Longer, heavier, and more powerful than its predecessors, it was the Navy's principal destroyer torpedo when World War II began. During the early war years, three new classes of improved Navy destroyers having twin deck mounts of multiple torpedo tubes began entering the fleet. As is recounted in this booklet, salvos of Mk 15 torpedoes launched from those destroyer tubes proved decisive on several occasions in the Pacific campaign.*
NPS/DKL Location: MICROFORM ADA274999

Paris, Alfonso and Omeed Alaverdi. **Nonlinear Aerodynamic Model Development and Extraction from Flight Test Data for the S-3B Viking**. Mclean VA: Science Applications International Corp.; Springfield, VA: Available from National Technical Information Service, 2000. (ADA384281). 19 p.
Abstract: *This paper addresses applied procedures for nonlinear aerodynamic model development and extraction from flight data for the S-3B Viking aircraft. The entire analysis procedures, from dynamic flight test data management to final blending and validation of the upgraded aerodynamic model, was performed within the Integrated Data Evaluation and Analysis Systems (IDEAS) developed by SAIC. IDEAS is a powerful database management system and analysis software containing a full complement of flight data preprocessing, calibration, simulation, model estimation, model verification, and validation tools.*

Electronic access: http://handle.dtic.mil/100.2/ADA384281

Rich, H.L., R.L. Bort, E.T. Habib, R.E. Baker and W. E. Carr. **OPERATION HARDTACK. Project 3.3. Shock Loading in Ships from Underwater Bursts and Response of Shipboard Equipment**. Washington, DC: David Taylor Model Basin; Springfield, VA: Available from National Technical Information Service, 1985. (ADA995438; DOE-WT-1627(EX)). 209 p.
Extracted version of report dated 30 Sep 61.
Abstract: *The shock loading in ships and the response of shipboard machinery were measured during Shots Wahoo and Umbrella to: (1) determine safe- and shock-damage ranges, particularly with respect to shipboard machinery and equipment, for delivery of antisubmarine nuclear weapons by destroyers and submarines; (2) determine the intensity and character of the shock motions on a submarine and on a merchant ship under quasi-lethal attack by and underwater nuclear explosion; and (3) acquire shock-motion data and correlate such data with other measurements and with theory in order to extrapolate the results to other attack geometries. Conclusions include: (1) The shock damaging ranges for ships from underwater explosions depend greatly on the design and condition of the machinery and equipments as well as on charge size, burst depth, water depth, and the like. (2) Immobilization ranges for a destroyer are given as horizontal ranges from surface zero to the center of the ship. (3) Temperature gradients in the water increase or decrease the damage ranges.*
NPS/DKL Location: MICROFORM ADA995438

Russell, Jerry C. **Ultra and the Campaign Against the U-Boats in World War II**. Carlisle Barracks, PA: Army War College, 1980. (ADA089275). 45 p.
Abstract: *The problem addressed is the extent to which the United States Navy used Ultra, or Special Intelligence, in the campaign against the German U-boats. Information was gathered through published and unpublished sources. Through a chronological approach, United States Navy involvement is traced from entry into the war until its conclusion. Many factors are involved in the final outcome of the war and Ultra is only one. The Battle of the Atlantic was long and gruesome rather than short and spectacular. The United States Navy used Ultra along with technology, tactics, brilliant leadership and courageous men at sea to win the Battle of the Atlantic in World War II. The lessons for the future are clear. If the United States intends to oppose the Soviet submarine force at sea anywhere in the world, then we must maintain the lead in intelligence, tactics and technology. Further, and most importantly, we must strive to regain superiority of forces in those ocean areas where our interests are at stake.*
NPS/DKL Location: MICROFORM ADA089275

Shaw, R. C. **Nonlinear Dynamic Analysis of Frangible Nosecap for Vertical Launch Antisubmarine Rocket (VLA)**. San Diego, CA Naval Command Control and Ocean Surveillance Center RDT&E Div.; Springfield, VA: Available from National Technical Information Service, 1994. (ADA276835). 31 p.
Abstract: *Described here is the application of a nonlinear finite element analysis (FEA) technique to predict the structural behaviors for a class of brittle materials that shows near-complete brittleness when loaded in tension, but exhibits some ductility when compressed. An ABAQUS* constitutive model, consisting of an isotropically hardening yield surface, which is active when the stress state is dominantly compressive, and an independent crack detection surface to determine if a point in the material fails by cracking in tension, is employed to simulate the failure of the brittle material. The application of the technique to determine if a potential frangible nosecap design of the Vertical Launch Antisubmarine Rocket (VIA) would break up as intended upon water impact for a given entry condition is presented as an example. Frangible nosecaps, Large deformation, Non-linear structural analysis, Water entry*
NPS/DKL Location: MICROFORM: ADA276835

Tang, P. Y. **A Piecewise Quadratic Strength Tensor Theory for Composites**. San Diego, CA: Naval Ocean Systems Center; Springfield, VA: Available from National Technical Information Service, 1987. (ADA190929; NOSC/TR-1188). 60 p.

Abstract: *Develop the piecewise quadratic strength tensor theory for composite materials and demonstrate its applicability to the available biaxial fracture data on composites. The theory will have application to current composite structures of Naval Ocean Systems Center's interest such as transducers and future composite structures such as torpedo hull section, Vertical Launch ASROC (VLA) nosecaps, and tethered deep submergence structures. The theory can also be used with a wide variety of other NAVY structures such as aircraft and submarine substructures.*

NPS/DKL Location: MICROFORM ADA190929

Thompson, J. **Common USW Picture**. Washington, DC: Naval Sea Systems Command, 2001. (ADA393700). 15 p.
Proceedings from the Navy Interoperability Workshop, 30-31 May 2001; sponsored by NDIA. Viewgraphs only.

Abstract: *Presentation given at the Navy Interoperability Workshop, held on 30- 31 May, 2001, and sponsored by the National Defense Industrial Association.*

Electronic access: http://handle.dtic.mil/100.2/ADA393700

Tofil, John A. **Engineering Report for P-3/UWB Flight Portage, Maine 26-27 June 1995**. Ann Arbor, MI: Environmental Research Inst of Michigan; Springfield, VA: Available from National Technical Information Service, 1995. (ERIM-241300-53-T-VOL-1 ADA368196). 72 p.

Abstract: *The report consists of a summary of each pass with crew comments included. Users of the data should review the pass summaries to be alerted for anomalies that may have occurred during the collection or noted by processing or analysis people. The appendix contains plots of critical parameters as a function of time. This information is provided so that users can examine, in some detail the exact time and magnitude of observed anomalies. In addition users can go back to this data to determine if there is a correlation between funnies observed during processing and the engineering data. For example, if there was a blank strip in the image it might correlate with the transmitter power plot showing that the transmitter was off during that time. All of the data shown in the plots is also recorded on the HDDT's in the Aux. data block and could be reproduced by the user. These plots are included in the appendix as a convenience for users of the data. The appendix is available upon request and thus is not included with this set of data.*

Electronic access: http://handle.dtic.mil/100.2/ADA368196

_____. **Appendix for P-3/UWB Flight Portage, Maine 26-27 June 1995**, Volume 3. Ann Arbor, MI: Environmental Research Inst of Michigan; Springfield, VA: Available from National Technical Information Service, 1995. (ADA368195; ERIM-241300-53-T-VOL-3). 266 p.

Abstract: *This report is an appendix containing the results of the P-3 passes and radar imagery data.*

Electronic access: http://handle.dtic.mil/100.2/ADA368195

_____. **Engineering Report for P-3/UWB Flight Portage, Maine 27-28 June 1995**, Volume 1. Ann Arbor, MI: Environmental Research Inst of Michigan; Springfield, VA: Available from National Technical Information Service, 1995. (ERIM-241300-54-T-VOL-1; ADA368075). 72 p.

Abstract: *This report is written as an aid to users of the SAR data collected by the P-3/SAR on 27-28 June 1995. The data set generated consists of: (1) HDDT #E1439, (2) HDDT #E1440, (3) HDDT#E1441,*

(4) HDDT #E1442, (5) Mission Plan disk, data, etc., (6) Navigation disks, (7) Post Pass disks, and (8) Post Pass Summary Sheets. The report consists of a summary of each pass with crew comments included. Users of the data should review the pass summaries to be alerted for anomalies that may have occurred during the collection or noted by processing or analysis people. The appendix contains plots of critical parameters as a function of time. This information is provided so that users can examine, in some detail the exact time and magnitude of observed anomalies. In addition users can go back to this data to determine if there is a correlation between funnies observed during processing and the engineering data. For example, if there was a blank strip in the image it might correlate with the transmitter power plot showing that the transmitter was off during that time. All of the data shown in the plots is also recorded on the HDDT's in the Aux. data block and could be reproduced by the user. These plots are included in the appendix as a convenience for users of the data. The appendix is available upon request and thus is not included with this set of data.
Electronic access: http://handle.dtic.mil/100.2/ADA368075

_____. **Appendix for P-3/UWB Flight Portage, Maine 27-28 June 1995**, Volume 3. Ann Arbor, MI: Environmental Research Inst of Michigan; Springfield, VA: Available from National Technical Information Service, 1995. (ADA368074; ERIM-241300-54-T-VOL-3). 249 p.
Abstract: *This document contains the appendix for P-3/UWB Flight Portage, Maine 27-28 June 1995, Volume 3.*
Electronic access: http://handle.dtic.mil/100.2/ADA368074

_____. **Engineering Report for P-3/UWB Flight Portage, Maine 28-29 June 1995**, Volume 1. Ann Arbor, MI: Environmental Research Inst of Michigan; Springfield, VA: Available from National Technical Information Service, 1995. (ADA368198, ERIM-241300-55-T-VOL-1). 60 p.
Abstract: *The report consists of a summary of each pass with crew comments included. Users of the data should review the pass summaries to be alerted for anomalies that may have occurred during the collection or noted by processing or analysis people. The appendix contains plots of critical parameters as a function of time. This information is provided so that users can examine, in some detail the exact time and magnitude of observed anomalies. In addition users can go back to this data to determine if there is a correlation between funnies observed during processing and the engineering data. For example, if there was a blank strip in the image it might correlate with the transmitter power plot showing that the transmitter was off during that time. All of the data shown in the plots is also recorded on the HDDT's in the Aux. data block and could be reproduced by the user. These plots are included in the appendix as a convenience for users of the data. The appendix is available upon request and thus is not included with this set of data.*
Electronic access: http://handle.dtic.mil/100.2/ADA368198

_____. **Appendix for P-3/UWB Flight Portage, Maine 28-29 June 1995**, Volume 2. Ann Arbor, MI: Environmental Research Inst of Michigan; Springfield, VA: Available from National Technical Information Service, 1995. (ADA368199; ERIM-241300-55-T-VOL-2). 250 p.
Abstract: *This appendix contains the data recorded from the P-3 passes.*
Electronic access: http://handle.dtic.mil/100.2/ADA368199

Torres, M. T. **The Role of the Army Air Corps in Antisubmarine warfare in World War II**. Maxwell AFB, AL: Air Command and Staff College; Available from National Technical Information Service, Springfield, VA, 1985. 41 p.

Abstract: *This historical review traces the development of the role of the Army Air Corps in antisubmarine warfare. Pre-war plans exempted the Air Corps from this duty. Despite lack of training and equipment, the Air Corps contributed significantly to the defeat of the submarine threat. In defeating this threat, the Air Corps had to first battle the Navy's strategy of using airplanes to escort convoys. Before being relieved of antisubmarine warfare duty, the Air Corps had proved the necessity of using the airplane in an offensive role to search and destroy submarines.*
NPS/DKL Location: MICROFORM ADA157118

Trinca, Joseph J. **Defense of North America during a NATO-Warsaw Pact Conflict: Some Implications of the USSR's Power Projection Capabilities**. Fort Leavenworth, KS: Army Command and General Staff College, 1980. (ADA094983; SBI-ADE750063). 122 p.
Master's thesis.
Abstract: *This study attempts to determine whether or not the existing conventional military forces and defense systems in North America are adequate both to meet commitments to NATO in the event of a major European conflict and provide for continental security. Investigation reveals that Canada is weakly defended relative to the capabilities of the USSR to project forces onto her territory. Thus, should the USSR choose to exploit this vulnerability by executing rear area military operations on the North American flank at the outset of a NATO-Warsaw Pact conflict, she could succeed in diverting crucial U.S. and Canadian reinforcements away from their primary missions on the battlefields of Europe. (author)*
NPS/DKL Location: MICROFORM ADA094983

United States. Congress. House. Committee on Armed Services. Subcommittee on Seapower and Strategic and Critical Materials. **Advanced submarine technology and antisubmarine warfare**: hearing before the Seapower and Strategic and Critical Materials Subcommittee and the Research and Development Subcommittee of the Committee on Armed Services, House of Representatives, One Hundred First Congress, first session, hearing held, April 18, 1989. Washington: U.S. G.P.O.: For sale by the Supt. of Docs., Congressional Sales Office, U.S. G.P.O., 1990. 72 p.
NPS/DKL Location: FEDDOCS Y 4.AR 5/2 A:989-90/40

U.S. Navy Electronics Laboratory. **Milestones in the NEL Deep Submergence Program: Special progress report**. San Diego, CA: Navy Electronics Laboratory, 1965. (ADA074138). 9 p.
Abstract: *This report summarizes the scientific efforts of the Deep Submergence Program at the Navy Electronics laboratory, San Diego, California. Since its inception in 1958, the program has been vitally concerned with development of techniques, instrumentation, and vehicles to fulfill its assignment – research of the marine environment, from the continental shelf to the abyssal sea floor. While NEL's primary interests are in relating applied research data in marine acoustics, biology, geology, and physical oceanography to antisubmarine and submarine warfare projects, the Deep Submergence Program has also added significantly to man's basic knowledge of the ocean sciences.*
NPS/DKL Location: MICROFORM ADA074138

Werking, William E. and Jeffery A. Kuhlman. **S-3B Small-Scale Applique Coupon Flight Test Evaluation Results**. Patuxent River, MD: Naval Air Warfare Center Aircraft Div.; Springfield, VA: Available from National Technical Information Service, 2001. (ADA388188 NAWCADPAX/RTR-2000/97). 32 p.
Abstract: *An evaluation of FP500 and FP1500 paint replacement film (applique) small-scale coupons with 52-4 adhesive installed on the S-3B aircraft was conducted during 225 hr of laboratory tests and 2*

flights totaling 2.0 flight-hours to determine system suitability for large-scale coupon evaluation. Complete adhesion of the film to the surface of the aircraft during basic maneuvers is an enhancing characteristic that will promote reliable performance of FP500 and FP1500 applique material reducing corrosion and maintenance down time aircraft. Tattering of prepeeled (failed) sections of the applique was an enhancing characteristic that will prevent any in-flight failures from becoming catastrophic failures. The capability of the adhesive to adhere to the film and not the aircraft surface during applique removal is an enhancing characteristic that will facilitate the rapid removal of the applique. There are no deficiencies.
Electronic access: http://handle.dtic.mil/100.2/ADA388188

West, Bo. **Analytic Optimizations in Crisis Stability**. Los Alamos, NM: Los Alamos National Lab; Springfield, VA: Available from National Technical Information Service, 1991. (ADA344705; LA-11959-MS). 33 p.
Abstract: *Second strikes are dominated by submarine-launched missiles in the absence of defenses, but shift to aircraft at modest levels of defense. Defenses protect some retaliatory missiles, but not enough to retaliate strongly. With defenses, missiles should be vestigial and could be eliminated without penalty. Then aircraft could also be significantly reduced without impacting stability. The combination of parameters that maximizes cost effectiveness also maximizes midcourse effectiveness and crisis stability.*
Electronic access: http://handle.dtic.mil/100.2/ADA344705

Westerlund, K.E. **At Least Six Soviet Submarines Participated in the Horsfjaerd Operation (Mindestens Sechs Sowjetische U-Boote Nahmen an der 'Horsfjaerden-Operation' Teil)**. Washington, DC: Naval Intelligence Support Center Translation Div.; Springfield, VA: Available from National Technical Information Service, 1983. (ADA137392; NISC-TRANS-7233). 10 p.
Trans. of Marine-Rundschau (Germany, F.R.) n10 p. 454-459 1983.
No abstract available.
NPS/DKL Location: MICROFORM ADA074138

Whymark, Roy R. **High Power Magnetostrictors For Sonar Applications**. Chicago, IL: IIT Research Institute; Springfield, VA: Available from National Technical Information Service, 1965. (IITRI-N149-7;AD471004). 99 p.
Final report: 1 Jun 59-31 Jul 65, Most Project-4.
Abstract: *This report summarizes research on both metallic and ferrite magnetostrictive type transducers. The research was directed toward (1) establishing methods for calculating how a magnetostrictor responds to large amplitude excitation, (2) determining how magnetostrictive elements respond to uniaxial static compressive stresses, (3) investigating the effect of hydrostatic stress on the important magnetostrictive parameters lambda and mu, (4) extending existing theory to include the dissipative forces inside a vibrating magnetostrictor, (5) investigations of the coupled vibrational modes of a ferrite tube magnetostrictor, and (6) investigations of magnetostrictive ferrite transducers for high power sonar applications. The lack of a suitable nonlinear theory for the behavior of magnetostrictors under large amplitude excitations dictated the empirical approach taken. However, throughout the contract limited work was directed, based upon these empirical results, toward the possible formulation of a nonlinear theory. (Author)*
NPS/DKL Location: MICROFORM AD0471004

Will, Albert S. and Robert R. Wilson. **Missile Safety System for Assuring Minimum Safe Distance**. Washington DC: Department of The Navy, 1999. (ADD019417). 5 p.
Patent, Filed 28 Dec 64, patented 23 Mar 99 PATENT-5 886 284, supersedes PAT-APPL-423 640-64. Government-owned invention available for U.S. licensing and

possibly, for foreign licensing. Copy of patent available Commissioner of Patents, Washington, DC 20231.

Abstract: *The invention pertains to an arming and safing system for a missile having an acceleration responsive mechanism for actuating a timing device upon launching to insure arming only after the passage of predetermined period of time, and an omni-directional impact switch for activating a dudding switch in case of missile impact at a distance less than a minimum safe distance from the launching vehicle.*

Electronic access: http://patft.uspto.gov/netahtml/srchnum.htm

_____ and _____. **Variable Range Timer Impact Safety System**. Washington DC: Department of The Navy, 1999. (ADD019414). 7 p.
Patent, Filed 28 Dec 64, patented 23 Mar 99 PATENT-5 886 285, supersedes PAT-APPL-423 642-64. Government-owned invention available for U.S. licensing and possibly, for foreign licensing. Copy of patent available Commissioner of Patents, Washington, DC 20231.

Abstract: *The invention pertains to an arming system for a missile which prevents destruction of the missile outside of a specified area. The missile may be launched from a submarine, and follow a water-air-trajectory, and includes a variable-range timer acting in conjunction with an impact detection system. The timer drives arming switches to the armed condition after a predetermined time, and then opens the switches after a second predetermined time, which establishes a maximum range for detonation of the missile warhead.*

Electronic access: http://patft.uspto.gov/netahtml/srchnum.htm

_____, _____ and George F. Fortin. **Guidance Information Analyzer**. Washington DC: Department of The Navy, 1999. (ADD019367). 4 p.
Patent, Filed 26 May 65, patented 23 Mar 99, PATENT-5 886 287, supersedes PAT-APPL-459 131-65. Government-owned invention available for U.S. licensing and possibly, for foreign licensing. Copy of patent available from Commissioner of Patents, Washington, DC 20231.

Abstract: *A gating circuit for a missile guidance system having a capacitor for supplying the output signal together with plurality of silicon controlled rectifiers for applying or removing a supply voltage to the capacitor.*

Electronic access: http://patft.uspto.gov/netahtml/srchnum.htm

Williams, M.J. and A.M. Arney. **A Mathematical Model of the Sea King Mk.50 Helicopter Aerodynamics and Kinematics**. Melbourne, Victoria, Aus.: Aeronautical Research Labs; Springfield, VA: Available from National Technical Information Service, 1986. (ADA174029, ARL-AERO-TM-379). 66 p.

Abstract: *Details are given of the expression used to describe the aerodynamics and kinematics of the Sea King Mk.50 helicopter during steady flight and low rate maneuvers up to an advance ratio of 0.3. The aerodynamics/kinematics formulation is a major component of the Sea King mathematical model developed by Aeronautical Research Laboratories (ARL) for flight simulation of this Anti-Submarine Warfare helicopter.*

NPS/DKL Location: MICROFORM ADA174029

Young, G.A. and D.E. Phillips. **OPERATION DOMINIC, SHOT SWORD FISH. Project Officers Report-Project 1.2 Surface Phenomena**. White Oak, MD: Naval Ordnance Lab; Springfield, VA: Available from National Technical Information Service, 1985. (ADA995301; DASA-POR-2001(EX); DASA-WT-2001(EX)0. 140 p.

Extracted version of report dated 14 Aug 64.

Abstract: *Shot Sword Fish was an operational test of the ASROC antisubmarine weapon system. The general objectives of the project were (1) to record and measure the formation, growth, and dissipation of the visible surface phenomena, including slicks, spray domes, plumes, fallout, base surge, and foam patch resulting from the underwater detonation of an ASROC weapon; (2) to use the data obtained to estimate the actual depth of burst, position of burst, yield, and bubble period; (3) to determine the location of ships and platforms in the experimental array before, during and after the test; (4) to provide surface-phenomena time-of-arrival data at platforms and ships in the array for use by other projects; and 85) to make the results available for improving the surface-phenomena scaling and prediction techniques which are currently employed for establishing delivery and lethal ranges for fleet nuclear weapons. In general, there was good agreement between the observed dimensions of the Sword Fish phenomena and the predictions.*

NPS/DKL Location: MICROFORM ADA995301

CONFERENCE ON LIMITATION OF ARMAMENT-- WASHINGTON DC 1921-1922

Archimbaud, Léon. **La conférence de Washington**. Paris, Payot, 1923. 364 p.

Barcia Trelles, Camilo. **La política exterior norteamericana de la postguerra (hasta los acuerdos de Washington de 1922)**. [Valladolid?]: Talleres tipográficos "Cuesta" [c1924]. 199 p.

Boston Evening Transcript. **Review of the Conference on Limitation of Armament, in connection with the Pacific and Far Eastern questions**. [Boston, 1922]. 142 p.

Bouy, Raymond. **Le désarmement naval**. Paris, Éditions des Presses universitaires de France, [1931]. 284 p.

Braisted, William Reynolds. **The United States Navy in the Pacific, 1909-1922**. Austin, University of Texas Press, [1971]. 741 p.
NPS/DKL Location: GENERAL E182 .B83

Buell, Raymond Leslie. **The Washington conference**. New York, Russell & Russell [1970, c1922]. 461 p.
Originally published: New York, London, D. Appleton and company, 1922.
Thesis (Ph.D.), Princeton University, 1923.

Conference on the Limitation of Armament (1921-1922: Washington, D.C.). **Conference on the limitation of armament, Washington, November 12, 1921-February 6, 1922 = Conference de la limitation des armements, Washington, 12 novembre 1921-6 fevrier 1922**. Washington: Govt. print. off., 1922. 1757 p.
NPS/DKL Location: GENERAL JX1974.7 .C7 WASHINGTON

Conference on the Limitation of Armament (1921-1922: Washington, D.C.). **Treaties and resolutions of the Conference on the Limitation of Armament: as ratified by the United States Senate, facts and tables**. New York: Federal Trade Information Service, c1922. 60 p.

Goldstein, Erik and John Maurer, eds. **The Washington Conference, 1921-22: naval rivalry, East Asian stability and the road to Pearl Harbor**. Ilford, Essex, UK; Portland, OR: Frank Cass, 1994. 319 p.

Guilfoile, Joseph Vincent. **The Japanese press and the Washington Conference of 1921-1922**. Washington, 1949. 104 l.

Hoag, C. Leonard. **Preface to preparedness, the Washington disarmament conference and public opinion**. Washington, DC, American council on public affairs, [1941]. 205 p.

Ichihashi, Yamato. **The Washington Conference and after, a historical survey**. New York, AMS Press, [1969]. 443 p.
Originally published: Stanford University, CA, Stanford University Press, 1928.

Ishimaru, Tota. **Kore de mo sekai heiwa ka.** [1925]. 415 p.

_____. **Washington Kaigi no shinso.** 11 [1922]. 438 p.

Ito, Masanori. **Gunshuku.** [1929]. 330 p.

_____. **Kafu Kaigi to sonogo.** 11 [1922]. 470 p.

Jacquemart, André. **La Conférence de Washington**. Paris, Jouvre & cie, 1923. 144 p.

Japan. Gaimusho. OBeikyoku. Dai 3-ka. **Kafu kaigi hokoku.** 11 [1922]. 2 v.

_____. **Washington kaigi keika.** 11 [1922]. 181 p.

Jusserand, Jean Adrien Antoine Jules. **La France et les sous-marins,** in **Washington, DC Conference on the limitation of armament, 1921-1922**, Washington, Govt. print. off., 1922, pp. 818-829.

Katsuizumi, Sotokichi. **Critical observation on the Washington conference**. Ann Arbor, Mich., 1922. 48 p.

Kawakami, Kiyoshi Karl. **Japan's Pacific policy, especially in relation to China, the Far East, and the Washington conference**. New York, E.P. Dutton & company, [c1922]. 380 p.

Knox, Dudley Wright. **The eclipse of American sea power**. New York city, American army & navy journal, inc. [c1922]. 140 p.

Komatsu, Midori. **Washington kaigi no shinso.** 11 [1922]. 337 p.

Kowark, Hannsjörg. **Die französische Marinepolitik 1919-1924 und die Washingtoner Konferenz**. Stuttgart: Hochschulverlag, 1978. 291 p.
Originally presented as the author's thesis, Stuttgart.

LePore, Herbert P. **The politics and failure of naval disarmament, 1919-1939: the phantom peace**. Lewiston, NY: Edwin Mellen Press, 2003. (Studies in political science; v. 19). 348 p.

Miura, Takeyoshi, ed. **Washinton Kaigi keika. Dai 1-bu. Gunbi seigen ni kansuru mondai**. [Tokyo]: Obeikyoku Daisanka, Taisho 11 [1922]. 471 p.

Mochizuki, Kotaro. **Kafu kaigi no shinso**. 11 [1922]. 293

National Council for Prevention of War (U.S.). **The United States navy since the Washington conference; comparison with the navies of Great Britain, Japan, France, and Italy; the facts derived from United States naval sources**. Washington, DC, National council for prevention of war, 1927. 26 p.

NHK "Dokyumento Showa" Shuzaihan. **Orenji sakusen: gunshukuka no Nichi-Bei Taiheiyo senryaku**. Shohan. Tokyo: Kadokawa Shoten, Showa 61 [1986]. (Dokyumento Showa; 5). 225 p.
Originally telecast in 1986 as an NHK television program: Dokyumento Showa.

Scheer, Reinhardt. **Amerika und die abrüstung der seemächte**. Berlin, A. Scherl, g.m.b.h. [c1922]. 44 p.

Shimada, Saburo. **Nihon kaizoron**. Tokyo: Ryobundo: Hatsubaijo Ono Shoten, Taisho 10 [1921]. 346 p.

Simpson, Bertram Lenox. **An indiscreet chronicle from the Pacific**, by Putnam Weale [pseud.]. New York, Dodd, Mead and company, 1922. 310 p.

Skopdzki, Bernhard. **Die unterseebootfrage auf der Washingtoner abrüstungskonferenz 1921/22**. Berlin: F. Dümmler, 1925. 88 p.

Sullivan, Mark. **The great adventure at Washington, the story of the conference**. Garden City, N.Y., Doubleday, Page & company, 1922. 290 p.

Tarbell, Ida M. **Peacemakers, blessed and otherwise; observations, reflections, and irritations at an international conference**. New York, Macmillan Co., 1922. 227 p.

Toho Tsushinsha Chosabu. **Kafu Kaigi taikan**. Tokyo: Toho Tsushinsha, Taisho 11 [1922]. 467 p.

Tsuneda, Tsutomu. **Washington Kaigi to eikyu heiwa**. [11 i.e. 1922]. 320 p.

Vinson, John Chalmers. **The parchment peace: the United States Senate and the Washington Conference, 1921-1922**. Westport, CT: Greenwood Press, 1984, c1955. 259 p.
Originally published: Athens: University of Georgia Press, 1955.
NPS/DKL Location: BUCKLEY E785 .V7

Wells, Herbert George. **Washington and the hope of peace**. London, W. Collins, [1922]. 272 p.
American edition (New York, The Macmillan Co.) has title: Washington and the riddle of peace.

Yasutomi, Shozo. **Kaigun gunshuku mondai**. [4 i.e. 1929]. (Taiheiyo mondai sosho, 3). 89 p.

LONDON NAVAL CONFERENCE. (1930)

Arima, Seiho. **Kaigun gunshuku mondai to Nihon no shorai**. 10 [1935]. 97 p.

Atkinson, James David. **The London Naval Conference of 1930**. Washington, 1949. 329 l.

Cardona y Prieto, Pedro María. **La Conferencia y el Tratado marítimo-naval de Londres (1930), desde el punto de vista español**. Madrid, Imprenta del Ministerio de marina, 1931. (Publicaciones de la Liga marítima española). 148 p.

Hall, Christopher. **Britain, America, and arms control, 1921-37**. New York: St. Martin's Press, 1987. 295 p.
Originally presented as the author's thesis (doctoral--Oxford University, 1982).

Harada, Kumao. **Fragile victory; Prince Saionji and the 1930 London treaty issue, from the memoirs of Baron Harada Kumao**. Translated with an introd. and annotations by Thomas Francis Mayer-Oakes. Detroit, Wayne State University Press, 1968. 330 p.
English translation of v. 1 of Saionji Ko to seikyoku.
Originally published as the translator's thesis, University of Chicago, 1955, under title: Prince Saionji and the London Naval Conference.
NPS/DKL Location: GENERAL DS888.2 .H2

Honda, Kumataro. **Beikoku no datsubo**. 5 [1930]. 103 p.

_____. **Gunshuku Kaigi to Nihon**. 5 [1930]. 119 p.

Matsushita, Yoshio. **Gunshuku mondai to rekkyo no sohasen**. 7 [1932]. 256 p.

Meyer, Karl Ludwig. **Das deutsch-englische flottenabkommen von 1935**. Stuttgart, W. Kohlhammer, 1940. 100 p.

Minardi, Salvatore. **Italia e Francia alla Conferenza navale di Londra del 1930**. Caltanissetta: S. Sciascia, 1989, c1988. (Collezione "Viaggi e studi"). 407 p.

Morley, James William, ed. **Japan erupts: the London Naval Conference and the Manchurian Incident, 1928-1932: selected translations from Taiheiyo Senso e no michi, kaisen gaiko shi**. New York: Columbia University Press, 1984. 410 p.

National Council for Prevention of War (U.S.) **The background of the London Naval conference, with comparative statistics on the five navies illustrated by tables and charts**. Washington, DC, National council for prevention of war, 1930. 76 p.

Nihon gaiko bunsho, Rondon Kaigun Kaigi keika gaiyo. 54 [1979]. 846 p. Documents on Japanese foreign policy, summaries of the proceedings of the London Naval Conferences.

O'Connor, Raymond Gish. **Perilous equilibrium; the United States and the London Naval Conference of 1930**. Lawrence, University of Kansas Press, 1962. 188 p. Originally published as thesis, Stanford University, under title: The United States and the London Naval Conference of 1930.
NPS/DKL Location: GENERAL JX1974 .O2

Pelz, Stephen E. Race to Pearl Harbor; the failure of the Second London Naval Conference and the onset of World War II. Cambridge, MA: Harvard University Press, 1974. (Harvard studies in American-East Asian relations, 5). 268 p.
NPS/DKL Location: GENERAL D742.J3 P44 1974

Toyoda, Jo. **Saisho Wakatsuki Reijiro: Rondon Gunshuku Kaigi shuseki zenken**. Tokyo: Kodansha, 1990. 436 p.

United States. Congress. Senate. Committee on Foreign Relations. **Treaty on the limitation of naval armaments**. Hearings before the Committee on foreign relations, United States Senate, Seventy-first Congress, second session, on Treaty on the limitation of naval armaments. May 12-16, 19-23, 26-28, 1930. Washington, U.S. Govt. print. off., 1930. 366 p.

United States. Dept. of State. **London naval treaty of 1930**. Washington, U. S. Govt. print. off., 1930. 23 p.

Wakatsuki, Reijiro. **Oshu ni tsukaishite**. 6 [1931]. 143 p.

Watanabe, Yukio. **Gunshuku: Rondon joyaku to Nihon Kaigun = Armament reduction**. Tokyo: Peppu Shuppan, c1988. 233 P.

Yamakawa, Tadao. **Rondon Kaigun Gunshuku Kaigi no seika**. 5 [1930]. 124 p.

CONVOYS -- GENERAL

Apitz, Peter. **Convoys - outdated by technology?** Oslo: Institutt for Forsvarsstudier, 1991. (Forsvarsstudier; 1991:5). 43 p.

Bowling, Roland Alfred. **The negative influence of Mahan on the protection of shipping in wartime: the convoy controversy in the twentieth century**. Orono, ME: University of Maine, 1980. 685 p.
Thesis (Ph.D.), University of Maine at Orono, 1980.
NPS/DKL Location: GENERAL VK15 .B68

Browning, Robert M., Jr. **The eyes and ears of the convoy: development of the helicopter as an anti-submarine weapon**. [Washington, DC: Coast Guard Historian's Office, 1993?]. 16 p. Govt Doc No.: D 5.402:EY 3;0378-H-01 (MF)

Kemp, Paul. **Convoy protection: the defence of seaborne trade**. London: Arms and Armour, c1993. 124 p.

Poyer, Jason M. **The next convoy war: the American campaign against enemy shipping in the twenty-first century**. Newport, RI: Naval War College, [1999]. 19 p.
Electronic access: http://handle.dtic.mil/100.2/ADA363196

Smith, Kevin. **Conflict over convoys: Anglo-American logistics diplomacy in the Second World War**. Cambridge [England]; New York, NY, USA: Cambridge University Press, 1996. 318 p.

Vidaud, Jean. **Les navires de commerce armés pour leur défense**. Paris, Picart, 1936. 163 p.
Issued also as Thése - Univ. de Paris.

Winton, John. **Convoy: the defence of sea trade, 1890-1990**. London: M. Joseph, 1983. 378 p.

CONVOYS -- WWI

Bowling, Roland Alfred. **Convoy in World War I: the influence of Admiral William S. Sims, U.S. Navy**. San Diego, CA: SDSU, 1975. 366 l.
Thesis (M.A.), San Diego State University, 1975.

Munro, Donald John. **Convoys, blockades and mystery towers**. London, S. Low, Marston & co., ltd. [1932]. 208 p.

CONVOYS -- WWII -- GENERAL

Bennett, William Edward. **The red duster at war**. London, Gollancz, 1942. 192 p.
by Warren Armstrong [pseud.]

Convoys in World War II. Washington, DC: Navy Dept. Library, [1993]. (World War II commemorative bibliography; no. 4). 14 p.

Campbell, A. B. (Archibald Bruce). **Salute the red duster**. London, C. Johnson, [1952]. 207 p.

Creighton, Kenelm. **Convoy commodore**. [2d ed.]. London, W. Kimber [1956]. 205 p.

Elphick, Peter. **Life line: the Merchant Navy at war, 1939-1945**. London: Chatham, 1999. 224 p.

Hague, Arnold. **The allied convoy system, 1939-1945: its organization, defence and operation**. Annapolis, MD: Naval Institute Press, c2000. 208 p.
Originally published: Canada: Vanwell Publ. Ltd., St. Catharines, Ontario, 2000; London: Chatham Publ., 2000.
NPS/DKL Location: GENERAL V182 .H38 2000

Hope, Stanton. **Ocean odyssey; a record of the fighting merchant navy**. London, Eyre & Spottiswoode, 1944. 220 p.

Johnston, Mac. **Corvettes Canada: convoy veterans of WWII tell their true stories**. Toronto: McGraw-Hill Ryerson, 1994. 310 p.

Lane, Tony. **The merchant seamen's war**. Manchester; New York: Manchester University Press; Distributed exclusively in the USA and Canada by St. Martin's Press, c1990. 287 p.

Lewis, William J. **Under the red duster: the merchant navy in World War II**. Shrewsbury: Airlife, 2003. 184 p.
NPS/DKL Location: GENERAL D771 .L49 2003

Nesbit, Roy Conyers. **The strike wings: special anti-shipping squadrons, 1942-1945**. London: W. Kimber, 1984. 288 p.

O'Flaherty, Ferocious. **Abandoned convoy; the U.S. merchant marine in World War II**. New York, Exposition Press, [1970]. 87 p.

Slader, John. **The fourth service: merchantmen at war, 1939-1945**. Wimborne Minster: New Guild, 1995. 347 p.

Originally published: London: Hale, 1994.

_____. **The Red Duster at war: a history of the merchant navy during the Second World War**. London: W. Kimber, 1988. 352 p.

Woddis, Jack. **Under the red duster: a study of Britain's merchant navy**. London: Senior Press, [1947]. 159 p.

CONVOYS -- WWII -- ARCTIC

Alotta, Robert I. and Donald R. Foxvog. **The last voyage of the SS Henry Bacon**. 1st ed. St. Paul, MN: Paragon House, 2001. 288 p.

Antier, Jean Jacques. **La bataille des convois de Mourmansk**. Paris: Presses de la Cite, c1981. (Troupes de choc). 271 p.

Barashkov, IUrii Anatol'evich. **Arkticheskie konvoi v nastroenii Glenna Millera: opyt kollektivnoi pamiati**. Arkhangel'sk: M'Art, 2000. 181 p.
Added TP Title: Arctic convoys in the mood for Glen Miller

Bártl, Stanislav. **Konvoj PQ 17: válecné drama v arktických morích**. Praha: Paseka, 1995. 192 p.

Blond, Georges. **Les Cargos massacrés: convois vers l'U.R.S.S**. Paris, Presses de la Cité, 1972. 251 p.

_____. **Convois vers l'U. R. S. S**. Paris, Fayard, [1950]. 277 p.

_____. **Convois vers l'U. R. S. S**. [Paris]: le Livre de poche, 1966. 256 p.

_____. **Kurs Murmansk: Die Schicksalsfahrten d. alliierten Eismeer-Konvois** [übers. von Karl Hellwig u. Cajus Bekker]. Oldenburg (Oldb); Hamburg: Stalling, 1957. 224 p.
German translation of Convoi vers l'URSS.

_____. **Kurs Murmansk: Die Schicksalsfahrten d. alliierten Eismeer-Konvois** [übers. von Karl Hellwig u. Cajus Bekker]. Bergisch Gladbach: Lübbe, 1981. 242 p.

_____. **Ordeal below zero; the heroic story of the Arctic convoys in World War II**. [1st English ed.]. London, Souvenir Press, 1956. 199 p.
English translation of Convoi vers l'URSS.

_____. **Ordeal below Zero**. London: Transworld Publishers, 1957. 224 p.

_____. **Ordeal below Zero**. London: Mayflower Books, 1965. 189 p.

_____. **Ordeal below zero**. London: Mayflower Books, 1968. 189 p.

_____. **Ordeal below zero**. Bath: Firecrest, 1987. 189 p.

Brookes, Ewart. **The Gates of Hell**. London: Jarrolds, 1960. 143 p.
On convoys to Murmansk during the 1939-45 war.

_____. **The Gates of Hell**. London: Arrow Books, 1962. 190 p.

_____. **The gates of hell**. New ed. London, Arrow Books, 1973. 192 p.

Bryzgalov, V. V. et al. **Konvoi: issledovaniia, vospominaniia, bibliografiia, dokumenty**. Arkhangel'sk: Arkhangel'skii tsentr Russkogo geogr. ob-va, 1995. (Istoriia i kul'tura Russkogo Severa). 317 p.

Campbell, Ian, Sir and Donald Macintyre. **The Kola run; a record of Arctic convoys, 1941-1945**. London, Muller, [c1958]. 254 p.

Carse, Robert. **A cold corner of hell; the story of the Murmansk convoys, 1941-45**. [1st ed.]. Garden City, NY: Doubleday, 1969. 268 p.

_____. **Lifeline; the ships and men of our merchant marine at war**. New York, W. Morrow and company, 1943. 189 p.

_____. **There go the ships**. New York, W. Morrow, 1942. 156 p.

Edwards, Bernard. **The road to Russia: Arctic convoys 1942**. Annapolis, MD: Naval Institute Press, 2002. 210 p.
Originally published: London: Leo Cooper, 2002.
NPS/DKL Location: GENERAL D770 .E394 2002

Evans, Mark Llewellyn. **Great World War II battles in the Arctic**. Westport, CT: Greenwood Press, 1999. (Contributions in military studies no. 172). 165 p.

Herman, Frederick Sawyer. **Dynamite cargo; convoy to Russia**. New York, Vanguard, [1943]. 158 p.

Heye, August Wilhelm. **"Z 13" von Kiel bis Narvik, kriegserleben einer zerstorerbesatzung**. Berlin, E. S. Mittler & sohn, 1942. 242 p.

Hughes, Robert. **Flagship to Murmansk; a gunnery officer in HMS 'Scylla', 1942-43**. London, Futura Publications, 1975. 191 p.
Originally published: London: Kimber, 1956 as 'Through the waters'.
Extracts and photos at http://www.world-war.co.uk/scylla_story.php3

Irving, David John Cawdell. **The destruction of convoy PQ.17**. New York, Simon and Schuster, [1969, c1968]. 337 p.
Originally published: London, Cassell, 1968.
NPS/DKL Location: GENERAL D771 .I7

Kemp, Paul. **Convoy!: drama in arctic waters**. London: Brockhampton Press, c1993. 256 p.

_____. **The Russian convoys, 1941-1945**. Poole, Dorset: Arms and Armour Press; New York, NY, U.S.A.: Sterling Pub. Co., [1987]. (Warships illustrated; no. 9). 64 p.

Kerslake, Sidney A. **Coxswain in the northern convoys**. London: W. Kimber, 1984. 191 p.

Kimata, Jiro. **Kyokuhoku no kaisen**. Shohan. Tokyo: Asahi Sonorama, Showa 60 [1985]. (Koku senshi shirizu; 53). 411 p.

Lund, Paul and Harry Ludlam. **Die Nacht der U-Boote: die Vernichtung des britischen Geleitzugs SC 7** [Dt. Übers. von Klaus Kamberger]. 9. Aufl. München: Heyne, 1994. 222 p.
Originally published: München: Heyne, 1983. 222 p.
German translation of Night of the U-boats

_____ and _____. **Night of the U-boats**. London, New York, Foulsham, 1973. 204 p.

_____ and _____. **Night of the U-boats**. London: New English Library, 1974. 173 p.

_____ and _____. **PQ 17--convoy to hell: the survivors' story**. London, New York [etc.], Foulsham, 1968. 240 p.

_____ and _____. **PQ 17--convoy to hell: the survivors' story**. London: New English Library, 1969. 192 p.

Macintyre, Donald G. F. W. **The battle of the Atlantic**. New York, Macmillan, [1961]. 208 p.
NPS/DKL Location: GENERAL D770 .M2

Mouton, Patrick. **L'or de Staline: 5 tonnes par 260 mètres de fond**. Rueil-Malmaison: Editions du Pen Duick, c1984. 196 p.

Pearce, Frank. **Last call for HMS Edinburgh: a story of the Russian convoys**. 1st American ed. New York: Atheneum, 1982. 200 p.
Originally published: London: Collins, 1982.

Pearson, Michael. **Red sky in the morning: the battle of the Barents Sea, 31 December 1942**. Shrewsbury: Airlife, 2002. 154 p.
NPS/DKL Location: GENERAL D771 .P43 2002

Penrose, Barrie. **Stalin's gold: the story of HMS Edinburgh and its treasure**. 1st American ed. Boston: Little, Brown, c1982. 223 p.
Originally published: London; New York: Granada, 1982. 223

Pertek, Jerzy. **Bitwy konwojowe na arktycznej trasie**. Wyd. 1. Pozna´n: Wydawn. Pozna´nskie, 1982. 197 p.

Platonov, A. V. **Poliarnye konvoi, 1941-1945**. Sankt-Peterburg: Galeia Print, 1999. (Arctic Allied Convoys in the World War II). 73 p.

Pope, Dudley. **73 North: the battle of the Barents Sea**. Annapolis, Md.: Naval Institute Press, [1988?]. 320 p.
NPS/DKL Location: BUCKLEY D771 .P6 1988

Schofield, Brian Betham. **The Arctic convoys**. London: Macdonald and Jane's, 1977. 198 p.

_____. **The Russian convoys**. London, B. T. Batsford, [1964]. (British battles series). 224 p.
NPS/DKL Location: GENERAL D771 .S3

_____. **The Russian convoys**. London: Pan Books, 1971. (British battles series). 237 p.

Smith, Peter Charles. **Arctic victory: the story of convoy PQ 18**. London: Kimber, 1975. 238 p.

_____. **Geleitzug nach Russland: d. Geschichte d. Konvois PQ 18** [Die Übertr. ins Dt. besorgte Hans Renker]. 1. Aufl.. Stuttgart: Motorbuch-Verlag, 1980. 284 p.
German translation of Arctic victory.

Suprun, M. N. **Lend-liz i severnye konvoi: 1941-1945 gg**. Moskva: Andreevskii flag, 1997. 363 p.

_____, V.V. Bryzgalov and V.A. Liubimov. **Severnye konvoi: issledovaniia, vospominaniia, dokumenty**. Arkhangel'sk: Arkhangel'skii filial Geograficheskogo obshchestva SSSR, 1991-1994. 2 v.

Taylor, J. E. **Northern escort**. London, G. Allen & Unwin, [1945]. 127 p.

Taylor, Theodore. **Battle in the arctic seas: the story of convoy PQ 17**. New York: Crowell, c1976. 151 p.

Wharton, Ric. **The Salvage of the Century**. Flagstaff, AZ: Best Publishing Company, 2000. 198 p.
Convoy QP-11; Recovery of HMS Edinburgh's cargo of gold.

Woodman, Richard. **The Arctic convoys, 1941-1945**. Pbk. ed. London: John Murray, 1995, c1994. 532 p.
Originally published: London: John Murray, 1994

CONVOYS -- WWII -- ATLANTIC

Bailey, Chris Howard. **The Royal Naval Museum book of the Battle of the Atlantic: the corvettes and their crews: an oral history**. Annapolis, Md.: Naval Institute Press, 1994. 156 p.

Bertrand, Michel. **Les escorteurs de la France libre**. Paris: Presses de la Cite, c1984. (Troupes de choc). 231 p.

Broome, John Egerton. **Convoy is to scatter**. London, Kimber, 1972. 232 p.

Burn, Alan. **The fighting commodores: the convoy commanders in the Second World War**. Annapolis, MD: Naval Institute Press, c1999. 262 p.
Originally published: London: Leo Cooper, 1999.
NPS/DKL Location: GENERAL V182 .B87 1999

Dennis, Owen. **The rest go on**. London, J. Crowther Ltd., 1942]. 132 p.

Edwards, Bernard. **Attack and sink!: the battle for convoy SC42**. Wimborne Minster: New Guild, c1995. 199 p.

Essex, James W. **Victory in the St. Lawrence: Canada's unknown war**. Erin, Ontario: Boston Mills Press, c1984. 159 p.

Gibson, Charles Dana, II. **The ordeal of Convoy NY 119; a detailed accounting of one of the strangest World War II convoys ever to cross the North Atlantic**. New York, South Street Seaport Museum, [1973]. 178 p.

_____. **The ordeal of Convoy NY 119; a detailed accounting of one of the strangest World War II convoys ever to cross the North Atlantic**. 2nd ed. Camden, ME: Ensign Press, c1992. 180 p.

Gretton, Peter, Sir. **Convoy escort commander**. London, Cassell, [1964]. 223 p.

_____. **Crisis convoy; the story of HX231**. London, P. Davies, 1974. 182 p.
NPS/DKL Location: GENERAL D770 .G76

_____. **Crisis convoy: the story of HX231**. Annapolis: Naval Institute Press, c1974. 182 p.

Halstead, Ivor. **Heroes of the Atlantic; the British merchant navy carries on!** New York, E. P. Dutton & Co., inc., 1942. 235 p.

Hartmark, Arne. **Atlanterhavsslaget og de norske korvettene**. Oslo: Norsk maritimt forlag, c1990. 123 p.

Haskell, Winthrop A. and Jürgen Rohwer. **Shadows on the horizon: the battle of Convoy HX-233**. Annapolis, MD: Naval Institute Press, 1998. 192 p.
Originally published: London: Chatham Publ., 1998.
NPS/DKL Location: GENERAL V182 .H37 1998

Kaplan, Philip and Jack Currie. **Convoy: merchant sailors at war, 1939-1945**. Annapolis, MD: Naval Institute Press, c1998. 224 p.
Originally published: London: Aurum Press, 1998.
NPS/DKL Location: BUCKLEY D810.T8 K37 1998

Kemp, Peter Kemp. **Decision at sea: the convoy escorts**. 1st ed. New York: Elsevier-Dutton, c1978. (Men and battle). 84 p.

Knox, Collie. **Atlantic battle**. London, Methuen & co., ltd., [1941]. 103 p.

Lund, Paul and Harry Ludlam. **Nightmare convoy: the story of the lost Wrens**. London. Foulsham. c1987. 128 p.

Middlebrook, Martin. **Convoy**. New York: Morrow, c1976. 350 p.

_____. **Convoy: the battle for convoys SC.122 and HX.229**. London: Allen Lane, 1976. 378 p.

_____. **Konvoi: dt. U-Boote jagen allierte Geleitzüge** [Aus d. Engl. von Erwin Duncker]. Rastatt: Moewig, 1984. 320 p.
German translation of Convoy.

Milner, Marc. **North Atlantic run: the Royal Canadian Navy and the battle for the convoys**. Annapolis, MD: Naval Institute Press, c1985. 326 p.

_____. **North Atlantic run: the Royal Canadian Navy and the battle for the convoys**. Toronto; Buffalo: University of Toronto Press, c1985. 326 p.

Mumford, J. Gordon. **The black pit-- and beyond**. Burnstown, Ont.: General Store Pub. House, c2000. 138 p.

Noli, Jean. **The admiral's wolf pack** [translated by J. F. Bernard]. [1st ed.]. Garden City, NY, Doubleday, 1974. 396 p.
English translation of Les loups de l'amiral.

_____. **Les loups de l'amiral**. [Paris], Fayard, [1970]. (Grands documents contemporains). 471 p.

O'Brien, David. **HX 72: the first convoy to die: the Wolfpack attack that woke up the admiralty**. Halifax: Nimbus Pub., 1998. 168 p.
NPS/DKL Location: GENERAL D771 .O27 1998

Parker, Mike. **Running the gauntlet: an oral history of Canadian merchant seamen in World War II**. Halifax, NS: Nimbus, c1994. 344 p.

Rayner, Denys Arthur, Stephen Wentworth and Evan Davies. **Escort: the Battle of the Atlantic**. Annapolis, MD: Naval Institute Press, c1999. (Classics of naval literature). 258 p.
Originally published: London: W. Kimber, 1955.
NPS/DKL Location: GENERAL D770 .R2 1999

Reid, Max. **D.E.M.S. and the Battle of the Atlantic, 1939-1945**. Ottawa: Commoners' Pub. Society, c1990. 100 p.

Revely, Henry. **The convoy that nearly died: the story of ONS 154**. London: Kimber, 1979. 222 p.

Reynolds, Quentin James. **Convoy**. New York, Random House, [1942]. 303 p.

Rohwer, Jürgen. **The critical convoy battles of March 1943: the battle for HX.299/SC122** [translated from the German by Derek Masters]. London: I. Allan, 1977. 256 p.
English translation with revisions of Geleitzugschlachten im Marz 1943.

Rutter, Owen. **Red ensign, a history of convoy**. London, R. Hale limited, 1942. 214 p.

Sawicki, Jan Kazimierz. **Bezbronne konwoje**. Wyd. 1. Gdynia: Wyzsza Szkola Morska, 1993. (Ksiegi Floty Ojczystej, 1230-7092; t. 3). 261 p.

Schofield, William G. **Eastward the convoys**. Chicago, Rand McNally [1965]. 239 p.

Seth, Ronald and P.W. Grelton. **The fiercest battle: the story of North Atlantic convoy ONS 5, 22nd April - 7th May 1943**. [1st American ed.]. New York: Norton, [1962, c1961]. 208 p.

Originally published: London: Hutchinson, 1961.

Smyth, Denis. **Battle of St. Patrick's Day, March 1943: the story of the HX.229 and SC.122 convoys-North Atlantic.** [Belfast?]: North Belfast History Workshop, [196-?]. [32] p.

Tennent, Alan J. **British and Commonwealth merchant ship losses to axis submarines, 1939-1945.** Stroud, Gloucestershire: Sutton Pub., 2001. 326 p.

Thomas, David Arthur. **The Atlantic Star, 1939-45**. London: W. H. Allen, 1990. 312 p.

Waters, John M. **Bloody winter**. Princeton, NJ: Van Nostrand, [1967]. 279 p.

_____. **Bloody winter**. Rev. ed. Annapolis, MD: Naval Institute Press, c1984. 285 p.

Watt, Frederick B. **In all respects ready: the Merchant Navy and the Battle of the Atlantic, 1940-1945.** Scarborough, Ont.; Englewood Cliffs, NJ: Prentice-Hall, c1985. 222 p.

Woon, Basil Dillon. **Atlantic front, the merchant navy in the war.** London, P. Davies [1941]. 323 p.

CONVOYS -- WWII -- MALTA/MEDITERRANEAN

Cameron, Ian. **Red duster, white ensign; the story of Malta and the Malta convoys.** Garden City, NY, Doubleday, 1960. 260 p.

Jurkovic, Ivo. **Oni sa saveznickih konvoja**. Rijeka: Visa pomorska skola, 1974. 151 p.

Kemp, Paul. **Malta convoys, 1940-1943**. London; New York: Arms & Armour Press, [1988]. (Warships illustrated; no. 14). 64 p.

Nassigh, Riccardo. **La battaglia dei convogli: Mediterraneo, 1940-1943.** Roma: Ufficio storia della marina militare, 1994. 233 p.
Contents: Part 1 -- La storia / Riccardo Nassigh -- La battaglia dei convogli -- Documentazione fotografica; Part 2 consists of papers presented at a conference held Mar. 22, 1993, Naples, Italy.

Payne, Donald Gordon. **Red duster, white ensign; the story of Malta and the malta convoys.** [1st ed.]. Garden City, NY, Doubleday, 1960. 260 p.
Originally published: London, Muller [1959].

Schembri, George and Mark Anthony Vella. **Operation Pedestal: the convoy that saved Malta, August 1942 = il-konvoj ta' Santa Marija, Awissu 1942**. St. Julians, Malta: M.A. Marketing, c1999. 1 v.

Shankland, Peter and Anthony Hunter. **Malta convoy**. London: Collins, 1961. 256 p.
Originally published: London: Collins, 1961.
NPS/DKL Location: GENERAL D763.M3 S5

_____ and _____. **Malta convoy**. London & Glasgow: Collins, 1963. 192 p.

_____. and _____. **Malta convoy**. London & Glasgow, Collins: 1972. 192 p.

_____ and _____. **Malta convoy**. Junior edition.. London: Collins, 1965. 256 p.

Smith, Peter Charles. **Pedestal: the Malta convoy of August, 1942**. London, Kimber, 1970. 208 p.

Thomas, David Arthur. **Malta convoys 1940-42: the struggle at sea**. London: Leo Cooper, 1999. 234 p.

Woodman, Richard. **Malta convoys, 1940-1943**. London: J. Murray, 2000. 532 p.

CONVOYS -- WWII -- PACIFIC

McAulay, Lex. **Battle of the Bismarck Sea**. 1st ed. New York: St. Martin's, 1991. 226 p.

Komamiya, Shinshichiro. **Senji yuso sendanshi**. Tokyo: Shuppan Kyodosha, Showa 62 [1987]. 408 p.

Matsubara, Shigeo and Akira Endo. **Rikugun senpaku senso**. Musashimurayama-shi: Senshi Kankokai; Tokyo: Hatsubai Seiunsha, Heisei 8 [1996]. 339 p.

Ota, Tsuneya. **Tatakau yusosen**. 19 [1944]. 178 p.

Wallace, Robert. **The secret battle, 1942-1944: the convoy battle off the East Coast of Australia during World War II**. Ringwood, Vic.: Lamont Publishing, 1995. 96 p.

FALKLAND ISLANDS WAR, 1982

Brown, David. **The Royal Navy and the Falklands war**. Annapolis, MD: Naval Institute Press, c1987. 384 p.
NPS/DKL Location: GENERAL F3031.5 .B76 1987

Barrella, Miguel. **Hundan al Belgrano: no mientas Margaret**. Bs. As. [i.e. Buenos Aires] Argentina: Ediciones Letra Buena, c1992. 139 p.

Craig, Chris, Captain. **Call for fire: sea combat in the Falklands and the Gulf War**. London: J. Murray, 1995. 300 p.

Dalyell, Tam. **Thatcher's torpedo**. London: Woolf, 1983. 80 p.

Gavshon, Arthur L. and Desmond Rice. **El hundimiento del Belgrano** [traduccion Fernando Estrada]. Buenos Aires, Argentina: Emece, 1984. 270 p.
English translation of: The sinking of the Belgrano.

_____ and _____. **The sinking of the Belgrano**. London: Secker & Warburg, 1984. 218 p.

Gould, Diana. **On the spot: the sinking of the "Belgrano"**. London: C. Woolf, 1984. 80 p.

Koburger, Charles W., Jr. **Sea power in the Falklands**. New York: Praeger, 1983. 186 p.

Waispek, Carlos Alberto. **Balsa 44: relato de un sobreviviente del crucero A.R.A. General Belgrano**. Buenos Aires: Editorial Vinciguerra, c1994. 158 p.

JUTLAND

Bacon, Reginald Hugh Spencer, Sir. **The Jutland scandal**. 2nd ed. London: Hutchinson & co., [1925]. 159 p.
NPS/DKL Location: BUCKLEY D582.J8 B2

Bellairs, Carlyon Wilfroy. **The battle of Jutland: the sowing and the reaping**. London: Hodder and Stoughton limited, [1920]. 312 p.
NPS/DKL Location: GENERAL D582.J8 B37/BUCKLEY D582.J8 B36

Bennett, Geoffrey Martin. **The Battle of Jutland**. London, B. T. Batsford [c1964]. 208 p.
NPS/DKL Location: GENERAL D582.J8 B4

_____. **The Battle of Jutland**. Philadelphia, Dufour Editions, 1964. 208 p.

_____. **The Battle of Jutland**. Newton Abbot: David and Charles, 1972. 208 p.

_____. **The Battle of Jutland**. New edition. Ware, Hertfordshire: Wordsworth Editions, 1999. 208 p.

Busch, Fritz Otto. **Die schlacht am Skagerrak**. Leipzig, Franz Schneider b.h., [c1933]. 78 p.

Campbell, N. J. M. **Jutland: an analysis of the fighting**. Annapolis, MD: Naval Institute Press, c1986. 439 p.
NPS/DKL Location: GENERAL D582.J8 C35 1986

_____. **Jutland: an analysis of the fighting**. New York: Lyons Press, 1998. 439 p.

Claxton, Bernard D., John H. Gurtcheff, and Jeffrey J. Polles. **Trafalgar and Jutland: a study in the principles of war**. Montgomery, AL: Air Command and Staff College, Maxwell Air Force Base, 1985. (Military history monograph series; 85-2). 86 p.
NPS/DKL Location: GENERAL DA88.5 1805 C52 1985

Corbino, Epicarmo. **La battaglia dello Jutland: vista da un economista**. 2nd ed. Milano: G. Colombi, 1935, c1933. 384 p.

Costello, John and Terry Hughes. **Jutland, 1916**. New York: Holt, Rinehart and Winston, 1977, c1976. 230 p.
Originally published: London: Weidenfeld & Nicolson, c1976.

Fawcett, H. W. and G.W.W. Hooper, eds. **The fighting at Jutland: the personal experiences of sixty officers and men of the British fleet**. [New ed.]. Annapolis, MD: Naval Institute Press, 2001. 448 p.
Originally published: Glasgow: MacLure, Macdonald, 1921.

Frost, Holloway Halstead. **The Battle of Jutland**. Annapolis, MD, United States Naval Institute; B.F. Stevens & Brown, ltd, 1936. 571 p.
NPS/DKL Location: GENERAL D582.J8 F9

_____. **The Battle of Jutland**. Annapolis, MD, U.S. Naval Institute [c1934, 1970?]. 571 p.
NPS/DKL Location: GENERAL D582.J8 F9

_____. **The Battle of Jutland**. New York: Arno Press, 1980, c1936. 571 p.

_____. **Grand Fleet und Hochseeflotte im Weltkrieg**. Berlin, O. Schlegel, 1938. 568 p.
German translation of The Battle of Jutland.

_____. **La battaglia dello Jutland, tradotto a cura dell'Ufficio storico della R. Marina dal capitano di vascello Carlo Giartosio**. Roma, Ministero della marina, Tipo-litografia dell' Ufficio di gabinetto, 1940. 494 p.
Italian translation of The Battle of Jutland.

Frothingham, Thomas Goddard. **A true account of the battle of Jutland, May 31, 1916**. Cambridge: Bacon & Brown, 1920. 54 p.

George, S. C. **Jutland to junkyard: the raising of the scuttled German High Seas Fleet from Scapa Flow: the greatest salvage operation of all time**. Edinburgh: Birlinn, 1999. 150 p.
Originally published: Cambridge: Patrick Stephen, 1973.
NPS/DKL Location: BUCKLEY VK1491 .G46 1999

Gibson, Langhorne and John Ernest Troyte Harper. **Das Rastel um Jutland, eine authentische Beschreibung**. [Hamburg, 1934]. 298 p.
German translation of the Riddle of Jutland.

_____ and _____. **The riddle of Jutland; an authentic history**. New York, Coward-McCann, inc., 1934. 416 p.
NPS/DKL Location: GENERAL D582.J8 G4

Gill, Charles Clifford. **What happened at Jutland; the tactics of the battle**. New York, George H. Doran company, [c1921]. 187 p.
NPS/DKL Location: GENERAL D582.J8 G43

_____. **Naval power in the war (1914-1918)**. New York: George H. Doran, [c1919]. 302 p.
NPS/DKL Location: BUCKLEY D580 .G4

Gordon, Gilbert Andrew Hugh. **The rules of the game: Jutland and British naval command**. Annapolis, MD: Naval Institute Press, c1996. 708 p.
Originally published: London: John Murray, 1996.
NPS/DKL Location: GENERAL D582.J8 G68 1996

Great Britain. Admiralty. **Battle of Jutland: 30th May to 1st June, 1916 / official despatches with appendices**. H.M. Stationery off. [printed by Eyre and Spottiswoode, ltd., 1920]. 603 p.
NPS/DKL Location: GENERAL D582.J8 G7

Harper, John Ernest Troyte. **The truth about Jutland**. London, J. Murray [1927]. 200 p.
NPS/DKL Location: BUCKLEY D582.J8 H2

Hase, Georg von. **Kiel and Jutland**. London: Skeffington, 1921. 233 p.

Hough, Richard Alexander. **The Battle of Jutland**. London: Hamish Hamilton, 1964. (Hamish Hamilton monographs). 64 p.

Irving, John James Cawdell. **The smoke screen of Jutland**. New York, D. McKay Co., [1967]. 256 p.
Originally published: London, Kimber, 1966.

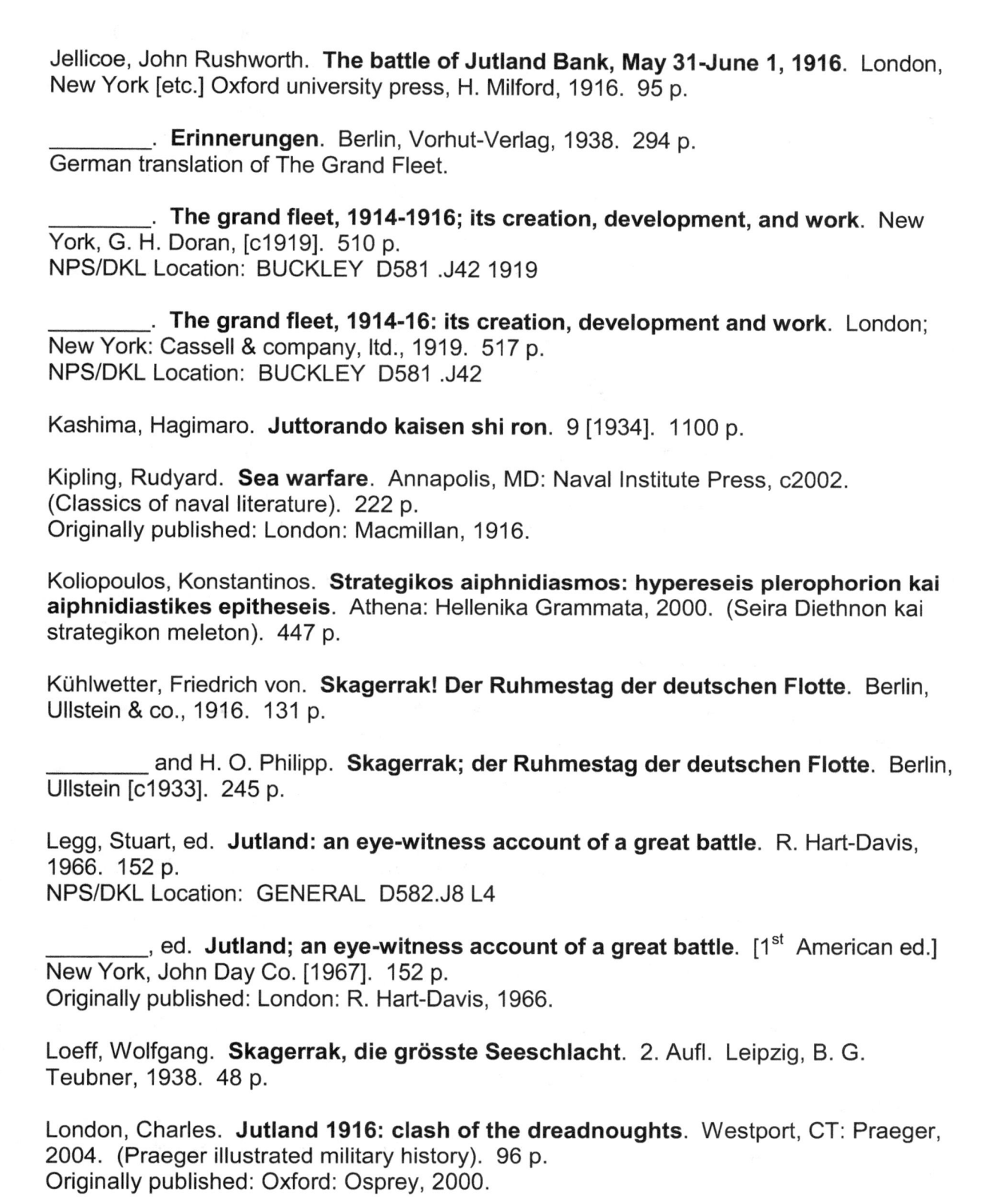

Jellicoe, John Rushworth. **The battle of Jutland Bank, May 31-June 1, 1916**. London, New York [etc.] Oxford university press, H. Milford, 1916. 95 p.

_____. **Erinnerungen**. Berlin, Vorhut-Verlag, 1938. 294 p.
German translation of The Grand Fleet.

_____. **The grand fleet, 1914-1916; its creation, development, and work**. New York, G. H. Doran, [c1919]. 510 p.
NPS/DKL Location: BUCKLEY D581 .J42 1919

_____. **The grand fleet, 1914-16: its creation, development and work**. London; New York: Cassell & company, ltd., 1919. 517 p.
NPS/DKL Location: BUCKLEY D581 .J42

Kashima, Hagimaro. **Juttorando kaisen shi ron**. 9 [1934]. 1100 p.

Kipling, Rudyard. **Sea warfare**. Annapolis, MD: Naval Institute Press, c2002. (Classics of naval literature). 222 p.
Originally published: London: Macmillan, 1916.

Koliopoulos, Konstantinos. **Strategikos aiphnidiasmos: hypereseis plerophorion kai aiphnidiastikes epitheseis**. Athena: Hellenika Grammata, 2000. (Seira Diethnon kai strategikon meleton). 447 p.

Kühlwetter, Friedrich von. **Skagerrak! Der Ruhmestag der deutschen Flotte**. Berlin, Ullstein & co., 1916. 131 p.

_____ and H. O. Philipp. **Skagerrak; der Ruhmestag der deutschen Flotte**. Berlin, Ullstein [c1933]. 245 p.

Legg, Stuart, ed. **Jutland: an eye-witness account of a great battle**. R. Hart-Davis, 1966. 152 p.
NPS/DKL Location: GENERAL D582.J8 L4

_____, ed. **Jutland; an eye-witness account of a great battle**. [1st American ed.] New York, John Day Co. [1967]. 152 p.
Originally published: London: R. Hart-Davis, 1966.

Loeff, Wolfgang. **Skagerrak, die grösste Seeschlacht**. 2. Aufl. Leipzig, B. G. Teubner, 1938. 48 p.

London, Charles. **Jutland 1916: clash of the dreadnoughts**. Westport, CT: Praeger, 2004. (Praeger illustrated military history). 96 p.
Originally published: Oxford: Osprey, 2000.

Lützow, Friedrich. **Der Nordseekrieg; Doggerbank-Skaggerrak**. Oldenburg i. O., Gerhard Stalling, 1931. (Einzeldarstellungen des Seekrieges 1914-1918, Bd. 1). 202 p.

Macintyre, Donald G. F. W. **Jutland**. London, Evans Bros. [1957]. 210 p.
NPS/DKL Location: BUCKLEY D582.J8 M2

_____. **Jutland**. [1st ed.]. New York, Norton, [1958]. 282 p.
Originally published: London, Evans Bros. [1957].

Maine, René. **Face à face, la bataille du Jutland**. Paris, Chassany & cie [1939]. 254 p.

Naval War College (U.S.). **The battle of Jutland, 31 May-I June 1916**. Newport, RI, 1920. (Naval War College Monograph; no. I). 148 p.
NPS/DKL Location: GENERAL D582.J8 U6 1920

_____. **Jutland**. 5th War College ed. Washington, Govt. print. off., 1927. 57 p.

Oakeshott, R. Ewart. **The blindfold game; the day at Jutland**. [1st ed.]. [Oxford, New York], Pergamon Press, [1969]. (The Pergamon English library). 128 p.

Parseval, Henri Louis Pie de. **La bataille navale du Jutland (31 mai 1916)**. Paris, Payot & cie, 1919. 190 p.

Pastfield, J. L. **New light on Jutland**. London: Heinemann, 1933. 30 p.

Pflugk-Harttung, Julius Albert Georg von. **Der kampf um die freiheit der meere, Trafalgar, Skagerrak**. Berlin, R. Eisenschmidt, 1917. 254 p.

Pollen, Anthony. **The great gunnery scandal: the mystery of Jutland**. London: Collins, 1980. 280 p.

Pollen, Arthur Joseph Hungerford. **The battle of Jutland, 31 May-I June 1916**. London, Chatto & Windus, 1918. 371 p.
American edition (Garden City, New York, Doubleday, Page & Company) published as The British Navy in battle.
NPS/DKL Location: BUCKLEY D581 .P75

_____. **The British navy in battle**. Garden City, New York, Doubleday, Page & company, 1919. 358 p.
London edition (Chatto & Windus) entitled: The navy in battle.
NPS/DKL Location: BUCKLEY D581 .P74

_____. **The Navy in battle**. London, Chatto & Windus, 1918. 371 p.
NPS/DKL Location: BUCKLEY D581 .P75

Rasor, Eugene L. **The battle of Jutland: a bibliography**. New York: Greenwood Press, 1992. (Bibliographies of battles and leaders, no. 7). 176 p.

Richelieu, Thomas de. **Jyllandsslaget 1916: historiens største søslag**. [København]: Aschehoug, c2002. 199 p.

Schoultz, Gustaf Johan Toivo von. **With the British battle fleet: war recollections of a Russian naval officer** [translated by Arthur Chambers]. London: Hutchinson & co., n.d. 360 p.
NPS/DKL Location: BUCKLEY D581 .S28

Steel, Nigel & Peter Hart. **Jutland, 1916: death in the grey wastes**. London: Cassell Military, 2003. 480 p.

Tarrant, V. E. **Jutland: the German perspective**. London: Cassell, 2001. 350 p. Originally published: London: Arms & Armour, 1995.

_____. **Jutland, the German perspective: a new view of the great battle, 31 May, 1916**. Annapolis, MD: Naval Institute Press, c1995. 318 p.
NPS/DKL Location: GENERAL D582.J8 T37 1995

United States. Office of Naval Intelligence. **The battle of the Skagerrak (Jutland) May 31, 1916: reports by Commander in Chief, German High Seas Fleet, and the Austria-Hungarian naval attache**. Washington, DC: U.S. Govt Print. Off., 1920. 58 p.
NPS/DKL Location: GENERAL D582.J8 U6

Wyllie, William Lionel et al. **More sea fights of the Great War: including the Battle of Jutland**. Cassell, 1919. 171 p.
NPS/DKL Location: BUCKLEY D581 .W9

Yates, Keith. **Flawed victory: Jutland, 1916**. Annapolis, MD: Naval Institute Press, c2000. 314 p.
NPS/DKL Location: GENERAL D582.J8 Y38 2000

MIDGET SUBMARINES, KAITEN (HUMAN TORPEDOS) & FROGMEN

Amar, Myriam S. **Operation Rimau, 11 September to 10 October, 1944: what went wrong?** Rev. ed. [Canberra]: Management Improvement and Manpower Policy Division, Administrative Service Branch, 1990. 1 v.

Barker, A. J. **Suicide weapon**. [New York, Ballantine Books, 1971]. (Ballantine's illustrated history of the violent century. Weapons book no. 22). 160 p.

Bekker, Cajus. **Einzelkämpfer auf See: Die dt. Torpedoreiter, Froschmänner u. Sprengbootpiloten im Zweiten Weltkrieg.** Oldenburg; Hamburg: Stalling, 1968. 210 p.

_____. **Einzelkämpfer auf See: Die dt. Torpedoreiter, Froschmänner u. Sprengbootpiloten im 2. Weltkrieg.** Herford: Koehler, 1978. 210 p.

_____. **Nemetskie morskie diversanty vo vmoroi mirovoi voine** [perevod s nemetskogo L. S. Azarkha, A. G. Bubnovskogo]. Moskva: Izdatel'stvo inostrannoi literatury, 1958. 231 p.
Russian translation of Und liebten doch das Leben.

_____. **und liebten doch das Leben: Die erregenden Abenteuer dt. Torpedoreiter, Froschmänner u. Sprengbootpiloten.** Hannover: Sponholtz, 1956. 236 p.

_____. **und liebten doch das Leben: Die erregenden Abenteuer dt. Torpedoreiter, Froschmänner u. Sprengbootpiloten.** München: Heyne, 1960. 172 p.

_____ and Hellmuth Heye. **K-men: the story of the German frogmen and midget submarines.** London: William Kimber, [1955]. 202 p.
[K = Kleinkampfmittel-Verband].
NPS/DKL Location: BUCKLEY D781 .B5

_____ and _____. **K-men: the story of German frogmen and midget submarines.** Kimber Pocket Ed. London: William Kimber, 1961. 157 p.

_____ and _____. **K-men: the story of the German frogmen and midget submarines** [translated from the German by George Malcolm]. Maidstone: George Mann Ltd, 1973. 202 p.
Reprint of 1955 edition.

Bertrand, Michel. **Commandos de la mer: sous-marins de poche, torpilles humaines, hommes-grenouilles (1940-1945).** [Paris]: Editions Maritimes & d'outre-mer, c1985. (Collection "Embruns de histoire"). 426 p.

Blassingame, Wyatt. **The U. S. frogmen of World War II.** New York, Random House [1964]. (Landmark books, 106). 171 p.

_____. **Underwater warriors, formerly called The U.S. frogmen of World War II.** New York: Random House, [1982] c1964. (Landmark books; 11). 151 p.
Originally published: New York: Random House, 1964 as The U.S. Frogmen of World War II.

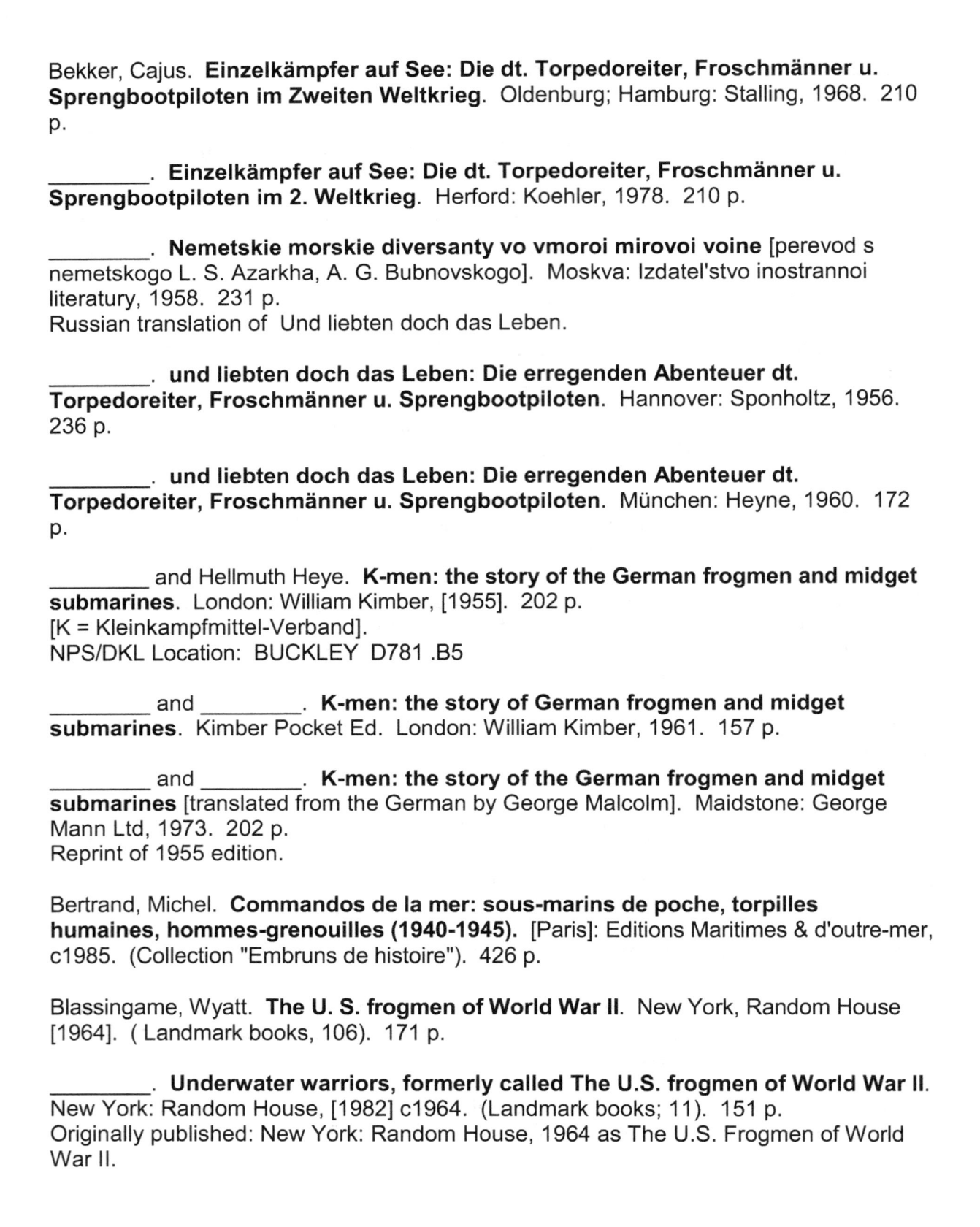

Borghese, Iunio Valerio. **Sea devils** [translated from the Italian Decima Flottiglia Mas by James Cleugh and adapted by the author]. Chicago, H. Regnery Co., 1954. 261 p.

Borotský, J. and A. Kunes. **Záludné miniponorky**. Vyd. 1. Praha: Nase vojsko, 1990. (Fakta a svedectví; sv. 106). 311 p.

Bracke, Gerhard. **Die Einzelkämpfer der Kriegsmarine: Einmanntorpedo- und Sprengbootfahrer im Einsatz**. 1. Aufl. Stuttgart: Motorbuch Verlag, 1981. 310 p.

Brou, Willy Charles. **Combat beneath the sea** [translated from the French by Edward Fitzgerald]. New York, Crowell, [1957]. 240 p.
Originally published: Paris: Documents du monde, [1955] as Nageurs de Combat.
NPS/DKL Location: GENERAL D780 .B8

_____. **Forceurs de rades**. Paris, editions de la Pense moderne, [1960]. 247 p.

Carruthers, Steven L. **Australia under siege: Japanese submarine raiders, 1942**. Sydney: Solus Books, 1982. 192 p.

Clarke, Hugh V. **Fire one!** [Sydney]: Angus & Robertson, 1978, c1966. 152 p.
Reprint of the 1966 ed. of To Sydney by stealth and of the 1965 ed. of Break-out published by Horwitz Publications.

Connell, Brian . **The return of the tiger**. London: Evans Brothers, 1960. 207 p.
An account of Operation Jaywick and Operation Rimau and the role played by Lieut.-Col. Ivan Lyon.

Date, John C. **Japanese "A" class midget submarines** / transcript by John C. Date. Garden Island, N.S.W.: Naval Historical Society of Australia, 1998. (Monograph / Naval Historical Society of Australia; 65). 22 p.
Alternate title: Japanese "A" class midget submarines and their daring attacks on Pearl Harbour 7 December 1941, Diego Suarez 30 May 1942, Sydney Harbour 31 May 1942.

_____. **Japanese midget submarine attack on Sydney Harbor, 31 May - 1 June** / transcript by John C. Date. Garden Island, N.S.W.: Naval Historical Society of Australia, 1993. (Monograph / Naval Historical Society of Australia; 27). 14 l.

Diomidov, Mikhail Nikolaevich and Aleksandr Nikolaevich Dmitriev. **Podvodnye apparaty**. 1966. 326 p.

Dmitriev, Vladimir Vasilevich. **Podvodnye "moskity."** (Po materialam inostr. pechati). Moskva, Voenizdat, 1969. 126 p.

Fane, Francis Douglas and Don Moore. **The naked warriors: the story of the U.S. Navy's Frogmen**. Annapolis, MD: Naval Institute Press, [1995]. (Naval Institute special warfare series). 308 p.

Originally published: New York: Appleton-Century-Crofts, 1956.
NPS/DKL Location: BUCKLEY D780 .F3 1995

Fell, William Richmond. **The sea our shield**. London, Cassell, 1966. 232 p.

Fleming, George. **Magennis V: the story of Northern Ireland's only winner of the Victoria Cross**. Dublin: History Ireland, 1998. 224 p.

Fock, Harald. **Marinekleinkampfmittel**. München, J. F. Lehmann, [1968]. (Wehrwissenschaftliche Berichte, Bd. 3). 155 p.

Frere-Cook, Gervis. **The attacks on the Tirpitz**. [Annapolis], Naval Institute Press, [1973]. (Sea battles in close-up, 8). 112 p.
Originally published: London, Allan, 1973.

Gagin, V. **Sverkhmalye podvodnye lodki i boevye plovtsy**. Voronezh: Poligraf, 1996. (Rossiia, prosnis´!). 52 p.

Gallagher, Thomas Michael. **Against all odds: midget submarines against the Tirpitz**. London: Macdonald, 1971. 170 p.

_____. **Senk "Tirpitz"!** [oversatt av Sigurd Valvatne]. Stavanger:Stabenfeldt, 1973. 171 p.
Swedish translation of The X-craft raid

_____. **The X-craft raid**. New York, Harcourt Brace Jovanovich [1971]. 170 p.
NPS/DKL Location: GENERAL D772.T5 G2

_____. **The X-craft raid**. New York [etc.]: Pinnacle Books, 1972. 186 p.

_____. **Zkaza bitevni lodi Tirpitz** [prelozil Robert Miller]. Praha: Mlada fronta, 1976. (Edice Archiv; 15). 156 p.
Czech translation of: The X-craft raid.

Grabatsch, Martin. **Torpedoreiter: Sturmschwimmer, Sprengbootfahrer: eine Geheimwaffe im Zweiten Weltkrieg**. Wels: Verl. Welsermuhl, c1979. 279 p.

Gregory, MacKenzie J. **"Under water warfare": the struggle against the submarine menace, 1939-1945**. Garden Island, N.S.W.: Naval Historical Society of Australia, Inc., 1997. 64 p.

Grosvenor, Joan and L. M. Bates. **Open the ports; the story of human minesweepers**. London, W. Kimber, [1956]. 199 p.

Hampshire, Arthur Cecil. **The secret navies**. London: W. Kimber, 1978. 272 p.

Higgins, Edward T. **Webfooted warriors; the story of a "frogman" in the Navy during World War II**. [1st ed.]. New York, Exposition Press, [1955]. 172 p.

Isono, Yasuko. **Ai to shi 768-jikan: ningen gyorai "Kaiten" tokubetsu kogeki taiin no memo**. Tokyo: Seishun Shuppansha, [Showa 60 i.e. 1985]. 253 p.
Based on Yamaguchi Hoso television program: Shishatachi no yuigon: Kaiten ni chitta gakutohei no kiseki.

Ito, Keiichi. **Kaiten**. [1968]. 226 p.

Jackson, Charles L. **On to Pearl Harbor and beyond**. Dixon, CA: Pacific Ship and Shore, c1982. 67 p.

Kemp, Paul. **Midget submarines**. London: Arms and Armour; New York: Distributed in the USA by Sterling Pub. Co., c1990. (Warships fotofax). 1 v.

_____. **Midget submarines of the Second World War**. London: Chatham, 1999. 125 p.

_____. **Underwater warriors**. Annapolis, MD: Naval Institute Press, c1996. 256 p.
Originally published: London: Arms and Armour, 1996.
NPS/DKL Location: GENERAL V857 .K36 1996

_____. **Underwater warriors: the fighting history of midget submarines**. London: Cassell, 2000. (Cassell military paperbacks). 256 p.
Originally published: London: Arms and Armour, 1996

Kent, Alexander. **Operation Monsun**. München: Cormoran, 2000. 382 p.

Kimata, Jiro. **Koto e no tokko**. 1975. 207 p.

Koryo Teicho Daijushichikikai. **Kisama to ore no seishunfu: Tokushu Senkotei Daijushichiki teicho no koseki**. Tokyo: Keiso Shobo, 1993. 303 p.

Kozu, Naoji. **Ningen gyorai kaiten**. Shohan. Tokyo: Tosho Shuppansha, 1989. 293 p.

Kurowski, Franz. **Fahrt ins Verderben: Einsatz d. Ein-Mann-Torpedos**. Balve/Westf.: Zimmermann, [1960]. 272 p.

Lau, Manfred. **Schiffssterben vor Algier: Kampfschwimmer, Torpedoreiter und Marine-einsatzkommandos im Mittelmeer 1942 - 1945**. 1. Aufl. Stuttgart: Motorbuch-Verl., 2001. 176 p.

Lind, L. J. **The midget submarine attack on Sydney**. Garden Island, NSW: Bellrope Press, c1990. 72 p.

_____. **Toku-tai: Japanese submarine operations in Australian waters.**
[Maryborough, Victoria?]: Kangaroo Press, 1992. 160 p.

Maeda, Masahiro. **Kaiten Kikusuitai no yonin: Kaigun chui Nishina Sekio no shogai.**
Tokyo: Kojinsha, 1989. 285 p.

McKie, Ronald. **The heroes.** Sydney: Angus & Robertson, 1960. 285 p.
The story of Operation Jaywick and Operation Rimau, two marine raids into Japanese-held Singapore from bases in Australia.

Merenov, Igor' Vladimirovich. **Sredstva peredvizheniia pod vodoi.** 1966. 78 p.

Mitchell, Pamela. **The tip of the spear: the submarine arm is the spearhead of the Royal Navy (Anon).** 2nd ed. Huddersfield: Richard Netherwood Ltd., 1995. 232 p.
Originally published: Huddersfield: Richard Netherwood Limited, 1993.
NPS/DKL Location: GENERAL   D784.G7 M57 1995

Nagasawa, Michio. **Kaitei no chinmoku: "Kaiten" hasshin seshi ya.** Tokyo: Nihon Hoso Shuppan Kyokai, 1999. 382 p.

Neuhauzer, A. **Bi-metsulot mifrats Si'am: parashat tivu`a ha-shayetet ha-Britit be-milhemet ha-`olam ha-sheniyah, be-10.12.41.** Tel-Aviv: Hotsa'at Yaron Golan, [1999]. 159 p.
Title on t.p. verso: Deep in Siam Bay

O'Neill, Richard. **Suicide squads: Axis and Allied special attack weapons of World War II: their development and their missions.** 1st Ballantine Books trade ed. New York: Ballantine Books, 1984, c1981. 296 p.
Originally published: London: Salamander, 1981.

_____. **Suicide squads: the men and machines of World War II special operations.** London: Salamander, c1999. (Classic conflicts). 272 p.
NPS/DKL Location: GENERAL   D744 .D54 1999

_____. **Suicide squads, W.W. II: Axis and Allied special attack weapons of World War II, their development and their missions.** 1st U.S. ed. New York: St. Martin's Press, c1981. 296 p.

Okazawa, Tadashi. **Kokuhakuteki "koku kagakusen" shimatsuki.** Tokyo: Kojinsha, 1992. (Shogen Showa no senso. Ribaibaru senki korekushon; [32]). 333 p.

Onda, Shigetaka. **Tokko.** Tokyo: Kodansha, 1988. 542 p.

Pöschel, Günther. **Froschmämmer.** Berlin, Deutscher Militärverlag, 1961. 303 p.

Potter, John Deane. **Yamamoto; the man who menaced America**. New York, Viking Press, [1965]. 332 p.
"Target A, Pearl Harbor's most secret weapon" p.72-115.
Originally published: London: Heinemann, 1965, as Admiral of the Pacific: the life of Yamamoto."
NPS/DKL Location: GENERAL DS890.Y25 P6 1965

_____. **Yamamoto: the man who menaced America**. New York: Paperback Library, 1965. 351 p.
Originally published: London: Heinemann, 1965 as Admiral of the Pacific, the life of Yamamoto.

Rastelli, Achille. **Caproni e il mare: progetti e realizzazioni per la guerra navale di un grande gruppo industriale milanese** [traduzione inglese di Gregory Alegi]. [Italy]: Museo aeronautico Gianni e Timina Caproni di Taliedo, [1999]. 131 p.

Rebikoff, Dimitri. **L'aviation sous-marine**. Paris, Flammarion, [1962]. 250 p.

_____. **En avion sous la mer**. Paris, P. Horay, [1956]. 222 p.

Schofield, William G. and and P.J. Carisella. **Frogmen: first battles**. Boston, MA: Branden Pub. Co., c1987. 187 p.

Shean, Max. **Corvette and Submarine**. Claremont, WA: M. Shean, 1992. 275 p.
Transfered from the corvette Bluebell to X-craft, he took part in the raids on Tirpitz, Bergen and in the Far East.

Sierra, Luis de la. **Buques suicidas; la historia de los submarinos de bolsillo, torpedos humanos y botes explosivos en el siglo XX**. [2. ed.]. Barcelona, L. de Caralt, [1963]. (La Vida vivida). 322 p.

_____. **Buques suicidas: la historia de los submarinos de bolsillo, torpedos humanos y botes explosivos en el siglo XX**. 2. ed. Editorial Juventud, 1976, c1972. 299 p.

Silver, Lynette Ramsay. **The heroes of Rimau: unravelling the mystery of one of World War II's most daring raids**. Birchgrove, N.S.W.: Sally Milner Publishing, 1990. 314 p.

_____. **The heroes of Rimau: unravelling the mystery of one of World War II's most daring raids**. Leo Cooper, 1991. 314 p.

_____. **The heroes of Rimau: unravelling the mystery of one of World War II's most daring raids**. Kuala Lumpur: S.A. Majeed, 1992. 314 p.

_____. **The heroes of Rimau: unravelling the mystery of one of World War II's most daring raids**. Singapore: Cultured Lotus, 2002. 320 p.

Skulski, Janusz. **The heavy cruiser Takao**. Annapolis, MD: Naval Institute Press, c1994. (Anatomy of the ship). 256 p.
Wrecked by British X-craft XE-1 and XE-3 in the Straight of Johore.

Thomson, Peter And Robert Macklin. **Kill the tiger: the truth about Operation Rimau**. Sydney: Hodder, 2002. 306 p.
These courageous men were part of Operation Rimau (Malay for tiger) which would use the latest one-man submarines - Sleeping Beauties - developed by Royal Navy scientists in Britain and yet to be tested in combat.

Torisu, Kennosuke. **Ningen gyorai: Tokko heiki "Kaiten" to wakodotachi**. Tokyo: Shinchosha, 1983. 329 p.

Toschi, Elios. **Tesei e i cavalieri subacquei**. Roma, G. Volpe, 1967. (Italiani in guerra, 5). 209 p.

Tsumura, Toshiyuki, pseud. **Ningen gyorai Kaiten**. 29 [1954]. 263 p.

Uehara, Mitsuharu. **"Kaiten" sono seishun gunzo: tokko senkotei no otokotachi**. Shohan. Tokyo: Shounsha, Heisei 12 [2000]. 410 p.

Waldron, Thomas John and James Gleeson. **The frogmen; the story of the wartime underwater operators**. London, Evans Bros., [1950]. 191 p.

_____. **The frogmen; the story of the wartime underwater operations**. Cadet ed. London, Evans Bros., 1966. 156 p.

_____. **The frogmen**. New ed. New York: Berkley Pub. Corp., 1963. 173 p.

_____. **The frogmen: the story of the wartime underwater operators**. [New ed.]. London: Pan Books Ltd., 1970. 195 p.

_____. **Midget submarine**. New York: Ballantine Books, 1975. (Ballantine's illustrated history of the violent century. Weapons book; no. 42). 159 p.

_____. **Submarinos enanos** [traductor: Juan Genova]. Madrid: Editorial San Martin, 1979. (Historia del siglo de la violencia. Armas; libro no. 33). 159 p.
Spanish translation of Midget Submarine.

Walker, Frank and Pamela Mellor. **The mystery of X-5: Lieutenant H. Henty-Creer's attack on the Tirpitz**. London: William Kimber, c1988. 239 p.

Warner, Denis Ashton, Peggy Warner and Sadao Seno. **The sacred warriors: Japan's suicide legions**. New York: Van Nostrand Reinhold, c1982. 370 p.

Warner, Peggy and Sadao Seno. **The coffin boats: Japanese midget submarine operations in the Second World War**. London: Leo Cooper in association with Secker & Warburg, 1986. 206 p.

Warren, Charles Esme Thornton and James Benson. **The midget raiders**. New York, W. Sloane Associates, 1954. 318 p.
NPS/DKL Location: BUCKLEY D784.G7 W2

_____, _____ and Sir George Creasy. **Above us the waves: the story of midget submarines and human torpedos**. London: Transworld, 1955. 332 p.
Originally published: London: George G. Harrap, 1953.

Watanabe, Yoshimitsu. **Seishun no wasurezaru hibi: kaiten tokko ichi taiin no senso**. [Aichiken Tokai-shi]: Watanabe Yoshimitsu, Heisei 2 [1990]. 446 p.

Wilkinson, Burke. **By sea and by stealth**. New York: Coward-McCann, c1956. 218 p.

Wright, Bruce S. **The frogmen of Burma: the story of the Sea Reconnaissance Unit**. Toronto, Vancouver, Clarke, Irwin, [c1968]. 152 p.

Yamaoka, Sohachi. **Kaitei senki**. [1943]. 218 p.

_____. **Kaitei senki**. Tokyo: Chuo Koron Shinsha, 2000. (Chuko bunko). 191 p.

Yokota, Yutaka. **Aa kaiten tokkotai**. 46 [1971]. ?? p.

_____. **Aa Kaiten Tokubetsu Kogekitai**. [1968]. 294 p.

_____ and Harrington, Joseph D. **Suicide submarine!** New York: Ballantine Books, [c1962]. (A Ballantine original; S600). 255 p.
Originally published: New York, Ballantine Books, [1962] as The kaiten weapon.

Q-SHIPS

Auten, Harold. **"Q" boat adventures, the exploits of the famous mystery ships**. [London], H. Jenkins limited, [1935]. 289 p.
NPS/DKL Location: BUCKLEY D581 .A9

Beyer, Kenneth M. **Q-ship versus U-boats: America's secret project**. Annapolis, MD: Naval Institute Press, 1999. 336 p.
NPS/DKL Location: GENERAL D783 .B49 1999

Bridgland, Tony. **Sea killers in disguise: the story of the Q-ships and decoy ships in the First World War**. Annapolis, MD: Naval Institute Press, c1999. 274 p.
Originally published: London: Leo Cooper, 1999
NPS/DKL Location: GENERAL D581 .B65 1999

Campbell, Gordon. **My mystery ships**. Garden City, NY, Doubleday, Doran & Company, inc., 1929. 318 p.
Originally published: London, Hodder & Stoughton, limited, [1928].

Chatterton, E. Keble. **Q-ships and their story**. [Annapolis], Naval Institute Press, 1972. 276 p.
Originally published: London: Sidgwick & Jackson, ltd., 1922; Boston, C. E. Lauriat, 1927.
NPS/DKL Location: BUCKLEY D581 .C44

_____. **Q-ships and their story**. New York: Arno Press, 1980, c1972. 276 p.
Reprint, originally published: Anapolis, Naval Institute Press, 1972.

Coder, Barbara J. **Q-Ships of the Great War**. Maxwell Air Force Base, AL: Air University Command and Staff College, 2000. 30 p.
Electronic access: http://www.au.af.mil/au/database/projects/ay2000/acsc/00-220.pdf

Coles, Alan. **Slaughter at sea: the truth behind a naval war crime**. London, Hale, 1986. 220 p.

Higgins, Alexander Pearce. **Defensively-armed merchant ships and submarine warfare**. London, Stevens, 1917. 56 p.

Noyes, Alfred. **Mystery ships: trapping the "U" boat**. London, Hodder & Stoughton, 1916. 181 p.

Ritchie, Carson I. A. **Q-ships**. Lavenham, Suffolk: Terence Dalton, 1985. 216 p.

SUBMARINES -- GENERAL

Arkin, William M. and Joshua Handler. **Naval accidents, 1945-1988**. Washington, DC (1436 U St., NW, Washington 20009: Greenpeace: Institute for Policy Studies, [1989]. (Neptune papers; no. 3). 83 p.

Bergelin, Jacques and Victor Alexandrov. **Guerre secrèté sous les océans**. Paris, Éditions maritimes it d'outremer, 1970. 198 p.

Bravetta, Ettore. **Sottomarini sommergibili e torpedini**. Milano: Fratelli Treves, Editori, 1915. 230 p.

Burcher, Roy and Louis Rydill. **Concepts in submarine design**. Cambridge [England]; New York: Cambridge University Press, 1998, 1994. (Cambridge ocean technology series; 2). 300 p.
NPS/DKL Location: GENERAL VM365 .B89 1998/VM365 .B89 1994

Burgess, Robert Forrest. **Ships beneath the sea: a history of subs and submersibles**. New York: McGraw-Hill, [1975]. 260 p.
NPS/DKL Location: GENERAL V857 .B9

Bynander, Fredrik, ed. **Ubåtsfrågan – ett symposium (The Submarine Issue – A Symposium)**. Stockholm: Swedish National Defence College, 2002. 132 p.

Cohen, P. **The realm of the submarine**. [New York], Macmillan, [1969]. 274 p.
NPS/DKL Location: GENERAL V857 .C6

Compton-Hall, Richard. **Submarine boats: the beginnings of underwater warfare**. New York: Arco Pub., 1984, c1983. 192 p.

_____. **The submarine pioneers**. Phoenix Mill, Thrupp, Stroud, Gloucestershire: Sutton Pub., c1999. 182 p.
NPS/DKL Location: BUCKLEY V857 .C652 1999

_____. **Submarine versus submarine: the tactics and technology of underwater confrontation**. 1st ed. New York: Orion Books, c1988. 191 p.
Originally published: Newton Abbot: David & Charles, 1988.
NPS/DKL Location: GENERAL V214.5 .C65 1988

_____. **Submarine warfare: monsters & midgets**. Poole, Dorset: Blandford Press; New York, NY: Distributed in the U.S. by Sterling Pub. Co., 1985. 160 p.

Cope, Harley Francis and Leland P. Lovette. **Serpent of the seas: the submarine**. New York; London: Funk & Wagnalls company, c1942. 252 p.

Douglas, Lawrence Henry. **Submarine disarmament: 1919-1936**. [Syracuse, NY]: c1970.] 276 p.
Thesis (Ph.D.), Syracuse University, 1970.

Dugan, James. **Man under the sea**. New York: Harper & Brothers, c1956. 332 p.
Originally published: London: Hamish Hamilton: London, 1956 as Man explores the sea, 366 p.

Dunmore, Spencer. **Lost subs: from the Hunley to the Kursk, the greatest submarines ever lost - and found**. Cambridge, MA: Da Capo, 2002. 178 p.
Contents: The silent service -- August 12, 2000: the final moments aboard the Russian Navy's nuclear submarine Kurst -- The birth of the submarine: A brief history of early subs, including the Turtle and the Nautilus -- The first underwater weapon of war: the

American Confederate sub H.L. Hunley -- The submarine comes of age: the influence of John Holland on submarine design -- Battle under the sea: the submarine has now become a formidable weapon, as Germany's U-boats prove during the First World War -- Lost in peace: two submarine tragedies, the US Navy's Squalus and the Royal Navy's Thetis, dominate peacetime headlines -- The U-boat war: Germany revolutionizes the way submarines are used in combat -- A casualty of the Cold War: a tiny piece of aluminum foil spells disaster for the US Navy's nuclear submarine Scorpion -- Disaster under the Barents Sea: the tragic story of the Kursk -- For those in peril.
NPS/DKL Location: GENERAL VB230 .D97 2002

Ekman, Per-Olof. **Havsvargar: ubatar och ubatskrig i Ostersjon**. [Helsinki]: Schildt, c1983. 348 p.

Ellacott, S. E. **Ships under the sea**. London: Hutchinson, 1961. 142 p.

Ferm, Anders. **Perspektiv på ubåtsfrågan. Hanteringen av ubåtsfrågan politiskt och militärt**. SOU 2001:85. Stockholm: Ministry of Defence, 2001.
http://forsvar.regeringen.se/propositionermm/sou/.
Summary in English: http://forsvar.regeringen.se/propositionermm/sou/pdf/sou2001_85d.pdf

Forsen, Bjorn och Annette Forsen. **Tysklands och Finlands hemliga ubatssamarbete**. [Helsingfors]: Soderstrom, c1999. 334 p.

Francis, Timothy L. **Submarines: leviathans of the deep**. New York: MetroBooks, c1997. 144 p.

Friedman, Norman. **Submarine design and development**. Annapolis, MD: Naval Institute Press, c1984. 192 p.
NPS/DKL Location: GENERAL V857 .F75 1984

Fyfe, Herbert C. **Submarine warfare: past, present and future**. London: Grant Richards, 1902. 332 p.

_____. **Submarine warfare past, present and future**. 2nd ed. London: Grant Richards, 1903. 332 p.
NPS/DKL Location: GENERAL V210 .F9 MISSING

_____. **Submarine warfare: past and present**. 2nd ed. Revised. London: E. Grant Richards, 1907. 302 p.

Gabler, Ulrich. **Submarine design**. Koblenz: Bernard & Graefe, c1986. 140 p.
English translation of Unterseebootbau.

_____. **Unterseebootsbau**. Darmstadt, Wehr und Wissen, [1964]. 157 p.

Gardner, W. J. R. **Anti-submarine warfare**. 1st English ed. London; Washington: Brassey's, 1996. (Brassey's sea power; v. 11). 160 p.

NPS/DKL Location: GENERAL V214 .G37 1996

Garrett, Richard. **Submarines**. 1st American ed. Boston: Little, Brown, c1977. 143 p.
NPS/DKL Location: GENERAL V857 .G24

Gates, P. J. and N. M. Lynn. **Ships, submarines, and the sea**. 1st ed. London; Washington: Brassey's, 1990. (Brassey's sea power; v. 2). 178 p.
NPS/DKL Location: GENERAL V396 .G38 1990

Gray, Edwyn. **Few survived: a comprehensive survey of submarine accidents and disasters**. 2nd ed. London: Leo Cooper, 1996. 274 p.
NPS/DKL Location: GENERAL VK1265 .G73 1996

_____. **Submarine warriors**. New York: Bantam Books, 1990, c1988. 324 p. Originally published: Novato, CA: Presidio, 1988.

Guierre, Maurice Casimir Lucien. **Aux postes de plongee**. 3rd ed. [Paris]: Gallimard, c1959. 477 p.

_____. **Sous mariniers**. Paris: Flammarion, 1948. 303 p.

Gunston, Bill. **Submarines in color**. New York: Arco Pub. Co., 1977, c1976. 200 p. Originally published: Poole: Blandford Press, 1976 as Submarines in colour.

Hervey, John B. **Submarines**. 1st English ed. London; New York: Brassey's; New York: Distributed by Macmillan, 1994. Brassey's sea power; v. 7). 289 p.

Herzog, Bodo. **60 Jahre deutsche Uboote 1906-1966**. München, J. F. Lehmann, [1968]. 324 p.

Hezlet, Arthur Richard. **The submarine and sea power**. London, P. Davies; New York, Stein & Day, 1967. 278 p.

Horton, Edward. **The illustrated history of the submarine**. Garden City, NY: Doubleday, [1974]. 160 p.
NPS/DKL Location: GENERAL V210 .H8

_____. **The illustrated history of the submarine**. London: Sidgwick & Jackson, c1974. 160 p.

Hutchinson, Robert. **Jane's Submarines: War Beneath the Waves from 1776 to the Present Day**. New York: HarperCollins, 2002. 223 p.

Ireland, Bernard. **Warships of the world: submarines & fast attack craft**. London: I. Allan, 1980. 160 p.

Jackson, George Gibbard. **The romance of the submarine**. Philadelphia, J.B. Lippincott company, [1930]. 244 p.
Originally published: London: Sampson Low, Marston, 1930.

Jackson, Robert. **Submarines of the World**. Etobicoke, Ontario: Prospero, 2000. 320 p.

_____. **Unterseeboote**. Bindlach: Gondrom-Vlg., 2001. 320 p.
German translation of Submarines of the World

Jameson, William. **The most formidable thing; the story of the submarine from its earliest days to the end of World War I**. London, R. Hart-Davis, 1965. 280 p.

Jane's Information Group. **Jane's underwater warfare systems**. Coulsdon, Surrey, UK: Jane's Information Group, annual.
NPS/DKL Location: REFERENCE V214 .J36 YY/YY

Jones, Eric R. **The proliferation of conventionally-powered submarines: balancing U.S. Cruise missile diplomacy? The cases of India and Iran**. Monterey, CA: Naval Postgraduate School; Springfield, VA: Available from National Technical Information Service, 1997. (ADA333352). 98 p.
Thesis (M.A. in National Security Affairs)--Naval Postgraduate School, 1997.
Abstract: *The end of the Cold War has left the United States as the world's sole superpower. The ability of the United States to strike deep into the territories of most nations with impunity represents a new security threat to many nations. Defeating the U.S. military is not feasible in mostcases, but balancing the United States may be possible, especially with weapons of mass destruction (WMD). Although WMD might provide a formidable deterrent, their technical, political, and economic costs preclude most nations from pursuing them. On the other hand, modern conventionally powered submarines are easier to obtain and operate and could present a significant deterrent to U.S. military force. This thesis assesses whether the perceived threat posed by the United States has emerged as a motivation for acquiring conventionally powered submarines since the end of the Cold War. After examining the motivations behind the recent submarine acquisitions of India and Iran, this thesis presents an economic model to predict when developing nations will be able to afford submarines if they choose to acquire them.*
Electronic access: http://handle.dtic.mil/100.2/ADA333352
NPS/DKL Location: THESIS J7115

Khiiainen, Lev Petrovich. **Razvitie zarubezhnykh podvodnykh lodok i ikh taktiki**. Moskva: Voenizdat, 1979. 150 p.

Khvoshch, V. A. **Taktika podvodnykh lodok**. Moskva: Voen. izd-vo, 1989. 262 p.

Kolesnik, Eugene M. **NATO and Warsaw Pact submarines since 1955**. Poole, England; New York: Blandford, 1987. [160] p.
NPS/DKL Location: GENERAL V857 .K65 1987

Kvitnitskii, Aleksei Alekseevich. **Bor´ba s podvodnymi lodkami (po inostrannym dannym)**. 125 p.

Lewis, David R. **The fight for the sea: the past, present and future of submarine warfare in the Atlantic.** Cleveland, World Pub. Co., [1961]. 350 p.
NPS/DKL Location: GENERAL V210 .L6

Lipscomb, Frank Woodgate and Malcolm McGregor. **Historic submarines.** New York, Prager, [1970]. 35 p.
Originally published: London: Evelyn, 1970. 39 p.
NPS/DKL Location: FOLIO V857 .L7

Mallison, W. Thomas. **Studies in the law of naval warfare: submarines in general and limited wars.** Washington: U.S. Govt. Print. Off., 1968. (International law studies; 58). 230 p.
NPS/DKL Location: FEDDOCS D 208.207:58

Mattila, Tapani and Olavi Vitikka. **Uhka lännestä: Suomen meripuolustus autonomian aikana.** Helsinki: Suomi merellä -säätiö, 1996. 222 p.

Middleton, Drew. **Submarine, the ultimate naval weapon: its past, present & future.** 1st ed. Chicago: Playboy Press, c1976. 256 p.
NPS/DKL Location: GENERAL V210 .M52 1976

Miller, David. **An illustrated guide to modern sub hunters.** Arco Pub., c1984. 160 p.
NPS/DKL Location: GENERAL V214 .M55 1984

_____. **An illustrated guide to modern submarines: the undersea weapons that rule the oceans today.** New York: Arco Pub., c1982. 159 p.
NPS/DKL Location: GENERAL V857 .M54 1982

_____. **Submarines of the world.** 1st American ed. New York: Orion Books, c1991. 189 p.
NPS/DKL Location: GENERAL V857 .M54 1991

_____ and John Jordan. **Modern submarine warfare.** London; New York: Salamander Books Ltd., c1987. 208 p.

Moore, John Evelyn, ed. **Jane's pocket book of submarine development.** New York: Collier Books, 1976. 240 p.
NPS/DKL Location: REFERENCE V857 .J3

_____ and Richard Compton-Hall. **Submarine warfare: today and tomorrow.** Bethesda, MD: Adler & Adler, 1987, c1986. 308 p.
Originally published: London: Joseph, 1986.
NPS/DKL Location: GENERAL V210 .M66 1987

Preston, Antony. **Navies of World War 3.** New York: Military Press: Distributed by Crown Publishers, c1984. 192 p.

NPS/DKL Location: GENERAL VA41 .P75 1984

_____. **Submarine warfare: an illustrated history**. San Diego, CA: Thunder Bay Press, 1998. 160 p.
NPS/DKL Location: GENERAL V210 .P67 1998

_____. **Submarines**. Greenwich, CT: Bison. 1982. 64 p.

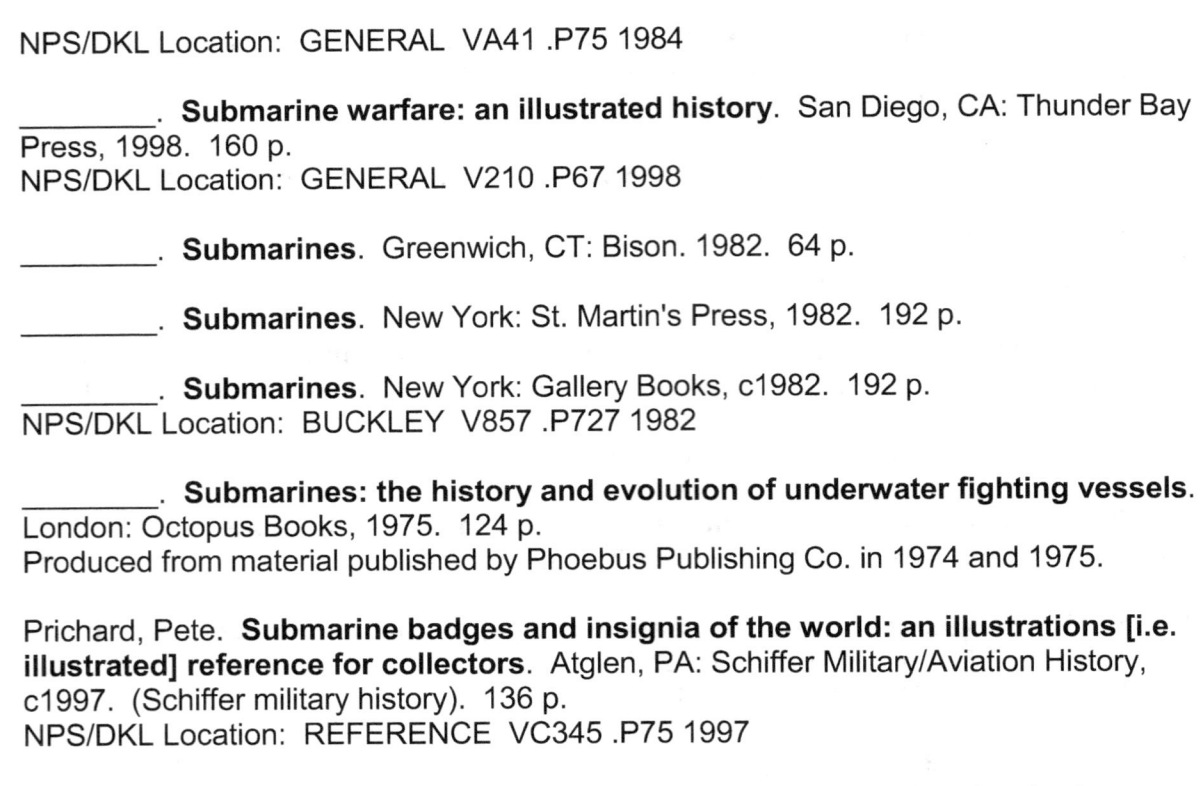

_____. **Submarines**. New York: St. Martin's Press, 1982. 192 p.

_____. **Submarines**. New York: Gallery Books, c1982. 192 p.
NPS/DKL Location: BUCKLEY V857 .P727 1982

_____. **Submarines: the history and evolution of underwater fighting vessels**. London: Octopus Books, 1975. 124 p.
Produced from material published by Phoebus Publishing Co. in 1974 and 1975.

Prichard, Pete. **Submarine badges and insignia of the world: an illustrations [i.e. illustrated] reference for collectors**. Atglen, PA: Schiffer Military/Aviation History, c1997. (Schiffer military history). 136 p.
NPS/DKL Location: REFERENCE VC345 .P75 1997

Rush, Charles W., W. C. Chambliss and H. J. Gimpel. **The complete book of submarines**. Cleveland: World Pub. Co., c1958. 159 p.

Selwyn, Philip A., Glenn R Spalding and Frank M. Lev. **Department of the Navy Exploratory Development (6.2) Investment Strategy**. Arlington, VA: Office of the Chief of Naval Research, 1990. (ADA229337). 24 p.
Abstract: *This document presents the Department of the Navy (DON) Exploratory Development (6.2) Investment Strategy, which establishes the focus and major thrusts of the 6.2 Program. The Navy's investment strategy for its exploratory development activities is derived from a national security strategy that mandates continued fulfillment of U.S. responsibilities in both Europe and Asia. In addition, the changing world environment will increase our responsibilities with respect to Third World nations. Successful implementation of U.S. strategy presumes that our naval forces will maintain technical superiority in their weaponry and platforms. As such, our warfighting strategy is achievable only if our nation maintains its current technological leadership. The Soviet Union's quantitative advantages and qualitative advances in space, submarine warfare, cruise missiles, and electronic warfare are tangible evidence that our ability to maintain a technological edge clearly is being challenged. Proliferation of high technology will continue to expand on the world arms market and will become prevalent in the Third World. Therefore, the requirement to maintain technological superiority remains absolutely essential.*
NPS/DKL Location: MICROFORM ADA229337

Siebe, Michael A. **Force Protection of Sea Based Logistics, A Historical Perspective**. Newport, RI: Naval War College, 1991. (ADA370875). 21 p.
Abstract: *The United States is heavily reliant on sea based logistics shipping. This shipping gives the United States great flexibility but it is also a critical vulnerability. During WW II Japan was a nation dependent on maritime shipping and Japan failed to provide adequate resources to protect that shipping. The results were disastrous. The U.S. and Great Britain also experienced attacks on maritime shipping in WW I and II. Resources were allocated and tactics developed to counter the German submarine threat.*

Current U.S. doctrine addresses protection of maritime shipping, but without a credible threat and with ever decreasing Naval resources, it is doubtful that the issue of maritime force protection will be seriously addressed until disaster strikes.
Electronic access: http://handle.dtic.mil/100.2/ADA370875

Stafford, Edward Peary. **The far and the deep**. New York, Putnam, [1967]. 384 p.
NPS/DKL Location: GENERAL V210 .S7

_____. **The far and the deep**. London: Barker, 1968. 382 p.

Thornton, W. M. and Gustavo Conde. **Submarine insignia & submarine services of the world**. Annapolis, MD: Naval Institute Press, c1997. 155 p.
NPS/DKL Location: GENERAL VC345 .T48 1997

Time-Life Books, eds. **Hunters of the deep**. Alexandria, VA: Time-Life Books; Morristown, NJ: School & library distribution, Silver Burdett, c1992. (The New face of war). 160 p.

Tunander, Ola. **Hårsfjärden: det hemliga ubåtskriget mot Sverige; översättning från engelska av Johan Erséus**. Stockholm: Norstedts, 2001. 405 p.

_____. **The secret war against Sweden: US and British submarine deception and political control in the 1980s**. Portland, OR: Frank Cass, 2003. (Cass series-- naval policy and history, 1366-9478; 21). 380 p.

United States. Office of Naval Intelligence. **Worldwide submarine challenges**. [Washington, DC?]: The Office, 1997. 31 p.
NPS/DKL Location: FEDDOCS D 201.2:W 89

_____. **Worldwide submarine challenges**. [Washington, DC?]: The Office, 1996. 32 p.
NPS/DKL Location: FEDDOCS D 5.202:W 89/2

Van der Vat, Dan. **Stealth at sea: the history of the submarine**. Boston: Houghton Mifflin, 1995, c1994. 374 p.
Originally published: London: Weidenfeld & Nicholson, 1994
NPS/DKL Location: GENERAL V857 .V35 1995

Wahlbäck, Krister. **Submarine Incursions in Swedish Waters, 1989-1992: A Comment on the Report of the Latest Official Investigation and the Debate It Brought About**. The Parallel History Project on NATO and the Warsaw Pact, by permission of the Center for Security Studies at ETH Zurich and the National Security Archive at the George Washington University on behalf of the PHP network. 30 October 2002.
http://www.isn.ethz.ch/php/research/AreaStudies/Wahlback.pdf

Watts, Anthony John. **Non-nuclear submarines--the world market**. Coulsdon, Surrey, UK; Alexandria, VA: Jane's Information Group, 1996. (Jane's special report). 167 p.

Weybrew, Benjamin B. **The ABC's of stress: a submarine psychologist's perspective**. Westport, CT: Praeger, 1992. 220 p.

SUBMARINES -- GENERAL -- BIBLIOGRAPHIES

Anderson, Frank J., comp. & ed. **Submarines, diving, and the underwater world: a bibliography**. Hamden, CT: Archon Books, 1975. 238 p.
Originally published in 1963 as Submarines, submariners, submarining.
NPS/DKL Location: REFERENCE VM365 .Z9

Beer, Albert. **Auswahl-Literaturverzeichnis zur Geschichte der Seekriegführung**. [Ellwangen, Kottenwiesen 7]: A. Beer, [1984 ?]. 72 p.

Bryce, Barbara Ann, comp. **An annotated literature survey of submarines, torpedoes, anti-submarine warfare, undersea weapon systems, and oceanography: 1941 to January 1962**. [California?]: Autonetics, [1962]. 1 v.

Jameson, Mary Ethel, comp. **Submarines; a list of references in the New York Public Library**. New York: New York Public Library, 1918. 97 p.

Martini, Ron. **Hot, Straight and Normal: A submarine bibliography**. Available from rontini@wavecom.net, 1997. 108 p.

National Research Council (U.S.). Committee on Undersea warfare. **An annotated bibliography of submarine technical literature, 1557 to 1953**. Washington, 1954. 261 p.
NPS/DKL Location: REFERENCE V857 .A26

Paine, Thomas O, comp. **Submarining: three thousand books and articles**. Santa Barbara, CA: General Electric Co.-TEMPO, Center for Advanced Studies, [1971]. 414 p.

Schlemm, Jürgen. **Der U-Boot-Krieg 1939-1945 in der Literatur: eine kommentierte Bibliographie**. Hamburg: Elbe-Spree-Verlag, c2000. 212 p.
NPS/DKL Location: REFERENCE D780 .S35 2000

SUBMARINES -- GENERAL -- FICTION (SELECTED AUTHORS)

Ballard, Robert D. and Tony Chiu. **Bright shark**. New York, NY: Delacorte Press, 1992. 483 p.
DKL Location: LEISURE BAL

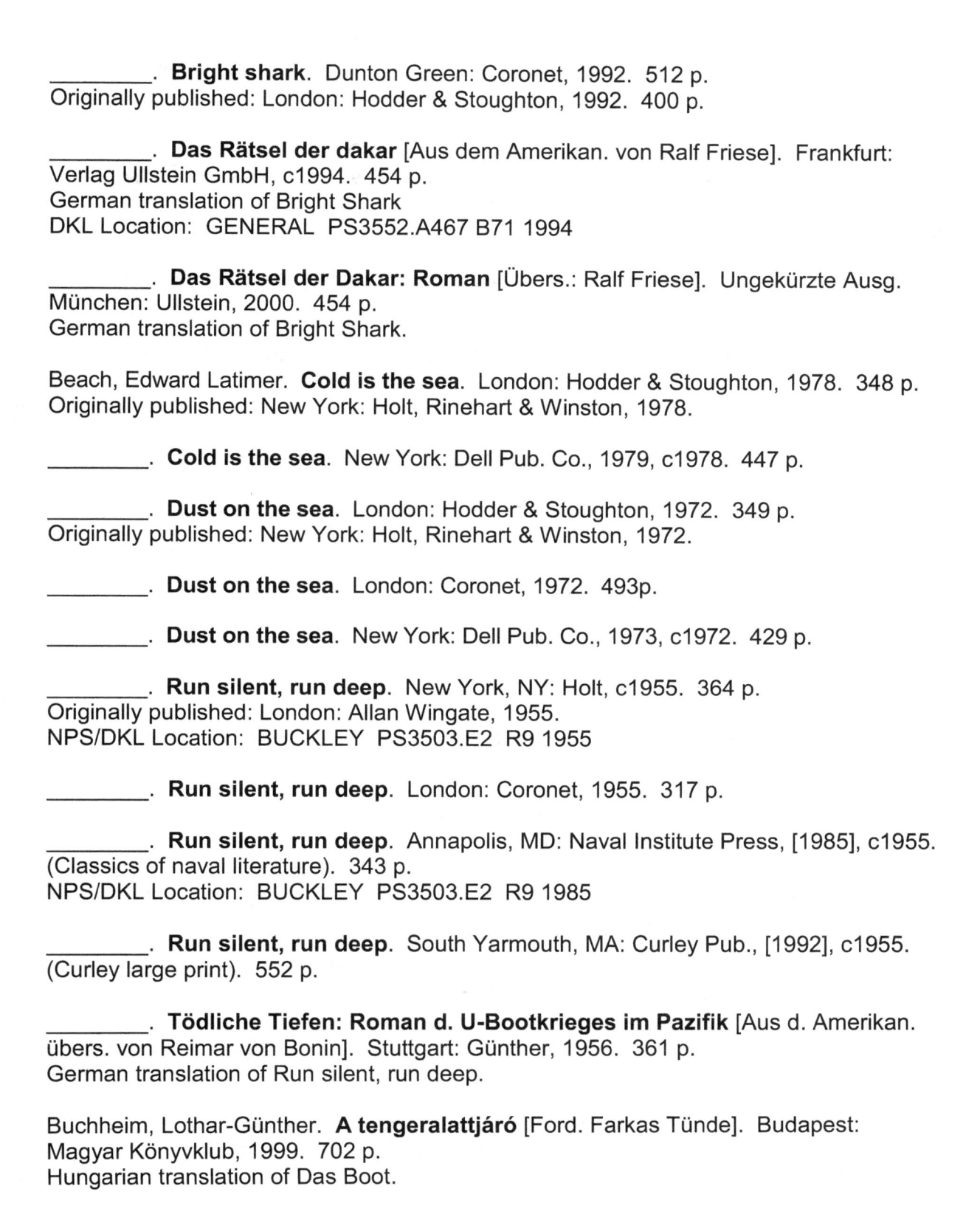

_____. **Bright shark**. Dunton Green: Coronet, 1992. 512 p.
Originally published: London: Hodder & Stoughton, 1992. 400 p.

_____. **Das Rätsel der dakar** [Aus dem Amerikan. von Ralf Friese]. Frankfurt: Verlag Ullstein GmbH, c1994. 454 p.
German translation of Bright Shark
DKL Location: GENERAL PS3552.A467 B71 1994

_____. **Das Rätsel der Dakar: Roman** [Übers.: Ralf Friese]. Ungekürzte Ausg. München: Ullstein, 2000. 454 p.
German translation of Bright Shark.

Beach, Edward Latimer. **Cold is the sea**. London: Hodder & Stoughton, 1978. 348 p.
Originally published: New York: Holt, Rinehart & Winston, 1978.

_____. **Cold is the sea**. New York: Dell Pub. Co., 1979, c1978. 447 p.

_____. **Dust on the sea**. London: Hodder & Stoughton, 1972. 349 p.
Originally published: New York: Holt, Rinehart & Winston, 1972.

_____. **Dust on the sea**. London: Coronet, 1972. 493p.

_____. **Dust on the sea**. New York: Dell Pub. Co., 1973, c1972. 429 p.

_____. **Run silent, run deep**. New York, NY: Holt, c1955. 364 p.
Originally published: London: Allan Wingate, 1955.
NPS/DKL Location: BUCKLEY PS3503.E2 R9 1955

_____. **Run silent, run deep**. London: Coronet, 1955. 317 p.

_____. **Run silent, run deep**. Annapolis, MD: Naval Institute Press, [1985], c1955. (Classics of naval literature). 343 p.
NPS/DKL Location: BUCKLEY PS3503.E2 R9 1985

_____. **Run silent, run deep**. South Yarmouth, MA: Curley Pub., [1992], c1955. (Curley large print). 552 p.

_____. **Tödliche Tiefen: Roman d. U-Bootkrieges im Pazifik** [Aus d. Amerikan. übers. von Reimar von Bonin]. Stuttgart: Günther, 1956. 361 p.
German translation of Run silent, run deep.

Buchheim, Lothar-Günther. **A tengeralattjáró** [Ford. Farkas Tünde]. Budapest: Magyar Könyvklub, 1999. 702 p.
Hungarian translation of Das Boot.

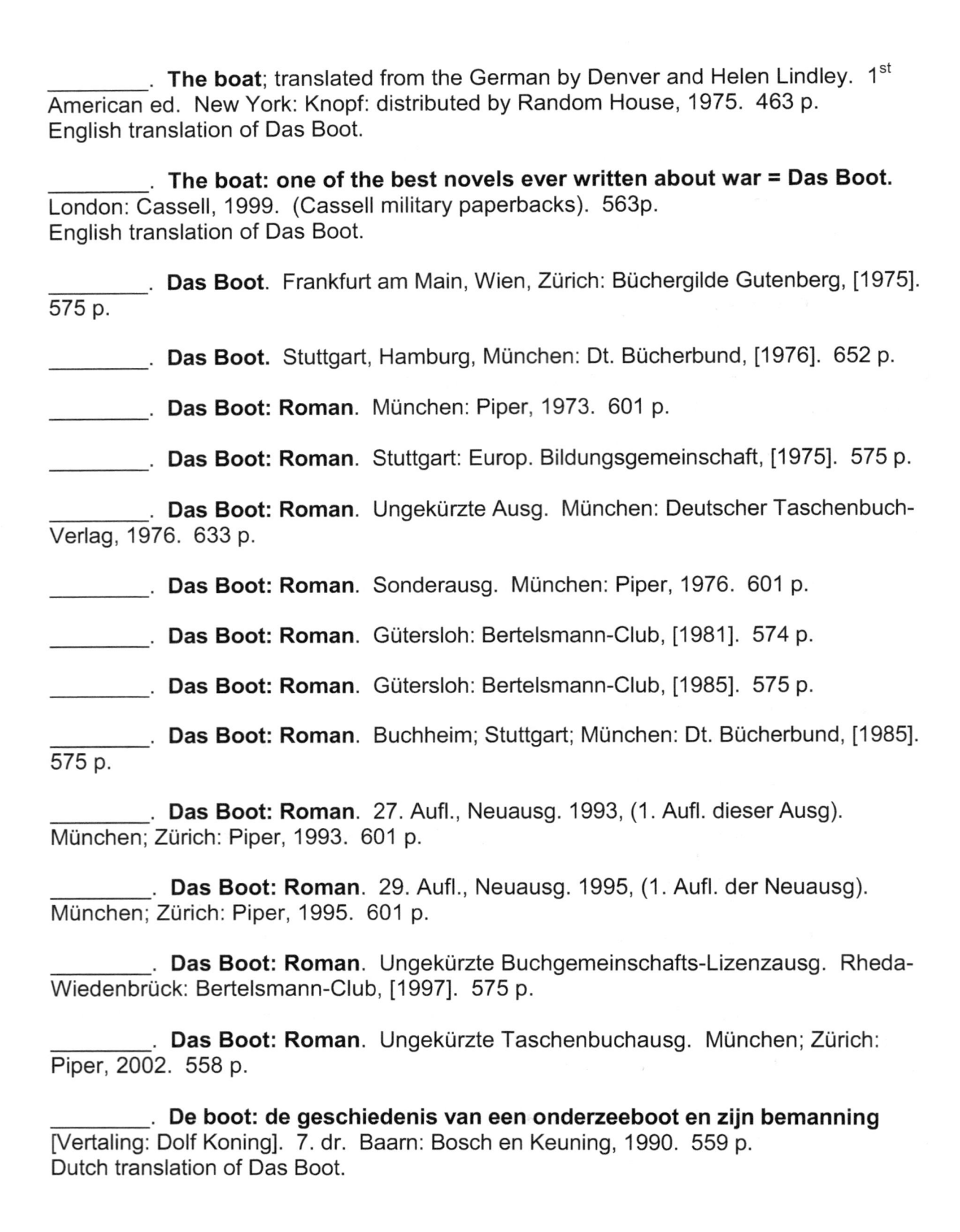

_____. **The boat**; translated from the German by Denver and Helen Lindley. 1st American ed. New York: Knopf: distributed by Random House, 1975. 463 p.
English translation of Das Boot.

_____. **The boat: one of the best novels ever written about war = Das Boot.** London: Cassell, 1999. (Cassell military paperbacks). 563p.
English translation of Das Boot.

_____. **Das Boot**. Frankfurt am Main, Wien, Zürich: Büchergilde Gutenberg, [1975]. 575 p.

_____. **Das Boot**. Stuttgart, Hamburg, München: Dt. Bücherbund, [1976]. 652 p.

_____. **Das Boot: Roman**. München: Piper, 1973. 601 p.

_____. **Das Boot: Roman**. Stuttgart: Europ. Bildungsgemeinschaft, [1975]. 575 p.

_____. **Das Boot: Roman**. Ungekürzte Ausg. München: Deutscher Taschenbuch-Verlag, 1976. 633 p.

_____. **Das Boot: Roman**. Sonderausg. München: Piper, 1976. 601 p.

_____. **Das Boot: Roman**. Gütersloh: Bertelsmann-Club, [1981]. 574 p.

_____. **Das Boot: Roman**. Gütersloh: Bertelsmann-Club, [1985]. 575 p.

_____. **Das Boot: Roman**. Buchheim; Stuttgart; München: Dt. Bücherbund, [1985]. 575 p.

_____. **Das Boot: Roman**. 27. Aufl., Neuausg. 1993, (1. Aufl. dieser Ausg). München; Zürich: Piper, 1993. 601 p.

_____. **Das Boot: Roman**. 29. Aufl., Neuausg. 1995, (1. Aufl. der Neuausg). München; Zürich: Piper, 1995. 601 p.

_____. **Das Boot: Roman**. Ungekürzte Buchgemeinschafts-Lizenzausg. Rheda-Wiedenbrück: Bertelsmann-Club, [1997]. 575 p.

_____. **Das Boot: Roman**. Ungekürzte Taschenbuchausg. München; Zürich: Piper, 2002. 558 p.

_____. **De boot: de geschiedenis van een onderzeeboot en zijn bemanning** [Vertaling: Dolf Koning]. 7. dr. Baarn: Bosch en Keuning, 1990. 559 p.
Dutch translation of Das Boot.

_____. **Okret** [Przel. Adam Kaska]. Wyd. 3. Warszawa: Wydawn. Bellona, 1996. 614 p.
Polish translation of Das Boot.

_____. **Okret** [Przel. Adam Kaska]. Wyd. 5. Warszawa: Dom Wydawn. Bellona, 2000. 614 p.
Polish translation of Das Boot.

_____. **Podmornica**. Zalozba Lipa, 1975. 680 p.
Slovenia translation of Das Boot.

_____. **Ponorka**. Praha: Nase vojsko, knizni obchod, 1991. 496 p.
Czech translation of Das Boot.

_____. **Le styx: roman** [traduit de l'allemand par Bernard Kreiss]. Paris: A. Michel, c1977. 496 p.
French translation of Das Boot.

_____. **Submarino**. Madrid: Ultramar, 1975. 416 p.
Spanish translation of Das Boot.

_____. **Sukellusvene**. Helsinki: Wilin & Göös, 1975. 459 p.
Finnish translation of Das Boot.

_____. **U-Båd** [Overs. af Hans Chr. Dahlerup Koch]. - 2. udg., 3. opl.. - [København]: Forum, 1992. 423 p.
Danish translation of Das Boot.

_____. **Ubåt** [övers. från tyska av Britta Hagwall; teknisk fackgranskning av Wilhelm Hagwall]. Forum, 1973. 470 p.
Swedish translation of Das Boot.

_____. **U-boat**; translated by J. Maxwell Brownjohn. London: Collins, 1974. 480 p.
English translation of Das Boot.

_____. **U-Boot: [Atlantico, 1941: la grande avventura di un sommergibile tedesco; romanzo]** [Trad. di Ursula Olmini Soergel]. 1. ed. Milano: Net, 2002 514 p.
Italian translation of Das Boot.

_____. **U-boto** [Rotaru-Gyunta Bufuhaimu]. Tokyo: Hayakawa Shobo, 1977. 503 p.
Japanese translation of Das Boot.

_____. **U 96: Szenen aus d. Seekrieg; e. Film**. Hamburg: Knaus, 1981. 348 p.

_____. **U 96: Szenen aus d. Seekrieg; e. Film**. Ungekürzte Ausg., 1. Aufl. München: Goldmann, [1985]. 347 p.

_____. **Der Film "Das Boot": e. Journal**. Orig.-Ausg., 1. Aufl., 1. München: Goldmann, 1981. 253 p.

_____. **Der Film Das Boot: e. Journal**. Gütersloh: Bertelsmann-Club, [1982] 253 p.

Clancy, Tom. **The hunt for Red October**. New York: Berkley Books, 1985, c1984. 469 p.
Originally published: Annapolis, MD: Naval Institute Press, 1984.

_____. **Jagd auf Roter Oktober:** Roman [Einzig berecht. Übers. aus d. Amerikan. von Hardo Wichmann]. 1. Aufl. Bern; München; Wien: Scherz, 1986. 383 p.
German translation of The hunt for Red October.

_____. **Jagd auf Roter Oktober: Roman** [Einzig berecht. Übers. aus d. Amerikan. von Hardo Wichmann]. Gütersloh: Bertelsmann-Club, [1987]. 383 p.
German translation of The Hunt for Red October.

_____. **Jagd auf "Roter Oktober": Roman** [Einzig berecht. Übertr. aus d. Amerikan. von Hardo Wichmann]. 1. Aufl. [München]: Goldmann, [1988]. 383 p.

_____. **Jagd auf Roter Oktober** [Übers. aus dem Amerikan. von Hardo Wichmann]. Stuttgart; München: Dt. Bücherbund, [1989]. 383 p.

_____. **Jagd auf Roter Oktober: Roman**; [Einzig berecht. Übertr. aus dem Amerikan. von Hardo Wichmann]. Klagenfurt: Kaiser, [1992]. 317 p.

_____. **Jagd auf Roter Oktober: Roman** [Einzig berechtigte Übers. aus dem Amerikan. von Hardo Wichmann]. Wien: Tosa, 1996. 383 p.
German translation of the Hunt for Red October.

_____. **Jagd auf Roter Oktober: Roman** [Aus dem Amerikan. von Hardo Wichmann]. Taschenbuchausg. München: Heyne, 2001. 493 p.
German translation of the Hunt for Red October.

_____. **Octobre rouge: roman** [traduit de l'américain par Marianne Véron avec la collab. de Jean Sabbagh]. Paris: Albin michel, 1986. 485 p.
Franch translation of The Hunt for Red October.

_____. **Okhota za "Krasnym Oktëiiabrem."** New York: Liberty Publishing House, 1986. 535 p.
Russian translation of the Hunt for Red October.

Grass, Günter. **Crabwalk**; translated from the German by Krishna Winston. 1st ed. Orlando: Harcourt, c2002. 234 p.
English translation of Im Krebsgang.

Recounts refugee family from the singking of the Wilhelm Gustloff.

_____. **Crabwalk**. London: Faber, 2002. 234 p.
Originally published as "Im Krebsgang". Gottingen: Steidl Verlag, 2002.
This English translation first published Orlando: Harcourt, 2002.

_____. **En crabe**. Paris: le Grand livre du mois, 2002. 263 p.
French translation of Im Krebsgang.

_____. **Hodom raka** [Prevela sa nemackog Aleksandra Gojkov Rajic]. Beograd: Narodna Knjiga, 2002. 204 p.
Serbian translation of Im Krebsgang.

_____. **Im Krebsgang: eine Novelle**. Göttingen: Steidl, 2002. 216 p.

_____. **Im Krebsgang: eine Novelle**. München: Saur, 2003. 300 p.

_____. **Im Krebsgang: Eine Novelle**. 1. Aufl. München: Deutscher Taschenbuch Verlag GmbH & Co. KG, 2004. 224 p.

Topol, Edward. **Chuzhoe litso**. Moskva: Egro-Press, 1994. (Oboima detektivov). 348 p.

_____. **Chuzhoe litso = Submarine U-137**. N'iu-Iork: Eduard Topol Ltd., 1987. 390 p.

_____. **Submarine U-137**. London: Corgi, 1985. 443 p.
Originally published: London, New York: Quartet Books, 1984.
English translation of Zagadka Y-137
In 1981 Soviet sub U-137 ran aground near the Swedish naval base at Karlskrona

_____. **U-137 oder Europa wird erschüttert** [Übers. aus d. Russ. von Nina Stein u. Hanns-Peter Pichl]. München: Roitman; Knaur: Droemer, 1983. 335 p.
German translation of Zagadka Y-137.

_____. **Ubåt 137: med uppdrag från KGB** [övers.: Jan Järnebrand]. 1985. 367 p.
Swedish translation of Zagadka Y-137

SUBMARINES -- GENERAL -- NUCLEAR

Beaver, Paul. **Nuclear powered submarines**. London; New York: Arms & Armour Press; New York, NY: Distributed in the U.S. by Sterling Pub. Co., 1986. (Warships illustrated; no. 5). 72 p.

Clancy, Tom. **Atom U Boot: Reise ins Innere eines Nuclear Warship**. München: Heyne, 1997. 375 p.
German translation of Submarine: a guided tour inside a nuclear warship.

_____. **Submarine: a guided tour inside a nuclear warship**. 1st ed. New York: Putnam, c1993. 328 p.
NPS/DKL Location: GENERAL V857.5 .C55 1993

_____ and John Gresham. **Submarine: a guided tour inside a nuclear warship**. Berkeley revised mass-market edition. New York: Berkeley Books, 2002. 323 p.

_____ and John Gresham. **Submarine: a guided tour inside a nuclear warship**. Berkeley revised trade papaerback edition. New York: Berkeley Books, 2003. 368 p.

Cote, Owen R. **The third battle: innovation in the U.S. Navy's silent Cold War struggle with Soviet submarines**. Newport, RI: Naval War College, Center for Naval Warfare Studies, 2003. (Newport paper; no. 16). 104 p.
Electronic access: http://www.nwc.navy.mil/press/npapers/np16/NewportPaper16.pdf

Craven, John P. **The silent war: the Cold War battle beneath the sea**. New York: Simon & Schuster, c2001. 304 p.
NPS/DKL Location: GENERAL V63.C7 A3 2001

Crouch, Holmes F. **Nuclear ship propulsion**. Cambridge, MD, Cornell Maritime Press, 1960. 347 p.
NPS/DKL Location: GENERAL VM774 .C9

Eriksen, Viking Olver. **Sunken nuclear submarines: a threat to the environment?** Oslo: Norwegian University Press; Oxford; New York: Distributed world-wide excluding Scandinavia by Oxford University Press, c1990. 176 p.

Ford, Thom W. **Ballistic missile submarines of the United States and the Soviet Union: a comparison of systems and doctrine**. Monterey, CA: Naval Postgraduate School; Springfield, VA: Available from National Technical Information Service, 1982. (ADA127829). 125 l.
Thesis (M.A. in National Security Affairs)--Naval Postgraduate School, 1982.
Also published in microfilm: Frederick, MD: University Publications of America, Inc., c1986. (Soviet Union special studies, 1982-1985; 2).
Abstract: *This thesis compares the development of fleet ballistic missile systems in the United States and Soviet Union and their contribution to the achievement of national security objectives of each nation. To this end, submarine and missile technologies, elements of operational practices and support, and general strategic doctrine, are traced. A comparative assessment of weapon system effectiveness and potential in achieving stated objectives is derived from capabilities, peacetime employment, and wartime plans as stated in open doctrinal documents.*
NPS/DKL Location: THESIS F5797

Fürst, Andreas, Volker Heise, and Steven E. Miller, eds. **Europe and naval arms control in the Gorbachev era**. Oxford; New York: Oxford University Press, 1992. 341 p.
NPS/DKL Location: GENERAL JX1974.7 .E85 1992

Gardiner, Robert and Norman Friedman, eds. **Navies in the nuclear age: warships since 1945**. Annapolis, MD: Naval Institute Press, c1993. (Conway's history of the ship). 224 p.
Originally published: London: Conway Maritime Press, 1993.
NPS/DKL Location: REFERENCE VM317 .N38 1993

Great Britain. Defence Radiological Protection Service. **Marine environmental radioactivity surveys at nuclear submarine berths in the UK, 1988**. London: H.M.S.O., 1989. 112 p.

Great Britain. Working Group on Marine Reactor Research. **Nuclear power for ship propulsion; report**. London, H.M. Stationery Off., 1964. (Papers by command, comnd. 2358). 21 p.

Hori, Motoyoshi. **Mei Su hai di zhan lue: Riben hai shang di Mei Su qian ting zhan**. Xianggang: Tian di tu shu gong si, 1978. 185 p.
Chinese translation of Nihonkai kyofu no sensuikan senso

_____. **Sensuikan**. 1980. 409 p.

_____, Kensuke Ebata and Kesaharu Imai. **Nihonkai kyofu no sensuikan senso**. 1976. 262 p.

Ladonchamps, Jean de et Jean Jacques Verdeau. **Propulsion nucléaire**. Paris (15e), Centre d'édition et de documentation de l'E.N.S.T.A., 32 Brd Victor, 1970-. v.
Incomplete contents: 1. Théorie du réacteur. -- 2. Technologie des réacteurs à eau pressurisée. -- 3. Applications à la propulsion navale.

_____ and _____. **Réacteurs nucléaires à eau pressurisée, théorie, technologie et applications à la propulsion navale**. Paris, Masson et Cie, 1972. 230 p.

LeSage, L. G. and A.A. Sarkisov, eds. **Nuclear submarine decommissioning and related problems**. (Proceedings of the NATO Advanced Research Workshop on Nuclear Submarine Decommissioning and Related Problems, Moscow, Russia, June 19-22, 1995). Dordrecht; Boston: Kluwer, c1996. (NATO ASI series. Partnership sub-series 1, Disarmament technologies; 8). 343 p.

_____ and _____, eds.. **Remaining issues in the decommissioning of nuclear powered vessels**. (Proceedings of the NATO/Russian Advanced Research Workshop on Scientific Problems and Unresolved Issues Remaining in the

Decommissioning of Nuclear Powered Vessels and in the Environmental Remediation of Their Supporting Infrastructure: Moscow, Russia, April 22-24, 2002). Boston, MA: Kluwer Academic, 2003. (NATO science series. 4, Earth and environmental sciences; v. 22). 420 p.

Lewis, John Wilson and Xue Litai. **China's strategic seapower: the politics of force modernization in the nuclear age.** Stanford, CA: Stanford University Press, c1994. (Studies in international security and arms control. 393 p.

Miyake, Yasuo. **Kakuheiki to hoshano**. 1969. 206 p.

Moore, Richard. **The Royal Navy and nuclear weapons**. London; Portland, OR: Frank Cass, 2001. 243 p.

The Nuclear duel: Strategic bombers, nuclear submarines, missiles, and manpower. New York: Arco, 1985, c1983. (War today--East vs. West). 66 p.

Newhouse, John. **War and peace in the nuclear age**. 1st ed. New York: Knopf: Distributed by Random House, 1989, c1988. 486 p.
Accompanies the PBS series War and peace in the nuclear age.
NPS/DKL Location: GENERAL U263 .N48 1989

Peng, Ziqiang. **Qi jing shen long: Zhongguo he qian ting ji shi**. Di 1 ban. Beijing Shi: Zhong gong zhong yang dang xiao chu ban she: Xin hua shu dian jing xiao, 1995 (1996 printing). (Zhongguo guo fang ke ji bao gao wen xue cong shu). 436 p.

Pocock, Rowland F. **Nuclear ship propulsion**. London, Allan, 1970. 176 p.

Polmar, Norman and Kenneth J. Moore. **Cold War submarines: U.S. and Soviet design and construction**. 1st ed. Washington, DC: Brassey's, c2003. 407 p.
Contents: Genesis -- Advanced diesel submarines -- Closed-cycle submarines -- U.S. Nuclear-propelled submarines -- Soviet nuclear-propelled submarines -- Cruise missile submarines -- Ballistic missile submarines -- "Polaris-from out of the deep..." -- The quest for speed -- Second-generation nuclear submarines -- The ultimate weapon I -- The ultimate weapon II -- "Diesel boats forever" -- Unbuilt giants -- Aircraft-carrying submarines -- Midget, small, and flying submarines -- Third-generation nuclear submarines -- Submarine weapons -- Fourth-generation nuclear submarines -- Soviet versus U.S. submarines -- U.S. submarine construction, 1945-1991 -- Soviet submarine construction, 1945-1991 -- U.S. submarine reactor plants -- Soviet submarine design bureaus -- Soviet ballistic missile systems.

Reynolds, Guy B. **The nuclear-armed Tomahawk Cruise Missile: its potential utility on United States and United Kingdom attack submarines**. Monterey, CA: Naval Postgraduate School; Springfield, VA: Available from National Technical Information Service, 1998. (ADA359545). 70 p.
Thesis (M.A. in National Security Affairs)--Naval Postgraduate School, 1998.

Abstract: *In July 1998, Britain published its Strategic Defense Review(SDR). The SDR outlined significant changes for Britain's nuclear weapons program and formalized the policy of sub-strategic deterrence using the Trident missile. It is unprecedented for a nuclear power to haveconsolidated its strategic and sub- strategic nuclear forces into a single system. The benefits offered by the British choice might be enjoyed for only a short time. The British have slashed their nuclear forces and eliminated the range of options previously available to their national command authority. Dependence on a single delivery system could result in the inability to respond to crises, to act autonomously, or to negotiate effectively with other nuclear weapon states. This thesis analyzes the benefits that nuclear Tomahawk could provide the British. Since the United States owns the system, the future of the nuclear Tomahawk in the American arsenal is crucial to any British decision to adopt it or a similar system. An unmanned nuclear cruise missile weapon offers many advantages in today's security environment. The United States should retain nuclear Tomahawk and Britain, with its mature maritime force, should consider acquiring a similar capability. The elimination of nuclear Tomahawk from the U.S. arsenal would be a mistake.*

NPS/DKL Location: THESIS R3686
Electronic access: http://library.nps.navy.mil/uhtbin/hyperion-image/98Dec_Reynolds.pdf
Electronic access: http://handle.dtic.mil/100.2/ADA359545

Ring, Jim. **We come unseen: the story of Britain's Cold War submariners**. London: John Murray, 2001. 270 p.

Rohwer, Jürgen. **66 [i.e. Sechsundsechzig] Tage unter Wasser; Atom-U-Schiffe und Raketen**. Oldenburg, G. Stalling [1964? c1962]. 88 p.

Sarkisov, Ashot Arakelovich and Alain Tournyol Du Clos, eds. **Analysis of risks associated with nuclear submarine decommissioning, dismantling, and disposal**. Dordrecht; Boston: Kluwer, 1999. (NATO ASI series. Partnership sub-series 1, Disarmament technologies; 16). 443 p.
Proceedings of the NATO Advanced Research Workshop on Analysis of Risks Associated with Nuclear Submarine Decommissioning, Dismantling, a nd Disposal, Moscow, Russia, November 24-26, 1997.
NPS/DKL Location: GENERAL V857.5 .A53 1999

Schomaekers, Gunter. **U-Boote auf allen Weltmeeren: Fahrten uber u. unter Wasser, Atom-U-Boote heute**. Wels; Munchen: Verl. Welsermuhl, c1978. 218 p.

Shambroom, Paul. **Face to face with the bomb: nuclear reality after the Cold War**. Baltimore: Johns Hopkins University Press, 2003. (Creating the North American landscape). 121 p.
Contents: Bombers -- Intercontinental ballistic missiles -- Submarines -- Command, control, and communications (c3) -- The 21st century.

Shakai Shinposha. **Henshubu**. Sasebo, 1968-nen 5-gatsu: genshiryoku sensuikan hoshano osen no kyofu. 46 p.

Solmi, Angelo. **Il mistero dei sottomarini atomici**. 1a ed. Milano: A. Mondadori, 1981. 177 p.

Tsipis, Kosta, Anne H. Cahn [and] Bernard T. Feld, eds. **The Future of the sea-based deterrent.** Cambridge, MA: MIT Press, [1973]. 266 p.
Papers of a symposium held in Racine, Wis., Nov. 1972 organized by the Carnegie Endowment for International Peace and the American Academy of Arts and Sciences under the sponsorship of the Johnson Foundation and the Alfred P. Sloan Foundation.
NPS/DKL Location: GENERAL V993 .F9

Tsukudo, Tatsuo. **Genshiryoku sensuikan.** [1979]. (Jiji mondai kaisetsu; 238). 181 p.

Weiss, Stephanie. **Nuclear reactors on ships; a bibliography.** [Santa Monica, Calif., Rand Corp.], 1969. (P-4077). 47 p.

Whitestone, Nicholas. **The submarine: the ultimate weapon.** London, Davis-Poynter, 1973. 146 p.
NPS/DKL Location: GENERAL V210 .W53

_____. **U-Boote: Superwaffe der Zukunft?** [übersetzt von Karl-Friedrich Merten]. München: J.F. Lehmanns, c1975. 158 p.
English translation of: The submarine: the ultimate weapon

Yomiuri Shinbunsha. **Genshiryoku sensuikan: senryaku kogeki gensen no zenbo.** Tokyo: Yomiuri Shinbunsha, Showa 60 [1985]. (Heiki saisentan; 2). 204 p.

SUBMARINES -- GENERAL -- WWI

Abbot, Willis J. **Aircraft and submarines, the story of the invention, development, and present-day uses of war's newest weapons.** New York & London, G. P. Putnam's sons, 1918. 388 p.

Blanchon, Georges. **Les sous-marins et la guerre actuelle.** Paris: Bloud et Gay, 1915. (Pages actuelles; 1914-1915 no. 20). 38 p.

Dommett, William Erskine. **Submarine vessels: including mines, torpedoes, guns, steering, propelling, and navigating apparatus, and with notes on submarine offensive and defensive tactics, and exploits in the present war.** London; New York: Whittaker & Co., 1915. 106 p.

Domville-Fife, Charles William. **Submarines of the world's navies.** London: F. Griffiths; Philadelphia: J.B. Lippincott co., 1911. 150 p.
NPS/DKL Location: GENERAL VM365 .D6

_____. **Submarines, mines and torpedoes in the war.** London; New York: Hodder & Stoughton, 1914. 192 p.

_____. **Submarines and sea power.** New York: Macmillan Co., [1919?]. 250 p.

Originally published: London: G. Bell & Sons, Ltd., 1919.

Gayer, Albert. **General survey of the history of the submarine warfare in all theaters of war, 1914-1918**. Newport, RI: Naval War College, 1930. 36 l.

Gill, C. C. **Naval power in the war (1914-1918)** [Lectures delivered at the United States Naval Academy in the winter of 1915-16 to midshipmen]. New York, George H. Doran company, [c1919]. 302 p.
NPS/DKL Location: GENERAL D580 .G4

Groeling, Dorothy Trautwein. **Submarines, disarmament and modern warfare**. [New York], 1935. 199 l.
Thesis (Ph.D.), Columbia University, 1935.

Wilson, Michael and Paul Kemp. **Mediterranean submarines**. Wilmslow: Crecy, 1997. 219 p.

Woodbury, David Oakes. **What the citizen should know about submarine warfare**. New York, W.W. Norton & Company, Inc. [1942]. 231 p.

SUBMARINES -- GENERAL -- WWII

Alden, John Doughty, comp. **United States and Allied submarine successes in the Pacific and Far East during World War II: chronological listing**. Pleasantville, NY: [s.n.], 1999. 1 v.

Bagnasco, Erminio. **I sommergibili della seconda guerra mondiale**. Parma: E. Albertelli, 1973. (Documentari della storia e della tecnica). 335 p.

_____. **Submarines of World War Two**. Annapolis, MD: Naval Institute Press, c1977. 256 p.
English translation of I sommergibili della seconda guerra mondiale.
NPS/DKL Location: REFERENCE V857 .B25

_____. **Submarines of World War Two**. London: Cassell & Co.; New York: Distributed in the USA by Sterling Pub., 2000. 256 p.

_____. **Uboote im 2. [Zweiten] Weltkrieg: [Technik - Klassen - Typen; eine umfassende Enzyklopädie]** [Die Übertr. ins Dt. besorgte: Wolfram Schürer]. 1. Aufl. Stuttgart: Motorbuch-Verl., 1988. 294 p.
German translation of Submarines of World War Two.

_____ and Achille Rastelli. **Sommergibili in guerra: centosettantadue battelli italiani nel secondo conflitto mondiale**. [Italy]: Albertelli, c1989. 263 p.

Cairns, Lynne. **Fremantle's secret fleets: Allied submarines based in Western Australia during World War II**. Fremantle, W.A.: Western Australian Maritime Museum, c1995. (Maritime history series; no. 1). 87 p.

Castex, Raoul. **Synthese de la guerre sous-marine; de Pontchartrain a Tirpitz**. Paris, A. Challamel, 1920. 228 p.

Cope, Harley Francis and Walter Karig. **Battle submerged; submarine fighters of World War II**. [1st ed.]. New York, Norton, [1951]. 244 p.

Dingman, Roger. **Ghost of war: the sinking of the Awa Maru and Japanese-American relations, 1945-1995**. Annapolis, MD: Naval Institute Press, c1997. 373 p.
NPS/DKL Location: GENERAL D777.5.A92 D56 1997

Icenhower, Joseph Bryan, comp. **Submarines in combat**. New York, F. Watts, [1964]. (The Watts seapower library). 180 p.
NPS/DKL Location: GENERAL D780 .I2

Kemp, Paul. **Friend or foe: friendly fire at sea 1939-1945**. London: Leo Cooper, 1995. 198 p.

Krug, Hans-Joachim et al. **Reluctant allies: German-Japanese naval relations in World War II**. Annapolis, MD: Naval Institute Press, 2001. 456 p.

Laurens, A. **Le blocus et la guerre sous-marine, 1914-1918**. Paris: Librairie Armand Colin, 1924. 217 p.

Manson, Janet M. **Diplomatic ramifications of unrestricted submarine warfare, 1939-1941**. New York: Greenwood Press, 1990. (Contributions in military studies; no. 104). 215 p.

Padfield, Peter. **War beneath the sea: submarine conflict during World War II**. New York: John Wiley, [1998]. 560 p.
Originally published: London: John Murray, 1995 as War beneath the sea. submarine conflict, 1939-45.
NPS/DKL Location: GENERAL D780 .P33 1998

Padoan, Gianni. **La fossa della morte: gli U-Boote in guerra nell'Atlantico**. Bologna: Capitol, 1977. 228 p.

Pearsall, G. H. **The Effects of World War II Submarine Campaigns of Germany and the United States; A Comparative Analysis.** Newport, RI: Naval War College; Available from National Technical Information Service, Springfield, VA, , 1994. 43 p.
Abstract: *The effectiveness of the German and United States submarine campaigns during World War II is compared by analyzing the genesis of each campaign, the commitment to each and the effort to overcome the losses imposed by submarine warfare. This comparison highlights one aspect of the strategic and operational consequences of conflict with an adversary able to build and maintain a superior industrial base*

in support of the military effort. This analysis places primary focus on German U-boat efforts in the Battle of the Atlantic and the U.S. submarine efforts in the Western Pacific. Ultimately, the overriding factor in the outcomes of both campaigns was the ability of the United States to produce more ships than the Germans could sink, to build more submarines than the Japanese could sink, and to sink more Japanese ships than the Japanese could build. As a result, the United States was able to sustain its total military effort against Germany; Japan was not able to sustain its efforts in the Pacific.
NPS/DKL Location: MICROFORM ADA283407

Pegolotti, Beppe. **Uomini contro navi.** [Firenze] Vallecchi, [1959]. 266 p.

Peillard, Léonce. **Histoire générale de la guerre sous-marine, (1939-1945).** Paris, R. Laffont: 1970. 451 p.

_____. **Histoire générale de la guerre sous-marine, 1939-1945.** Genève, Edito-service: [Evreux], [diffusion le Cercle du bibliophile][1973]. 442 p.

Poolman, Kenneth. **Allied submarines of World War Two.** London: Arms & Armour; New York: Distributed in the USA by Sterling Pub. Co., 1990. 160 p.

Robertson, Terence J., Jerauld Wright and N.L.A. Jewell. **The ship with two captains.** E.P. Dutton, 1957. 256 p.
NPS/DKL Location: BUCKLEY D772.S4 R6

Rohwer, Jürgen. **Allied submarine attacks of World War Two: European theatre of operations, 1939-1945.** Annapolis, MD: Naval Institute Press, c1997. 252 p.
Companion volume to Rohwer's Axis submarine successes, 1939-1945.
Originally published: London: Greenhill Books, 1997.
NPS/DKL Location: GENERAL D780 .R56 1997

_____. **Axis submarine successes, 1939-1945.** Annapolis, MD: Naval Institute Press, c1983. 386 p.
Originally published: Cambridge, [Cambridgeshire]: Patrick Stephens, c1983.
English translation of Die U-Boot-Erfolge der Achsenmachte, 1939-1945, München, J. F. Lehmann, [1968].

_____. **Axis submarine successes of World War Two: German, Italian and Japanese submarine successes, 1939-1945.** London: Greenhill; Annapolis, MD: Naval Institute Press, 1999. 366 p.
Companion volume to Rohwer's Allied submarine attacks of World War Two: European theatre of operations, 1939-1945.
NPS/DKL Location: GENERAL D780 .R5713 1999

_____. **Die U-Boot-Erfolge der Achsenmachte, 1939-1945.** (Bibliothek fur Zeitgeschichte Bd. 1). München, J. F. Lehmann, [1968]. 376 p.

Tennent, Alan J. **British and Commonwealth merchant ship losses to axis submarines, 1939-1945.** Stroud, Gloucestershire: Sutton Pub., 2001. 326 p.

United States. Office of the Chief of Naval Operations. **German, Japanese, and Italian submarine losses. World War II**. Washington, DC: The Office, 1946. (OPNAV-P33-100 NEW 5-46). 28 p.
NPS/DKL Location: GENERAL D780 .U55 1946

Van der Vat, Dan. **The Atlantic campaign: the great struggle at sea 1939-1945**. Edinburgh: Birlinn, 2001. 576 p.

_____. **The Pacific campaign: World War II, the U.S.-Japanese naval war, 1941-1945**. New York: Simon & Schuster, c1991. 430 p.
NPS/DKL Location: GENERAL D767.9 .V36 1991

Whitehouse, Arthur George Joseph. **Subs and submariners**. London: F. Muller, 1963, c1961. 416 p.
Originally published: Garden City, NY, Doubleday, 1961.
NPS/DKL Location: GENERAL V210 .W5

Wilson, Michael. **A Submariners' War: The Indian Ocean 1939-45**. Gloucestershire: Tempus; Charleston, SC: Arcadia, 2000. 192 p.

SUBMARINES -- AMERICAN – GENERAL

Alden, John Doughty. **The fleet submarine in the U.S. Navy: a design and construction history**. Annapolis, MD: Naval Institute Press, c1979. 290 p.
NPS/DKL Location: GENERAL V858 .A48

_____. **The fleet submarine in the U.S. Navy: a design and construction history**. Shrewsbury: Airlife, 1989. 285 p.

Barrows, Nathaniel A. **Blow all ballast! The story of the Squalus**. New York, Dodd, Mead & Company, 1940. 298 p.
NPS/DKL Location: BUCKLEY VA65.S7 B2

Beach, Edward Latimer. **Around the world submerged: the voyage of the Triton**. New York: Holt, Rinehart & Winston, [1962]. 293 p.
NPS/DKL Location: GENERAL VA65.T48 B3

_____. **Around the world submerged: the voyage of the Triton**. Annapolis, MD:Naval Institute Press, [2001]. (Bluejacket books). 336 p.

Beckman, Philip J. **Scheduling attack submarine deployments**. Monterey, CA: Naval Postgraduate School; Springfield, VA: Available from National Technical Information Service, 1997. (ADA331771). 47 p.
Thesis (M.S. in Operations Research)--Naval Postgraduate School, 1997.

Abstract: *The Navy's peacetime mission is "to conduct forward presence operations to help shape the strategic environment by deterring conflict, building interoperability, and by responding, as necessary, to fast breaking crises with the demonstration and application of credible combat power." (OPNAV INSTRUCTION 3501.316, February 1995) The ability to carry out this mission hinges on the Navy's ability to maintain ships and submarines forward deployed in regions where such crises may occur. The end of the Cold War and current budget constraints have caused a drawdown in the number of ships and submarines with which to provide forward presence. Coupled with the continued requirement to maintain a certain level of forward presence, this drawdown creates shortfalls when attempting to deploy ships or submarines to fill certain mission requirements. To minimize these shortfalls, this thesis formulates the problem of scheduling attack submarine deployments as an integer program. Due to its size and complexity, heuristic algorithms are developed to provide near-optimal solutions in a reasonable amount of time. In addition to providing near-optimal deployment schedules, results from the algorithms are also useful in evaluating changes in maintenance and operational policies.*
NPS/DKL Location: THESIS B3355

Bentley, John. **The Thresher disaster; the most tragic dive in submarine history**. [1st ed.] Garden City, NY, Doubleday, 1974. 372 p.
NPS/DKL Location: GENERAL VA65.T7 B44

Birkler, John L. et al. **The U.S. submarine production base: an analysis of cost, schedule, and risk for selected force structures.** Santa Monica, CA: Rand, 1994. (MR-456-OSD);(ADA285646). 203 p.
NPS/DKL Location: GENERAL V858 .U2 1994

_____. **The U.S. submarine production base: an analysis of cost, schedule, and risk for selected force structures: executive summary**. Santa Monica, CA: Rand, 1994. (MR-456/1-OSD);(ADA288636). 27 p.
NPS/DKL Location: GENERAL V858 .U213 1994
Electronic access: http://handle.dtic.mil/100.2/ADA288636

Blair, Clay. **The atomic submarine and Admiral Rickover**. [1st ed.]. New York: Holt, [c1954]. 277 p.
NPS/DKL Location: BUCKLEY VA65.N3 B6

Cable, Frank T. **The birth and development of the American submarine**. NY; Lond.: Harper, 1924. 337 p.

Callahan, Vincent Francis et al, eds. Under**water defense handbook**. [2d ed.] Washington, Callahan Publications, 1963. 138 p.
NPS/DKL Location: GENERAL V396.3 .C3

Calvert, James. **Surface at the Pole; the extraordinary voyages of the USS Skate**. [1st ed.]. New York, McGraw-Hill, [1960]. 220 p.
NPS/DKL Location: GENERAL VA65.S6 C2

Christley, J. L. **United States Naval submarine force information book**. Marblehead, MA: Graphic Enterprises of Marblehead, c2000. 196 p.
NPS/DKL Location: REFERENCE V858 .C47 2000

Clark, James I. **Steel coffin at forty fathoms**. Milwaukee: Raintree Publishers, c1980. 47 p.
The Squalus rescue.

Colby, C. B. **Submarine: men and ships of the U. S. submarine fleet**. New York: Coward-McCann, c1953. 48 p.

_____. **Submarine warfare; men, weapons, and ships**. New York, Coward-McCann, [1967]. 48 p.

Cote, Owen. **Mobile targets from under the sea: new submarine missions in the new security environment.**
[Cambridge, MA]: MIT Security Studies Program, [1999]. 72 p.
Electronic access: http://web.mit.edu/ssp/Publications/confseries/mobtarg.pdf

_____. **Precision strike from the sea: new missions for a new Navy**. Cambridge, MA: Massachusetts Institute of Technology, 1998. 44 p.
NPS/DKL Location: GENERAL V210 .C683 1998
Electronic access: http://web.mit.edu/ssp/

Craven, John P. "**Ocean technology and submarine warfare**," *IN* Institute for Strategic Studies (London, England) The implications of military technology in the 1970s. New York: Institute for Strategic Studies, c1968, 38-46; (Adelphi papers; 46 (March 1968)).

_____. **The silent war: the Cold War battle beneath the sea**. New York; London: Simon & Schuster, c2001. 304 p.
NPS/DKL Location: GENERAL V63.C7 A3 2001

Cross, Wilbur. **Challengers of the deep; the story of submarines**. New York, W. Sloane Associates, 1959. 258 p.
NPS/DKL Location: GENERAL V210 .C9

_____. **The encyclopedia of American submarines**. New York: Facts on File, 2002. 304 p.
NPS/DKL Location: REFERENCE V858 .C76 2003

Friedman, Norman. **U.S. submarines since 1945: an illustrated design history**. Annapolis, MD: Naval Institute Press, c1994. 280 p.
NPS/DKL Location: BUCKLEY V858 .F75 1994

_____. **U.S. submarines through 1945: an illustrated design history**. Annapolis, MD: Naval Institute Press, c1995. 379 p.
NPS/DKL Location: BUCKLEY V858 .F75 1995

Genat, Robert and Robin Genat. **Modern U.S. Navy submarines**. Osceola, WI, USA: Motorbooks International, 1997. (Enthusiast color series). 96 p.

Gordon, Robert B. **Working for Admiral Rickover: memoir**. Washington, DC: Naval Historical Foundation Memoir program, 2000. (Memoir -- Naval Historical Foundation). 17 l.
NPS/DKL Location: BUCKLEY CT18.I52 G67 2000

Hastings, Scott A. **An Assessment of Future United States Naval Force Structure in the Pacific Theater**. Fort Leavenworth, KS: Army Command and General Staff College; Available from National Technical Information Service, Springfield, VA, 1994. 178 p. Master's thesis.
Abstract: *The naval force structure proposed by the 1993 Department of Defense 'Bottom Up Review' was analyzed in terms of three force planning cases built around illustrative scenarios using representative depictions of future threats. Each case included: regional analysis in terms of mission, forces, area, and command and control; development of military requirements; and comparison of requirements and capabilities, identification of shortfalls, and characterization of risk. A notional U.S. carrier battle group and air wing for the year 2000 were examined in scenarios involving a conventional global war with a reconstituted Russia, a major regional contingency on the Korean peninsula, and a lesser regional contingency involving a freedom of navigation dispute with Indonesia. The scenarios represent different levels from the spectrum of conflict. The future naval force was found insufficient to ensure victory in global conventional war, the scenario which involved the greatest risk to U.S. interests. The future force, optimized for blue-water operations, was also shown seriously deficient in countering mines and diesel submarines, another threat which entailed the potential for damage to U.S. interests.*
NPS/DKL Location: MICROFORM ADA284646

Hinkle, David, ed. **United States submarines**. Southport, CT: Hugh Lauter Levin; Lancaster: Gazelle, 2002. 352 p.
Published in association with the Naval Submarine League.
NPS/DKL Location: FOLIO V858 .U55 2002

Howes, Brian T. **Determining the Future of the US Submarine Force**. Monterey, CA: Naval Postgraduate School; Springfield, VA: Available from National Technical Information Service, 1992. (ADA260898). 130 p.
Thesis (M.A. in National Security Affairs)--Naval Postgraduate School, 1992.
Abstract: *The end of the Cold War has been the watershed event for changes in the international and national security environments that present tremendous implications for the US submarine force. These changes include calls for significant US defense cuts to reap a peace dividend, the increasing importance of economics as a determinant of defense spending, and the disintegration of the Soviet Union resulting in the absence of a clear tangible global threat to US national interests. What has resulted from these chances is the formulation of a new US national security strategy that focuses on regional contingencies, and the decision to cut US defense forces by at least 25% over five years including the cancellation of the Seawolf submarine program. This thesis addresses the implications of these tremendous changes on the US submarine force. Specifically, issues that are addressed include roles and missions, force structure, submarine design, and changing the institutional mindset of the submarine community. The issue of roles and missions involves demonstrating the applicability of the submarine to regional warfare. The issue of submarine force structure deals with both the short term and long term factors affecting submarine force reductions and ultimate submarine force size. The issue of submarine design addresses concerns over the submarine industrial base, the Centurion program, and design requirements for a regional warfighting submarine.*

NPS/DKL Location: THESIS H8253

Hyatt, A. M. J., ed. **Dreadnought to Polaris: maritime strategy since Mahan: papers from the Conference on Strategic Studies at the University of Western Ontario, March 1972**. Toronto: Copp Clark Pub. Co.; Annapolis, MD: Naval Institute Press, 1973. 125 p.
NPS/DKL Location: GENERAL V163 .C7

Jorgensen, Eric J., Joseph J. Fuller, Samuel C. Rainey and Theodore J. Post. **Potential Benefits to Navy Training Programs Resulting from Increased Use of Interactive Electronic Technical Manuals. Phase 1. Initial Evaluation of IETM Applicability to Schoolhouse and Worksite Training Functions: Interim report Mar-Oct 1995**. Bethesda, MD: Naval Surface Warfare Center Carderock Division, 1995. (ADA309992; NSWCCARDIV-TR-96/016). 118 p.
Abstract: *This report summarizes Phase I of a study entitled Training Benefit Analysis of the Accelerated Use of Interactive Electronic Technical Manuals (IETM's). An initial evaluation of the interactive, computer controlled display of technical information has been carried out by the Navy training community. Results indicate the use of IETM's, integrated with automated courseware, could significantly improve training processes. Forty-seven candidate projects covering surface, air and submarine warfare areas were identified. Fifteen IETM hypotheses and associated implementation scenarios were evaluated. Of these, twelve were supported by more than two-thirds of the participants in this study. Candidate projects were identified for business-case-analyses to be performed in Phase II. This report also identifies technical and administrative issues which must be addressed before the full potential of IETM's can be realized. Measures needed for greater integration, infrastructure support and standardization of IETM's in training are recommended. Phase II of the study will consist of a more detailed analysis of the selected candidate projects, particularly from the standpoint of return on investment. This will provide the Chief of Naval Operations with the basis for training input to the Program Objective Memorandum (POM) '98 preparation process.*
Electronic access: http://handle.dtic.mil/100.2/ADA309992

Kaufman, Steve and Yogi Kaufman. **Silent chase: submarines of the U.S. Navy**. Charlottesville, VA: Thomasson-Grant, c1989. 160 p.
NPS/DKL Location: FOLIO V858 .K38 1989

_____ and _____. **Silent chase: submarines of the U.S. Navy**. Shrewsbury, England: Airlife, 1989. 160 p.
NPS/DKL Location: FOLIO V858 .K38 1989

Kuenne, Robert E. **The attack submarine; a study in strategy**. New Haven, Yale University Press, 1965. 215 p.
NPS/DKL Location: GENERAL V210 .K9

Largess, Robert P. and James L. Mandelblatt. **U.S.S. Albacore: forerunner of the future**. Portsmouth, NH: P.E. Randall, c1999. (Publication (Portsmouth Marine Society) 25). 189 p.
NPS/DKL Location: BUCKLEY VA65.A25 L37 1999

Leary, William M. Under ice: **Waldo Lyon and the development of the Arctic submarine**. 1st ed. College Station: Texas A&M University Press, c1999. (Texas A & M University military history series; 62). 303 p.
NPS/DKL Location: GENERAL V857.5 .L43 1999

Lockwood, Charles A. **Down to the sea in subs: my life in the U.S. Navy**. [1st ed.]. New York: Norton, [1967]. 376 p.
NPS/DKL Location: GENERAL CT12.O23 A2

Maas, Peter. **The rescuer**. London, Collins, 1968. 256 p.
Originally published: New York: Harper & Row [1967]

_____. **The terrible hours: the man behind the greatest submarine rescue in history**. 1st ed. New York: HarperCollins Publishers, c1999. 259 p.
NPS/DKL Location: GENERAL VA65.S68 M33 1999

Mielcarz, Diane C. **Submarines - Who Needs Em?** Newport, RI: Naval War College; Available from National Technical Information Service, Springfield, VA, 1995. 23 p.
Abstract: *The current military/political climate of budget cuts, downsizing, the demise of the Soviet Union, and the establishment of a new world order has forced all services to reexamine and justify their existence. I will specifically examine what U.S. submarine has to offer in the defense of America in relation to the National Military Strategy, to include deterrence, forward presence, crisis response and reconstitution; enduring characteristics which include stealth, endurance and agility; critical roles such as peacetime engagement, surveillance, deterrence, regional sea denial, precision strike, task group support, ground warfare support and battlespace dominance; and contributions that the submarine can make to the post-Cold War era. While there is no possible way to predict where or when the next conflict will occur, the submarine possesses valuable attributes which can be successfully utilized at the operational level to enhance the mission and achieve success.*
NPS/DKL Location: MICROFORM ADA293371

Merrill, John and Lionel D. Wyld. **Meeting the submarine challenge: a short history of the Naval Underwater Systems Center**. 1st ed. [Washington, DC: Dept. of the Navy, 1997]. 372 p.
NPS/DKL Location: FEDDOCS D 201.2:SU 1

Moss, Lee O. **Operational Employment of Fast Attack Submarines in Littoral warfare-A Force Multiplier or a Force**. Newport, RI: Naval War College; Available from National Technical Information Service, Springfield, VA, 1996. 22 p.
Abstract: *With the Navy's shift in focus from blue water operations to littoral warfare, significant effort has been expended toward developing operational concepts that optimize the employment of naval forces, with the emphasis being placed on the Naval Expeditionary Task Force (Carrier Battle group and Amphibious Ready Group). Submarines play a key role in tactical operations within the battle group, but also have the potential to have a significant impact on the operational level of war, if properly utilized by the theater CINC. In order to get the most out of the submarine force, the CINC must balance operational tasks with the tactical needs of the CJTF. If the CINC allows the CJTF to control the SSNs to conduct littoral warfare tasks, he is forfeiting some of his operational flexibility. By viewing the JTF as a unit operating at the operational/tactical level, the CINC can better focus on tasks that need to be accomplished at the operational level. Employment of SSNs at this level of warfare can fill the gap that has been left by the Navy's concentration on littoral warfare. Emphasis on the development of doctrine for littoral warfare need*

to focus on the value of the SSN at all levels of warfare. Proper integration at all levels will result in maximum flexibility and will restore blue water Navy capabilities to the theater commander.
NPS/DKL Location: MICROFORM ADA312180

Navy times. **They fought under the sea; the saga of the submarine** [compiled by the editors of Navy times]. Harrisburg, PA, Stackpole Co., [1962]. 184 p.
NPS/DKL Location: GENERAL V210 .N3

Norris, David Thomas. **Strategic planning, Polaris, and Tomahawk: technological imperative hypotheses.** Monterey, CA: Naval Postgraduate School; Available from National Technical Information Service, Springfield, VA, 1987. (ADA193027). 126 p. Thesis (M.A. in National Security Affairs)--Naval Postgraduate School, 1987.
Abstract: *This thesis examines the force procurement element of the military strategic planning process and is comprised of two parts. First, models are constructed to depict ideal strategic planning. The initial step in each model is the formulation of the national interest. The national interest is defined in terms useful to strategic planners by creating a unique paradigm based on the Constitution. The technological imperative hypothesis is explored as an aberration to the ideal strategic planning process. Second, the technological imperative hypothesis is tested with case studies of the Polaris and the Tomahawk. Even though the hypothesis was disproved in each case, the case studies yielded useful relationships between technology, strategy, and doctrine.*
NPS/DKL Location: THESIS N885

Polmar, Norman. **The American submarine.** Annapolis, MD: Nautical & Aviation Pub. Co. of America, c1981. 172 p.

_____. **The American submarine.** 2nd ed. Annapolis, MD: Nautical & Aviation Pub. Co. of America, c1983. 171 p.
NPS/DKL Location: GENERAL V858 .P59 1983

_____. **Atomic submarines.** Princeton, NJ, Van Nostrand, [1963]. 286 p.
NPS/DKL Location: GENERAL V858 .P7

_____. **Death of the Thresher.** [1st ed.]. Philadelphia, Chilton Books, [1964]. 148 p.
NPS/DKL Location: GENERAL VA65.T7 P7

_____ and Thomas B. Allen. **Rickover.** New York: Simon & Schuster, c1982. 744 p.
NPS/DKL Location: GENERAL CT18.I52 P64 1982

Poluhowich, John. **Argonaut: the submarine legacy of Simon Lake.** College Station: Texas A&M University Press, c1999. West Texas A&M University series; no. 4). 181 p.
NPS/DKL Location: GENERAL VM1 .L25 P65 1999

Rees, Edwin. **The seas and the subs.** [1st ed.]. New York, Duell, Sloan & Pearce, [1961]. 233 p.
NPS/DKL Location: GENERAL V858 .R3

Rodengen, Jeffrey L. **Serving the silent service: the legend of Electric Boat**. Fort Lauderdale, FL: Write Stuff Syndicate, c1994. 176 p.
NPS/DKL Location: ON-ORDER

Sakitt, Mark. **Submarine warfare in the Arctic: option or illusion?** Stanford, CA: Stanford University, International Strategic Institute at Stanford, 1988. 93 p.
"An Occasional Paper of the Center for International Security and Arms Control."
NPS/DKL Location: GENERAL V210 .S24 1988

Smith, Carole J. **Small Fleet -- Big Risk**. Newport, RI: Naval War College; Available from National Technical Information Service, Springfield, VA, 1995. 21 p.
Abstract: *In any future conflict, the U.S. Navy will most likely enjoy a significant technological and numerical superiority over its adversary. A relatively small navy may, however, avoid decisive battle and influence events at sea indefinitely. Naval strategists have coined the terms fleet in being, fortress fleet, and risk fleet to describe strategies designed to use inferior forces to an advantage. Through an examination of these strategies as analyzed by both Alfred Thayer Mahan and Julian Corbett and study of historical examples of each, the relevance of these strategies to current naval thought can be determined. While fortress fleet and risk fleet have very limited value today, a diesel submarine fleet in being poses a significant threat that must be addressed in future U.S. naval strategy.*
NPS/DKL Location: MICROFORM ADA293409

Snider, Don M., ed. **Attack submarines in the post-Cold War era: the issues facing policymakers**. [Washington, DC]: Center for Strategic & International Studies, c1993. 19 p.

Sontag, Sherry and Christopher Drew. **Blind man's bluff: the untold story of American submarine espionage**. New York: Public Affairs, c1998. 352 p.
NPS/DKL Location: BUCKLEY VB231.U54 S65 1998

_____ and _____. **Blind man's bluff: the untold story of American submarine espionage**. Thorndike, ME: Thorndike Press, 1999. 728 p.
Originally published: New York: Public Affairs, 1998.

_____ and _____. **Jagd unter Wasser: Die wahre Geschichte der U-Boot-Spionage**. München: Goldmann, 2001. 542 p.
German translation of Blind man's bluff.

Soto, Alberto A. **The Flaming Datum problem with varying speed**. Monterey, CA: Naval Postgraduate School; Springfield, VA: Available from National Technical Information Service, 2000. (ADA379766). 49 p.
Thesis (M.S. in Operations Research)--Naval Postgraduate School, 2000.
Abstract: *The problem of detecting an enemy submarine whose possible position was revealed by the hit of a torpedo is known as the "Flaming Datum" problem. All previous studies devoted to this theme make unrealistic assumptions about the speed of the escaping target when dealing with a diesel-electric submarine. In this kind of submarine the constraint imposed by the remaining charge of its batteries determines that its behavior is essentially conservative in how fast it should escape. The objective of this thesis is to explore the idea of varying speed in the flaming datum problem. Two different approaches are considered. An analytical model is developed based on the relationship among some of the physical factors*

that could determine or constrain the behavior of a diesel submarine while escaping from the area of the flaming datum. The second approach considers a discrete event simulation using the Java-based Simkit package. Data analysis is used to determine a possible fit for the simulation results. Several tactics are explored to determine their effects on detection probability.
NPS/DKL Location: THESIS S66611635
Electronic access: http://handle.dtic.mil/100.2/ADA379766

Stambler, Irwin. **The battle for inner space; undersea warfare and weapons**. New York, St. Martin's Press, [1962]. 259 p.
NPS/DKL Location: GENERAL V210 .S695 1962

Steele, George P. **Seadragon: Northwest under the ice**. [1st ed.]. New York, Dutton, 1962. 255 p.
NPS/DKL Location: GENERAL VA65.S43 S7

Stern, Robert Cecil. **U.S. subs in action**. Carrollton, TX: Squadron/Signal Publications, c1979. (Warships; no. 2). 45 p.
NPS/DKL Location: BUCKLEY D783 .S73 1979

Stumpf, David K. **Regulus, the forgotten weapon: a complete guide to Chance Vought's Regulus I and II**. Paducah, KY: Turner Pub. Co., c1996. 192 p.
NPS/DKL Location: GENERAL UG1312.C7 S78 1996

Terzibaschitsch, Stefan. **Submarines of the US Navy** [translated by M.J. Shields]. London: Arms & Armour Press; New York: Distributed in the USA by Sterling, c1991. 214 p.
English translation of U-Boote der US Navy.

_____. **U-Boote der US Navy**. Herford: Koehler, 1990. 214 p.
German translation of Submarines of the US Navy.

Thompson, Roger. **Brown Shoes, Black Shoes, and Felt Slippers: Parochialism and the Evolution of the Post-War U.S. Navy**. Newport, RI: Naval War College; Available from National Technical Information Service, Springfield, VA, 1995. 70 p.
Abstract: *This report examines how intra-service parochialism has affected the United States Navy since the end of the Second World War. It traces the development of naval bureaucratic dominance from the prewar battleship admirals, through the rise of naval aviators to the eventual dominance by nuclear submariners. The author posits that the Navy may now have entered a new era once again dominated by surface warfare officers and wonders what the consequences of this change may be. The study argues for balance and urges naval leadership to rise above the natural tendency to square the past by primarily promoting the interests of the dominant warfare group.*
NPS/DKL Location: MICROFORM ADA299970

Tritten, James John. **The submarine's role in future naval warfare**. Monterey, CA: Naval Postgraduate School; Springfield, VA: Available from National Technical Information Service, [1992]. (NPS-NS-92-010; ADA252808). 43 p.
Abstract: *The three basic elements where we traditionally commence strategic planning have changed dramatically in the past two years. A new national security strategy recasts the roles and missions of the*

armed forces in new terms. The submarine force needs to be justified under the new grammar for warfare as a part of the four new mission areas under the new national military strategy. The submarine force alone can perform the strategic deterrence and defense missions. The submarine's role in presence involves a high/low mix choice. There should be increased emphasis on the submarine force for crisis response: (1) rapid response (2) shore bombardment and strike (3) as the initial leading maritime component for second major regional contingencies, and (4) initial and limited sea control. A European regional war evolving out of a major regional contingency is not the same thing as the old European-centered global war with the USSR. Decreased emphasis should be placed on strategic anti-submarine warfare. Reconstitution goals could be met with at-sea nuclear weapons.
NPS/DKL Location: FEDDOCS D 208.14/2:NPS-NS-92-010

United States. Defense Science Board. Task Force on Submarine of the Future. **Report of the Defense Science Board Task Force on Submarine of the Future**. Washington, DC: Office of the Under Secretary of Defense for Acquisition & Technology, 1998. 46 p.
NPS/DKL Location: GENERAL V857 .R46 1998

United States. General Accounting Office. **Navy acquisitions: improved littoral warfighting capabilities needed: report to the Chairman and ranking minority member, Subcommittee on Military Research and Development, Committee on Armed Services**, House. of Representatives. Washington, DC: The Office, [2001]. (GAO-01-493). 32 p.
NPS/DKL Location: FEDDOCS GA 1.13:01-493
Electronic access: http://purl.access.gpo.gov/GPO/LPS12498

United States. Naval History Division. **The submarine in the United States Navy**. Washington [For sale by the Supt. of Docs., U.S. Govt. Print. Off.], 1960. 30 p.

_____. **The submarine in the United States Navy**. 2d ed. Washington [For sale by the Supt. of Docs., U.S. Govt. Print. Off.], 19639. 20 p.

_____. **The submarine in the United States Navy**. 3d ed. Washington [For sale by the Supt. of Docs., U.S. Govt. Print. Off.], 1969. 24 p.
NPS/DKL Location: GENERAL V858 .U3

United States. Naval Torpedo Station, Keyport, Wash. **Undersea warfare fleet support**. Keyport, WA: Dept. of Defense, Dept. of the Navy, Naval Torpedo Station; Washington: for sale by the Supt. of Docs., U.S. Govt. Print. Off., [1976?]. 25 p.
NPS/DKL Location: GENERAL VA68.K4 U6

United States. Navy. Atlantic Fleet. Submarine Force. **United States Ship Thresher (SSN 593): in memoriam, April 10, 1963**. [New York?, 1964]. 146 p.
NPS/DKL Location: GENERAL VA65.T7 T5

United States. Navy. Pacific Fleet. Submarine Force. **Submarine search and rescue plan**. [Vallejo, CA, 1955]. (ComSubGru SFran; no. 1-55). 1 v.
NPS/DKL Location: GENERAL VK1265 .U6

Watson, Kenneth M. and Victor C. Anderson. **The Marine Physical Laboratory Multi-Disciplinary Ocean Science and Technology Program**. La Jolla, CA: Scripps Institution of Oceanography, 1989. (ADA220008, MPL-U-27/89). 35 p.
Abstract: *This report contains summaries of the research performed at the Marine Physical Laboratory. Brief descriptions of the research and bibliographies of the publications resulting therefrom are included. The general areas of our research include: Ocean Environmental Acoustics, Marine Physics, Marine Geophysics, Signal Processing, Ocean Technology, Platform Development and Support, and Technical Assistance and Technology Transfer.*
NPS/DKL Location: MICROFORM ADA220008

Weir, Gary E. **Building American submarines, 1914-1940**. Washington, DC: Naval Historical Center, 1991. 166 p.
NPS/DKL Location: GENERAL V858 .W45 1991

_____. **Forged in war: the naval-industrial complex and American submarine construction, 1940-1961**. Washington: Naval Historical Center, Dept. of the Navy: For sale by U.S. G.P.O., Supt. of Docs., 1993. 314 p.
NPS/DKL Location: GENERAL V858 .W46 1993

Williams, Marion D. **Submarines under Ice: The U.S. Navy's Polar Operations**. Annapolis, MD: Naval Institute Press, 1998. 256 p.
NPS/DKL Location: GENERAL V858 .W55 1998

Zierke, William C. and D. A. Boger, eds. **A physics-based means of computing the flow around a maneuvering underwater vehicle**. State College, Pa.: Pennsylvania State University, Applied Research Laboratory, [1997]. (TR 97-002). 276 p.
NPS/DKL Location: GENERAL QA911 .P49 1997

SUBMARINES -- AMERICAN -- KOREAN WAR

Billy, Gregory M. **An Operational Analysis of United States Submarine Employment in the Korean War**. Newport, RI: Naval War College; Available from National Technical Information Service, Springfield, VA, 1994. 29 p.
Abstract: *United States submarine operations during the Korean War are critically analyzed from an operational perspective. The Korean War represented a prototype for future Major Regional Conflicts (MRCs). Examining the Operational Commander's use of submarines against a relatively weak naval power, in a conflict dominated by land battle, provides lessons which may be applicable to future MRCs. Brief historical and operational overviews are followed by operational analyses of submarine command and control, operational reconnaissance missions, and the war's impact on the submarine force. Conclusions discuss lessons learned for present and future operational planning. Compared to their significant contribution during World War II, U.S. submarines did not play a major role in Korea. Their employment was mostly directed towards training and reconnaissance operations. Korean War operational reconnaissance set the stage for submarine operations throughout the Cold War. Submarine employment in the Korean War was affected by three key issues: difficulty in preventing blue-on-blue engagements, communications limitations which inhibited rapid, reliable submarine operational tasking, and defensive mining of the littoral region. These three issues will continue to challenge operational Commanders when employing submarines in future MRCS.*
NPS/DKL Location: MICROFORM ADA279727

Cagle, Malcolm W. and Frank A. Manson. **The sea war in Korea**. Annapolis, [MD]: United States Naval Institute, c1957. 555 p.
NPS/DKL Location: GENERAL DS920.A3 C2

Field, James A. **History of United States naval operations Korea**. [U.S. Naval History Division], 1962. 499 p.
NPS/DKL Location: GENERAL DS920.A3 F4

Lederer, William J. **The last cruise; the story of the sinking of the submarine, U.S.S. Cochino**. New York, Sloane, [1950]. 110 p.
A shorter version appeared in the Saturday Evening Post as *Miracle under the Arctic sea.*"
NPS/DKL Location: BUCKLEY V858 .L4

Naval Institute Press and Sonalysts, Inc. **The sea services in the Korean War, 1950-1953** [computer file] / produced by the Naval Institute Press and Sonalysts, Inc. in conjunction with the historical offices of the U.S. Navy, Marine Corps and Coast Guard. 50th Anniversary of the Korean War Commemoration ed. Annapolis, MD: U.S. Naval Institute Press, 2000, c1957. 1 computer optical disc.
NPS/DKL Location: CIRCDESK DS920.A2 C3 2000

Schratz, Paul R. **Submarine commander: a story of World War II and Korea**. Lexington, KY: University Press of Kentucky, c1988. 322 p.
NPS/DKL Location: GENERAL V63.S38 A3 1988

SUBMARINES -- AMERICAN – NUCLEAR

Adcock, Al. **U.S. ballistic missile subs in action**. Carrollton, TX: Squadron/Signal Publications, c1993. (Warships; no. 6). 50 p.

Anderson, William R. and Clay Blair, Jr. **Nautilus 90 north**. [1st ed.]. Cleveland, World Pub. Co., [1959]. 251 p.
NPS/DKL Location: GENERAL VA65.N3 A6

_____ and _____. **Nautilus 90 north**. 1st TAB ed. Blue Ridge Summit, PA: Tab Books, [1989], c1959. (Military classics series). 251 p.

Baar, James and William E. Howard. **Polaris!** [1st ed.]. New York, Harcourt, Brace, [1960]. 245 p.
NPS/DKL Location: GENERAL VG93 .B2

Bivens, Arthur Clark. **Of nukes and nose cones: a submarine story**. Baltimore, MD: Gateway Press, 1996. 125 p.

Duncan, Francis. **Rickover and the nuclear navy: the discipline of technology**.
Annapolis, MD: Naval Institute Press, 1990. 374 p.
NPS/DKL Location: GENERAL CT18 .I522 D86 1990

_____. **Rickover: the struggle for excellence**. Annapolis, MD: Naval Institute
Press, 2001. 416 p.
NPS/DKL Location: GENERAL V63.R63 D86 2001

Duthie, Richard E. C. and Billy D. Donald. **Nuclear powered submarines**. [Oak Ridge,
TN] U. S. Atomic Energy Commission, Division of Technical Information, [1964]. 1 v.

Govan, Dale R. **The Submarine Contribution to Operational Protection**. Newport, RI:
Naval War College, 1997; Available from National Technical Information Service,
Springfield, VA. 24 p.
Abstract: *How can nuclear powered submarines (SSNs) contribute to joint force protection? Are these submarines essential to a joint force commander's concept of operations? Would their absence significantly alter his branch and sequel plans? Although SSNs represent a significant combat capability, do they possess the necessary range of capabilities to enhance operational protection in a given theater? SSNs can be a force multiplier in the right scenario. 'The modern attack submarine is a versatile multi-mission warship that is more survivable than any other naval vessel in history.' However, just as the special operating forces complement ground troops, SSNs complement the naval forces. SSNs can not accomplish all tasks all the time, but the capabilities they bring to joint force operations can free other forces to act in contributing areas to accomplish the overall mission. This is their forte. The principle missions submarines can perform have grown tremendously from the pre-World War II tasks. These tasks included covert strike warfare, surface warfare, undersea warfare, intelligence collection and surveillance, covert indication and warning, electronic warfare, special warfare, covert mine warfare, and battlegroup support. With so many capabilities available, the operational commander must rely on doctrine to incorporate these tasks into his concept of operations. This paper will attempt to articulate the fundamental principles to guide the use of SSNs in warfare. Just as air superiority against an adversary requires phasing of operations, so does undersea superiority. Submarines can best combine time and space with stealth to help prepare the littoral battlespace for future operations.*
NPS/DKL Location: MICROFORM ADA328103

Kuenne, Robert E. **The Polaris missile strike; a general economic systems analysis**.
[Columbus]: Ohio State University Press, [1967, c1966]. 434 p.
The third volume to emerge from the General Economic Systems Project at Princeton
University.
NPS/DKL Location: GENERAL V993 .K9

Lawliss, Chuck. **The submarine book: a portrait of nuclear submarines and the men who sail them**. New York: Thames & Hudson, 1991. 160 p.
NPS/DKL Location: GENERAL V857 .L29 1991

_____. **The submarine book**. Short Hills, NJ: Burford Books, c2000. 204 p.
Originally published: New York: Thames & Hudson, 1991.

LaCroix, F. W. et al. **A concept of operations for a new deep-diving submarine**.
Santa Monica, CA: Rand, 2001. (MR-1395). 153 p.

Abstract: *By 2012, the reactor on the U.S. Navy's only deep-diving research submarine will be exhausted, making it necessary to either refuel the reactor or replace the submarine. If the Navy opts for a new submarine, what capabilities should it retain and what capabilities should be added? What would be its most important missions and what would be required for it to perform those missions? In this report, the authors worked with panels of qualified scientists, defense experts, and naval officers to develop a concept of operation for a possible replacement platform, analyzing which military and scientific missions should have the highest priorities. The authors conclude by offering a list of the highest-priority missions and two design concepts that would best be able to achieve them.*
NPS/DKL Location: GENERAL V857.5 .C65 200
Electronic access: http://www.rand.org/publications/MR/MR1395/

Leary, William M. **Under ice: Waldo Lyon and the development of the Arctic submarine**. 1st ed. College Station: Texas A&M University Press, c1999. (Texas A & M University military history series; 62). 303 p.

Lightbody, Andy and Joe Poyer. **Submarines: hunter-killers & boomers**. Lincolnwood, IL: Publications International, c.1990. 192 p.

Polmar, Norman. **Atomic submarines**. Princeton, NJ, Van Nostrand, [1963]. 286 p.
NPS/DKL Location: GENERAL V858 .P7

_____. **Death of the Thresher**. [1st ed.]. Philadelphia, Chilton Books, [1964]. 148 p.
NPS/DKL Location: GENERAL VA65.T7 P7

Rees, Ed. **The seas and the subs**. [1st ed.]. New York, Duell, Sloan & Pearce, [1961]. 233 p.
NPS/DKL Location: GENERAL V858 .R3

Rockwell, Theodore. **The Rickover effect: how one man made a difference**. Annapolis, MD: Naval Institute Press, c1992. 411 p.
NPS/DKL Location: GENERAL CT18 .I52 R62 1992

_____. **The Rickover effect: the inside story of how Adm. Hyman Rickover built the nuclear Navy**. New York: J. Wiley, 1995. 411 p.
Originally published: Annapolis, MD: Naval Institute Press, c1992.

Rodengen, Jeffrey L. **Serving the silent service: the legend of Electric Boat**. Fort Lauderdale, FL: Write Stuff Syndicate, c1994. 176 p.

Sapolsky, Harvey M. **The Polaris system development; bureaucratic and programmatic success in government**. Cambridge, MA: Harvard University Press, 1972. 261 p.
NPS/DKL Location: GENERAL V993 .S2

Scharfenberg, Horst. **Nautilus 90 Grad Nord; Atom-U-Boote erobern die Meere**. [1. Aufl.]. Stuttgart, K. Thienemanns Verlag, [1959]. 190 p.

German translation of Nautilus 90 north.

Spinardi, Graham. **From Polaris to Trident: the development of US Fleet ballistic missile technology.** Cambridge [England]; New York: Cambridge University Press, 1994. (Cambridge studies in international relations; 30). 253 p.
Revision of author's thesis (Ph.D.) -- University of Edinburgh.
NPS/DKL Location: GENERAL V993 .S65 1994

Steele, George P. and Herbert J. Gimpel. **Nuclear submarine skippers and what they do**. New York: F. Watts, 1962. 140 p.

Tritten, James J. A New Case for Naval Arms Control: Interim report Sep-Dec 1992. Monterey, CA: Naval Postgraduate School Department of National Security; Springfield, VA: Available from National Technical Information Service, 1992. (ADA259759)1 v.
Abstract: *This paper opens with an examination of existing legal restraints on naval forces and arms control agreements and concludes that the U.S. is already heavily engaged in naval arms control. Given the new international security environment and the new U.S. regionally-oriented national security and military strategies, the author then recommends a series of additional naval arms control measures that should be taken: exchanges of data, transparency, INCSEA, cooperative measures, an agreement on the laws of submarine warfare, abolishing NCND, no first tactical nuclear use at sea, NWFZs, advanced notification of operational-level exercises, environmental protection measures, controls over maritime technologies, armed escorts of nuclear shipments, new Roes, PALs, the resolution of outstanding political issues at sea, deep cuts in nuclear forces, CFE follow-on, limits on specific types of naval forces, geographic limits, expanded standing naval forces, and a re negotiation of the ABM Treaty. The paper then addresses verification and compliance issues. Author concludes that since the U.S. Navy has already managed to avoid major arms control while balanced on the precarious 'slippery slope', there is no reason to continue its stonewalling policies.*
NPS/DKL Location: FEDDOCS D 208.14/2:NPS-NS-92-016

United States. Congress. House. Committee on Merchant Marine and Fisheries. **Disposal of decommissioned nuclear submarines**: hearing before the Committee on Merchant Marine and Fisheries, House of Representatives, Ninety-seventh Congress, second session, on oversight of the Ocean Dumping Act and National Ocean Pollution Planning Act and the disposal of defueled, decommissioned nuclear submarines, October 19, 1982--Manteo, N.C. Washington: U.S. G.P.O., 1983. 166 p.
NPS/DKL Location: FEDDOCS Y 4.M 53:97-47

United States. Congress. House. Committee on National Security. Subcommittee on Military Procurement. **New attack submarine**: hearing before the Military Procurement Subcommittee of the Committee on National Security, House of Representatives, One Hundred Fourth Congress, first session, hearing held September 7, 1995. Washington: U.S. G.P.O.: For sale by the U.S. G.P.O., Supt. of Docs., Congressional Sales Office, 1996. 127 p.
NPS/DKL Location: FEDDOCS Y 4.SE 2/1 A:995-96/16

United States. Congress. Joint Committee on Atomic Energy. **Nuclear submarines of advanced design**. Hearing, Ninetieth Congress, second session ... June 21, 1968. Washington, U.S. Govt. Print. Off., 1968. 131 p.

United States. Congress. Senate. Committee on Armed Services. Subcommittee on Nuclear Deterrence, Arms Control, and Defense Intelligence. **Shipment of spent nuclear fuel from U.S. Navy ships and submarines to the Idaho National Engineering Laboratory (INEL)**: hearing before the Subcommittee on Nuclear Deterrence, Arms Control, and Defense Intelligence of the Committee on Armed Services, United States Senate, One Hundred Third Congress, first session, July 28, 1993. Washington: U.S. G.P.O.: For sale by the U.S. G.P.O., Supt. of Docs., Congressional Sales Office, 1994. 163 p.
NPS/DKL Location: GENERAL KF26 .S20 V.103 NO.4

United States. Congressional Budget Office. **Increasing the Mission Capability of the Attack Submarine Force**. [Washington, D.C.]: The Office, 2002. 35 p.
NPS/DKL Location: FICHEDOCS Y 10.2:AT 8
Electronic access: http://www.cbo.gov/showdoc.cfm?index=3312&sequence=0&from=1

United States. General Accounting Office. **Attack submarines: alternatives for a more affordable SSN force structure**: report to Congressional requesters. Washington, DC: The Office; Gaithersburg, MD (P.O. Box 6015, Gaithersburg 20884-6015): The Office [distributor, 1994]. 45 p.
NPS/DKL Location: FEDDOCS GA 1.13:NSIAD-95-16

United States. Naval Material Command. Trident System Project Office. **Trident system**. Washington, DC: Trident System Project Office, Navy Dept., 1977. 13 p.
NPS/DKL Location: GENERAL V993 .U63

United States. Navy Dept. Strategic Systems Programs. **Facts/chronology: Polaris-Poseidon-Trident**. Arlington, VA: Strategic Systems Programs, Navy Department, 1996. 95 p.
NPS/DKL Location: GENERAL V993 .F35 1996

_____. **FBM facts/chronology: Polaris, Poseidon, Trident**. Washington, DC: Strategic Systems Project Office, Navy Department, 1982. 59 p.
NPS/DKL Location: GENERAL V993 .F35 1982

_____. **FBM facts/chronology: Polaris, Poseidon, Trident**. Washington, DC: Strategic Systems Programs, Navy Department, 1990. 75 p.
NPS/DKL Location: GENERAL V993 .F35 1990

Waller, Douglas C. **Big Red: three months on board a Trident nuclear submarine**. New York: HarperCollins Publishers, c2001. 336 p.
NPS/DKL Location: GENERAL VA65.N35 W35 2001

Williams, Gorfon C. **Employment of Fast Attack Submarines by the Operational Commander**. Newport, RI: Naval War College, 1998. (ADA348604). 27 p.

Abstract: *For operational commanders, the use of military force today requires flexibility, efficiency, and careful risk management among joint forces. In light of these requirements, this paper examines the influence of the fast attack nuclear submarine (SSN) and the Joint Force Commander's JFC) employment of SSNs as an operational-level asset. The history of submarine warfare provides many lessons regarding submarine employment. Although the JFC can use submarines to accomplish or support a broad scope of missions, optimum employment requires analysis of the characteristics, capabilities, and expertise of submarines. The intrinsic and enduring characteristics of the SSN are stealth, mobility, endurance, and flexibility. The SSN also possesses diverse capabilities that allow it to perform a number of missions in support of the operational commander. These include theater ISR, support of ground components and operations ashore, and attacks on sea-based threats and objectives. Matching the submarine's characteristics with these capabilities yields several employment principles. For the JFC, SSNs are flexible assets that are best employed operationally deep where autonomous, survivable, or enabling forces are required. As an operational asset, the SSN contributes directly to several of the JFC's operational functions. First, submarines are important to the JFC because of their role in operational intelligence. Because of their stealth and forward positions, SSNs are able to conduct ISR directed at the enemy's operational weaknesses and centers of gravity. Second, the SSN's mobility and endurance enable it to execute operational maneuver. Third, through their USW, SUW and MIW capabilities, SSNs provide superb operational protection of naval forces. Finally, the SSN's extensive operational reach and autonomy make it particularly effective at conducting operational fires. Effective employment of SSNs will help the JFC achieveleverage and freedom of action in the theater.*
Electronic access: http://handle.dtic.mil/100.2/ADA348604

Vyborny, Lee and Don Davis. **Dark waters: an insider's account of the NR-1, the Cold War's undercover nuclear sub**. New York: New American Library, 2003. 243 p.

SUBMARINES -- AMERICAN – PERSIAN GULF WAR

Brown, Ron A. and Robert J. Schneller, comps. **United States Naval Forces in Desert Shield and Desert Storm: a select bibliography**. Washington, D.C.: Naval Historical Center, 1993. 50 p.
NPS/DKL Location: REFERENCE DS79.72 .B76 1993

Marolda, Edward J. and Robert J. Schneller, Jr. **Shield and sword: the United States Navy and the Persian Gulf War**. Washington: Naval Historical Center, Dept. of the Navy: For sale by the U.S. G.P.O., Supt. of Docs., 1998. 517 p.
NPS/DKL Location: FEDDOCS D 221.2:G 95

Meisner, Arnold. **Desert Storm sea war**. Osceola, WI, USA: Motorbooks International, 1991. 128 p.

Pokrant, Marvin. **Desert Storm at sea: what the Navy really did**. Westport, Ct: Greenwood Press, 1999. (Contributions in military studies, no. 175). 329 p.
NPS/DKL Location: GENERAL DS79.744 .N38 P65 1999

United States. Office of the Chief of Naval Operations. **The United States Navy in "Desert Shield" "Desert Storm."** Washington, DC: The Office, [1991]. 1 v.
NPS/DKL Location: GENERAL DS79.72 .U55 1991

SUBMARINES -- AMERICAN -- WWI

Bass, Herbert J., ed. **America's entry into World War I: submarines, sentiment, or security?** New York, Holt, Rinehart & Winston, [1964]. 122 p.
NPS/DKL Location: GENERAL D619 .B3

Battey, George Magruder. **70,000 miles on a submarine destroyer; or, The Reid boat in the world war.** Atlanta, The Webb & Vary company, 1919. 384 p.
NPS/DKL Location: BUCKLEY D589.U6 B3 1919

Leighton, John Langdon. **Simsadus: London; The American navy in Europe.** New York, H. Holt & Company, 1920. 169 p.
NPS/DKL Location: GENERAL D589.U6 L5

Millholland, Ray. **The splinter fleet of the Otranto barrage.** New York, The Bobbs-Merrill company, [c1936]. 307 p.

Moffat, Alexander White. **Maverick Navy.** 1st ed. Middletown, CT: Wesleyan University Press, c1976. 157 p.

Silverstone, Paul H. **U.S. warships of World War I.** Garden City, NY, Doubleday, 1970. 304 p.
Simultaneously published: Shepperton: Allan, 1970.

Sims, William Sowden and Burton J. Hendrick. **The victory at sea.** Garden City, NY: Doubleday, Page & Company, 1920. 410 p.
NPS/DKL Location: BUCKLEY D589.U6 S6 1920

_____, _____ and David F. Trask. **The victory at sea.** Annapolis, MD: Naval Institute Press, c1984. (Classics of naval literature). 420 p.
Originally published: Garden City, NY: Doubleday, Page & Co., 1920.
NPS/DKL Location: GENERAL D589.U6 S6 1984

Thompson, Terry Brewster. **Take her down, wartime adventures of U.S. submarine L-9.** [New York] Sheridan house, [1942]. 313 p.
Originally published: New York: Sheridan House, [c1937] as Take her down, a submarine portrait.
NPS/DKL Location: GENERAL D595.U3 T4

United States. Office of Naval Records and Library. **American ship casualties of the world war including naval vessels, merchant ships, sailing vessels, and fishing craft.** Comp. by Historical section. Cor. to April 1, 1923. Washington, Govt print. off., 1923. 24 p.
NPS/DKL Location: GENERAL D589 .U545 1923

SUBMARINES -- AMERICAN -- WWII

Alden, John D. **U.S. submarine attacks during World War II: including Allied submarine attacks in the Pacific theater.** Annapolis, MD: Naval Institute Press, c1989. 285 p.
NPS/DKL Location: GENERAL D783 .A54 1989

Banning, Kendall. **Submarine! The story of undersea fighters.** Random House, 1942. 51 p.
NPS/DKL Location: BUCKLEY VM365 .B2

Barnes, Robert Hatfield. **United States submarines.** [1st ed.] New Haven, CT, H.F. Morse Associates, [1944]. 195 p.
Reprinted mostly from articles published in the Connecticut Circle magazine.

_____. **United States submarines.** 2nd Edition. New Haven, CT, H. F. Morse associates, inc. [1945]. 221 p.
Reprinted mostly from articles published in the Connecticut Circle magazine.

_____. **United States submarines.** 3d ed. New Haven, CT: H.F. Morse, 1946. 221 p.
Reprinted mostly from articles published in the Connecticut Circle magazine.

Beach, Edward Latimer. **Salt and steel: reflections of a submariner.** Annapolis, MD: Naval Institute Press, 1999. 299 p.
NPS/DKL Location: GENERAL CT2.E44 A3 1999

_____. **Submarine!** New York, Holt, [1952]. 301 p.
NPS/DKL Location: BUCKLEY D783.5.T7 B3

_____. **Submarine!** London: W. Heinemann, 1953. 274 p.

_____. **Submarine** [Traduction par J. Jouan]. Paris: les Presses de la Cité, 1953. 316 p.
French translation of Submarine

_____. **Submarine!** London: Brown, Watson, 1958. 318 p.

_____. **Submarine** [Traduction de R. [René] Jouan]. Paris: Presses pocket, 1966. 317 p.
French translation of Submarine

_____. **Submarine!** London: Coronet, 1974. 301 p.

_____. **Submarine!** Annapolis, MD:Naval Institute Press, [2003]. (Bluejacket books). 301 p.

_____. **Ubåt!: amerikanske undervannsbåter i kamp mot Japan** [til norsk ved Herbrand Lavik; forord av S. Valvatne]. Bergen: J.W. Eides, 1953. 326 p.
Norwegian translation of Submarine!

Bell, Art. **Peter Charlie: the cruise of the PC 477**. Woodland Hills, CA: Courtroom Compendiums, 1982. 384 p.

Benere, Daniel E. **A Critical Examination of the U.S. Navy's Use of Unrestricted Submarine Warfare in the Pacific Theater during WWII**. Newport, RI: Naval War College, Department of Operations, 1992. (ADA253241). 31 p.
Abstract: *This paper analyzes the United States use of unrestricted submarine warfare against the Japanese in World War II. Within the framework of the principles of war, the paper critically analyzes the strategy of the use of submarines during the war and how the operational strategy changed during the course of the war. This paper also critically surveys the use (or misuse) of the key tenets of modern, fundamental military thought. Recommendations and observations are offered which are considered applicable to modern warfare.*
NPS/DKL Location: MICROFORM ADA253241

Beynon, Robert P. **The Pearl Harbor avenger**. 1st ed. Deland, FL: Just Books 1, c2002. 298 p.
USS Bowfin

Blair, Clay. **Silent victory: the U.S. submarine war against Japan**. Philadelphia, Lippincott, [1975]. 1072 p.
NPS/DKL Location: GENERAL D783 .B6

_____. **Silent victory: the U.S. submarine war against Japan**. Annapolis, MD: Naval Institute Press, 2001. 1071 p.
Originally published: Philadelphia: Lippincott, 1975.

Bouslog, Dave. **Maru killer: the war patrols of the USS Seahorse**. Sarasota, Fla.: Seahorse Books, c1996. 210 p.

Boyd, Carl. **American command of the sea through carriers, codes, and the silent service: World War II and beyond**. Newport News, Va.: Mariners' Museum, c1995. (Mariners' Museum publication; no. 43). 79 p.

Briggs, Raymonde. **The super submarine, the only vessel now capable of safely carrying and servicing a sixteen inch gun**. [Bogota, NJ, The Dancey printing company, 1942]. 20 p.
Discussion of a proposed new submarine with sixteen inch gun, invented by N. M. Hopkins.

Calvert, James F. **Silent running: my years on a World War II attack submarine**. New York: J. Wiley, c1995. 282 p.
NPS/DKL Location: GENERAL D782.J33 C35 1995

Carmer, Carl Lamson. **The Jesse James of the Java Sea**. New York: Farrar & Rinehart, [1945]. 119 p.
NPS/DKL Location: GENERAL D783.5.S8 C2

Carr, Roland T. **To sea in Haste**. Washington: Acropolis Books, c1975. (The Acropolis Americana bicentennial series). 260 p.

Casey, Robert J. **Battle below: the war of the submarines**. Indianapolis: Bobbs-Merrill, [c1945]. 380 p.
NPS/DKL Location: BUCKLEY D783 .C3

Chambliss, William C. **The silent service**. [New York], New American Library, [1959]. 158p.

Christodoulou, Nicholas. **Saga of the submarine Scabby, U.S.S. Scabbardfish (SS-397)**. 1st ed. Nashville, TN: L.H. Cunningham, c1985. 208 p.

Cline, Rick. **Final dive: the gallant and tragic career of the WWII submarine, USS Snook**. Placentia, CA: R.A. Cline Pub., c2001. 235 p.

_____. **Submarine Grayback: the life & death of the WW II sub, USS Grayback**. 1st ed. Placentia, CA: R.A. Cline Pub., c1999. 252 p.
NPS/DKL Location: GENERAL D783.5 .G73 C55 1999

Conner, Claude C. **Nothing friendly in the vicinity: my patrols on the submarine USS Guardfish during WWII**. 1st ed. Mason City, IA: Savas Publ., c1999. 230 p.

Crowell, Charles D. **SHO-1 versus KING II - Victory at Leyte Gulf - Was it United States Luck or Japanese Mistakes?** Carlisle Barracks, PA: Army War College; Available from National Technical Information Service, Springfield, VA, 1989. 88 p.
Abstract: *The Battle for Leyte Gulf was the greatest naval battle of all time in terms of number of ships involved, losses of ships and aircraft and size of area over which the battle was fought. The American victory effectively marked the end of the Japanese Navy in World War Two. The battle was marked by furious surface, air and submarine action at sea and fierce fighting ashore on Leyte Island by US Army and Marine ground forces. While U.S. Navy dealt devastating losses to the Japanese fleet and claimed a resounding victory, the battle continues to be discussed for the significant operational, tactical and judgmental errors made by commanders of both sides. This study examines the errors made, the reasons for the errors and the effect the errors had toward deciding the outcome of this battle. It investigates the Japanese plan for the battle and the Japanese philosophy toward the war in 1944 and how these issues affected the outcome. It also considers the American chain of command in the Pacific theater and the problems caused by that unique setup. The paper discusses what we have learned, if anything from Leyte Gulf, and if in a similar situation would we make the same mistakes again. Finally the paper evaluates the composite effect of errors on both sides.*
NPS/DKL Location: MICROFORM ADA209582

DeRose, James F. and Roger W. Paine, Jr. **Unrestricted warfare: how a new breed of officers led the submarine force to victory in World War II**. New York: Wiley, c2000. 310 p.
NPS/DKL Location: GENERAL D783.5.W3 D39 2000

Dienesch, Robert M. **Submarine against the rising sun: the impact of radar on the American submarine war in 1943, the year of change**. [Fredericton]: University of New Brunswick. 1996. 227 l.
Thesis (M.A.), University of New Brunswick, Dept. of History 1996.

Dingman, Roger. **Ghost of war: the sinking of the Awa Maru and Japanese-American relations, 1945-1995**. Annapolis, MD: Naval Institute Press, c1997. 373 p.
NPS/DKL Location: GENERAL D777.5 .A92 D56 1997

Farago, Ladislas. **The Tenth Fleet**. New York, I. Obolensky, [1962]. 366 p.
NPS/DKL Location: GENERAL D783 .F22

_____. **The Tenth Fleet**. New York: Paperback Library, 1971, c1962. 366 p.

Felsen, Henry Gregor. **He's in submarines now**. New York: Robert M. McBride & Co., c1942. 175 p.

Fluckey, Eugene B. **Thunder below!: the USS Barb revolutionizes submarine warfare in World War II**. Urbana: University of Illinois Press, c1992. 444 p.
NPS/DKL Location: GENERAL D783.5 .B36 F58 1992

Frank, Gerold, James D. Horan and Joseph M. Eckberg **U.S.S. Seawolf, submarine raider of the Pacific**. New York, G. P. Putnam's sons, [1945]. 197 p.
NPS/DKL Location: GENERAL D783.5.S4 F7

Galantin, I. J. **Take her deep!: a submarine against Japan in World War II**. Chapel Hill, NC: Algonquin Books of Chapel Hill, 1987. 262 p.

Galvin, John R. and Frank Allnutt. **Salvation for a doomed zoomie: a true story**. Indian Hills, CO: Allnutt Pub., c1983. 272 p.

Gannon, Robert. **Hellions of the deep: the development of American torpedoes in World War II**. University Park, PA: Pennsylvania State University Press, 1996. 241 p.
NPS/DKL Location: GENERAL V850 .G36 1996

Germinsky, Robert A. **Submarines in World War II: the silent service**. [Washington, DC]: U.S. Navy Commemorative Committee, [1994?]. (WWII fact sheet). 1 sheet.
NPS/DKL Location: FEDDOCS D 201.39:SU 1

Grider, George and Lydel Sims. **War fish**. [1st ed.] Boston, Little, Brown, [1958]. 282 p.
NPS/DKL Location: GENERAL D783 .G8

Gugliotta, Bobette. **Pigboat 39: an American sub goes to war**. Lexington, KY: University Press of Kentucky, c1984. 224 p.

Hawkins, Maxwell. **Torpedoes away, sir! Our submarine navy in the Pacific**. New York, H. Holt & company, [1946]. 268 p.
NPS/DKL Location: BUCKLEY D783 .H3

Holmes, Harry. **The last patrol**. Shrewsbury, Eng.: Airlife Pub. Ltd., 1994. 212 p.
NPS/DKL Location: GENERAL D783 .H58 1994

Holmes, Wilfred Jay. **Double-edged secrets: U.S. naval intelligence operations in the Pacific during World War II**. Annapolis, MD: Naval Institute Press, c1979. 231 p.
NPS/DKL Location: GENERAL D810.S7 H74

_____. **Double-edged secrets: U.S. naval intelligence operations in the Pacific during World War II**. Annapolis, MD: Naval Institute Press, c1979 (1998 printing). (Bluejacket books). 231 p.

_____. **Undersea victory: the influence of submarine operations on the war in the Pacific**. Garden City, NY: Doubleday, 1966. 505 p.
NPS/DKL Location: GENERAL D783 .H7

Howarth, Stephen, ed. **Men of war: great naval leaders of World War II**. 1st U.S. ed. New York: St. Martin's Press, 1993. 602 p.
Originally published: London: Weidenfeld & Nicolson, 1992.
NPS/DKL Location: GENERAL CT32 .M461 1993

Hoyt, Edwin Palmer. **Bowfin**. Short Hills, NJ: Burford Books, [1998]. (Classics of war). 234 p.
Originally published: New York: Van Nostrand Reinhold, 1983.

_____. **The destroyer killer**. New York: Pocket Books, c1989. 222 p.
USS Harder.

_____. **Submarines at war: the history of the American silent service**. New York: Stein & Day, 1983. 329 p.
NPS/DKL Location: GENERAL V858 .H68 1983

_____. **Submarines at war: the history of the American silent service**. New York: Jove Books, 1992, c983. 316 p.

Hunnicutt, Thomas G. **The Operational Failure of U.S. Submarines at the Battle of Midway - and Implications for Today**. Newport, RI: Naval War College; Available from National Technical Information Service, Springfield, VA, 1996. 31 p.
Abstract: *U.S. submarine operational failure led to tactical insignificance at the Battle of Midway. This was a remarkable outcome since interwar U.S. policy, submarine design, and fleet exercises dictated fleet*

support by submarines. From today's view this failure is neither unique to a platform nor specific to an operation. It can and does cross all services. The operational failure at Midway resulted from the failure to abide by the operational art factors of synergy, simultaneity and depth, anticipation, and leverage. These were compounded by failure to provide adequate C31 system operational support. These failures were a consequence of the submarine force, and the Navy, not adequately addressing and training on operational art during the interwar years. Today, Navy doctrine and training still have not adequately addressed operational art though it is an essential part of joint warfare. The present use of exercises designed only to test and build tactical proficiency of air, land, or sea forces risk the same type of operational failure in future wars. Suggestions on developing operational art proficiency through innovation as a function of today's forces, budgets, and training technology are presented for consideration.
Electronic access: http://handle.dtic.mil/100.2/ADA311656

Jaffee, Walter W. **Steel shark in the Pacific: USS Pampanito SS-383**. 1st ed. Palo Alto: Glencannon Press, 2001. 208 p.

Kaufman, Yogi and Paul Stillwell. **Sharks of steel**. Annapolis, MD: Naval Institute Press, c1993. 107 p.
NPS/DKL Location: GENERAL V857 .K28 1993

Kiefer, Edward A., comp. **The Golden belly of the 202 U.S.S. Trout: a memorial edtion of the Second War Patrol during World War II**. [1st ed.]. [Fairlawn, Ohio]: [Kiefer Publications, c1993]. 1 v.

Kimble, David L. **Chronology of U.S. Navy submarine operations in the Pacific, 1939-42**. 2nd World War II monograph ed. Bennington, VT: Merriam Press, 1999. (World War II monograph series; no. 11). 60 p.
NPS/DKL Location: GENERAL D780 .K56 1999

Kimmett, Larry and Margaret Regis. **U.S. submarines in World War II: an illustrated history**. Seattle, WA: Navigator Pub., [c1996]. 159 p.
NPS/DKL Location: BUCKLEY D783 .K56 1996

LaVo, Carl. **Back from the deep: the strange story of the sister subs Squalus and Sculpin**. Annapolis, MD: Naval Institute Press, c1994. 226 p.
The Squalus is also indexed as Sailfish (Submarine).
NPS/DKL Location: GENERAL D783 .L38 1994

_____. **Slade Cutter, submarine warrior**. Annapolis, Md: Naval Institute Press, 2003. 265 p.
NPS/DKL Location: GENERAL V63.C88 L39 2003

Lanigan, Richard J. **Kangaroo express: the epic story of the submarine Growler: with recollections by "Skipper" Arnold Schade**. Laurel, FL: RJL Express Publications, c1998. 173 p.
NPS/DKL Location: GENERAL D782.G76 L36 1998

Lenton, H. T. **American submarines**. Garden City, NY, Doubleday [1973]. (Navies of the Second World War). 128 p.

Originally published: London, MacDonald, [1973].

Lockwood, Charles A. **Sink 'em all: submarine warfare in the Pacific**. 1st ed. New York: Dutton, 1951. 416 p.
NPS/DKL Location: BUCKLEY D783 .L6 1951

_____. **Topi ikh vsekh**. Perevod s angliiskogo Zheltova, K. M. Moskva, Voen. izd-vo, 1960. 398 p.
Russian translation of Sink 'em all.

_____. and Hans Christian Adamson. **Battles of the Philippine Sea**. New York, Crowell [1967]. 229 p.

_____ and _____. **Hell at 50 fathoms**. [1st ed.]. Philadelphia, Chilton Co., Book Division, [1962]. 299 p.
NPS/DKL Location: GENERAL V858 .L8

_____ and _____. **Hellcats of the sea**. New York, Chilton, [1955]. 335 p.

_____. and _____. **Hellcats of the sea**. New York: Greenberg, 1955. 335 p.
NPS/DKL Location: BUCKLEY D780 .L8

_____ and _____. **Through hell and deep water; the stirring story of the Navy's deadly submarine, the U. S. S. Harder, under the command of Sam Dealey, destroyer killer**. New York, Greenberg, [1956]. 317 p.
NPS/DKL Location: BUCKLEY D783.5.H3 L8

_____ and _____. **Zoomies, subs, and zeros**. New York: Greenberg, [1956]. 301 p.
NPS/DKL Location: BUCKLEY D769.45 .L8

Lowder, Hughston E. **The silent service: U.S. submarines in World War II**. 1st ed. Baltimore, MD: Silent Service Books, c1987. 492 p.

_____. and Jack Scott. **Batfish, the champion "submarine-killer" submarine of World War II**. Englewood Cliffs, NJ: Prentice-Hall, c1980. 232 p.

Matsunaga, Ichiro, Gordon J. Van Wylen and Kan Sugahara. **Encounter at sea and a heroic lifeboat journey**. Troy, MI: Sabre Press, c1994. 232 p.

McCants, William R. **War patrols of the USS Flasher: the true story of one of America's greatest submarines, officially credited with sinking the most Japanese shipping in World War II**. Chapel Hill, NC: Professional Press, c1994. 465 p.

Meigs, Montgomery C. **Slide rules and submarines: American scientists and subsurface warfare in World War II**. Washington, DC: National Defense University Press: Supt. of Docs., U.S. G.P.O. [distributor], 1990. 269 p.
NPS/DKL Location: GENERAL D783 .M45 1990

Mellor, William Bancroft. **Sank same**. New York: Howell, Soskin, c1944. 224 p.
NPS/DKL Location: BUCKLEY D790 .M5

Mendenhall, Corwin. **Submarine diary: The silent stalking of Japan**. Chapel Hill, NC: Algonquin Books of Chapel Hill, 1991. 290 p.

Michno, Gregory. **USS Pampanito: killer-angel**. Norman: University of Oklahoma Press, c2000. 445 p.
NPS/DKL Location: GENERAL D783.5.P4 M53 2000

Miller, Vernon J. **Analysis of Japanese submarine losses to Allied submarines**. Bennington, VT, U.S.A. (218 Beech St., Bennington 05201): Weapons & warfare Press, c1984. (Weapons and warfare special; no. SP 29). 32 p.

Milton, Keith M. **Subs against the rising sun**. Las Cruces, NM: Yucca Tree Press, c2000. 369 p.
NPS/DKL Location: GENERAL D783 .M55 2000

Morison, Samuel Eliot. **Coral Sea, Midway and submarine actions, May 1942-August 1942**. Little, Brown, 1947. 611 p.
NPS/DKL Location: RESERVE D773 .M8 V.4

Nolte, Carl. **USS Pampanito: a submarine and her crew**. 1st ed. San Francisco, CA: San Francisco Maritime National Park Association, 2001. 68 p.

O'Kane, Richard H. **Clear the bridge!: the war patrols of the U.S.S. Tang**. Chicago: Rand McNally, c1977. 480 p.
NPS/DKL Location: GENERAL D783.5 .T35 O43

_____. **Clear the bridge!: the war patrols of the U.S.S. Tang**. Novato, CA: Presidio Press, 1989, c1977. 480 p.
NPS/DKL Location: GENERAL CT15 .K36 1989

_____. **Wahoo: the patrols of America's most famous World War II submarine**. Novato, CA: Presidio Press, c1987. 345 p.

Polmar, Norman and Thomas B. Allen. **World War II: America at war, 1941-1945**. New York: Random House, c1991. 940 p.
NPS/DKL Location: GENERAL D743.5 .P56 1991

Roscoe, Theodore. **United States submarine operations in World War II**. Annapolis: United States Naval Institute, [1949]. 577 p.
NPS/DKL Location: BUCKLEY D783 .R7

Ruhe, William J. **Slow dance to Pearl Harbor: a tin can ensign in prewar America**. Washington: Brassey's, c1995. 210 p.
NPS/DKL Location: GENERAL D783 .R83 1995

_____. **War in the boats: my World War II submarine battles**. Washington [DC]: Brassey's, c1994. 303 p.
NPS/DKL Location: GENERAL CT18.U17 R83 1994

Russell, Dale. **Hell above, deep water below**. Tillamook, OR: Bayocean Enterpises, c1995. 212 p.

Sasgen, Peter T. **Red scorpion: the war patrols of the USS Rasher**. Annapolis, MD: Naval Institute Press, c1995. 366 p.
NPS/DKL Location: GENERAL D783.5.R37 S27 1995

Schratz, Paul R. **Submarine commander: a story of World War II and Korea**. Lexington, KY: University Press of Kentucky, c1988. 322 p.
NPS/DKL Location: GENERAL V63.S38 A3 1988

Sheridan, Martin. **Overdue and presumed lost: the story of the U.S.S. Bullhead**. Francestown, NH: M. Jones Co., [1947]. 143 p.
NPS/DKL Location: BUCKLEY D783.5.B8 S5

Silverstone, Paul H. **U.S. warships of World War II**. Annapolis, MD: Naval Institute Press, 1989, c1965. 444 p.
Originally published: London, Ian Allen; Garden City, NY, Doubleday, 1965.

_____. **US warships since 1945**. Annapolis, MD: Naval Institute Press, 1987. 240 p.
NPS/DKL Location: REFERENCE VA61 .S54 1987

Smith, Ron. **Torpedoman**. 2nd rev. ed. Nicholasville, KY: Appaloosa Press, 1998. 95 p.

Smith, Steven Trent. **The rescue: a true story of courage and survival in World War II**. New York: J. Wiley, 2001. 326 p.
NPS/DKL Location: GENERAL D783.5.C74 S63 2001

_____. **Wolf pack: the American submarine strategy that helped defeat Japan**. Hoboken, NJ: John Wiley, c2003. 312 p.
NPS/DKL Location: GENERAL D783 .S65 2003

Sterling, Forest J. **Wake of the Wahoo: the heroic story of America's most daring WWII submarine, USS Wahoo.** 4th ed. Placentia, CA: R. A. Cline Pub., 1999. 221 p. Originally published: Philadelphia: Chilton Co., Book Division, [1960].

Thompson, Lawrance Roger. **The Navy hunts the CGR 3070.** Garden City, NY: Doubleday, Doran & co., inc., 1944. 150 p.
Published in condensed form in Harper's magazine under the title: The strange cruise of the yawl Zaida.'"
NPS/DKL Location: BUCKLEY D783.5.C6 T4

Trumbull, Robert. **Silversides.** New York, H. Holt & company, [1945]. 217 p.
NPS/DKL Location: GENERAL D783.5.S5 T8

Tuohy, William. **The bravest man: the story of Richard O'Kane & U.S. submariners in the Pacific war.** Stroud: Sutton, 2001. 422 p.
NPS/DKL Location: GENERAL D783.O43 T86 2001

United States. Navy. Pacific Fleet. **U.S. submarine losses: World War II** / [prepared by the Commander Submarine Forces, U.S. Pacific Fleet. Fourteenth Naval District. Publications & Printing Office, 1946]. [174] p.
NPS/DKL Location: GENERAL D783 .U6

United States. Navy. Pacific Fleet. **U. S. submarine losses, World War II.** Navpers 15,784. 1949 issue. [Reprint]. [Washington, U. S. GoVT Print. Off., 1949, 1946]. 174 p.
NPS/DKL Location: BUCKLEY D783 .U6 1949

United States. Navy. Pacific Fleet. **United States submarine losses: World War II** / reissued with an appendix of Axis submarine losses, fully indexed by Naval History Division. [Rev. & augmented ed.]. Office of the Chief of Naval Operations, 1963. (NavPers 15784; OpNav-P33-100). 244 p.
NPS/DKL Location: GENERAL D783 .U62

United States Submarine Veterans of World War II. **United States Submarine Veterans of World War II: a history of the veterans of the United States Naval Submarine Fleet.** Dallas, TX: Taylor Pub. Co., <1987- >. v.

Walkowiak, Thomas F. **Fleet submarines of World War Two.** Missoula, MT: Pictorial Histories Publishing Co., 1988. 47 p.
History of the Gato, Balco and Trench class submarines

_____. **Fleet submarines of World War Two: the floating drydock.** Missoula, MT: Pictorial Histories Pub. Co., 1988. 47 p.

Wheeler, Keith and the editors of Time-Life Books. **War under the Pacific.** Alexandria, VA: Time-Life Books; Morristown, NJ: school & library distribution by Silver Burdett, c1980. 208 p.

NPS/DKL Location: GENERAL D767 .W47

Zim, Herbert Spencer. **Submarines, the story of undersea boats**. New York, Harcourt, Brace [c1942]. 306 p.
NPS/DKL Location: BUCKLEY VM365 .Z7

SUBMARINES -- AUSTRALIAN -- GENERAL

Ball, Desmond. **The new submarine combat information system and Australia's emerging information warfare architecture**. Canberra, ACT: Strategic and Defence Studies Centre, Australian National University, 2001. (Working paper (Australian National University. Strategic and Defence Studies Centre); no. 359). 23 p.

Garrisson, A. D. **The Australian submarine project: an introduction to some general issues**. Canberra: Strategic and Defence Studies Centre, Australian National University, 1987. (Working paper / Strategic and Defence Studies Centre, Research School of Pacific Studies, Australian National University; no. 142). 1v.

Gillett, Ross. **Australian & New Zealand warships, 1914-1945**. Sydney: Doubleday, 1983. 360 p.

_____. **Australian & New Zealand warships since 1946**. 1st ed. Brookvale, NSW, Australia: Child & Associates, 1988. 192 p.

_____. **Warships of Australia**. Adelaide: Rigby, 1977. 296 p.

McIntosh, Malcolm Kenneth and John B. Prescott. **Report to the Minister of Defence on the Collins Class Submarine and related matters**. [Canberra, A.C.T.]: Commonwealth of Australia, c1999. 1 v.

Topmill Pty. Ltd. **Australian Seapower: Submarines**. Marrickville, NSW: Topmill, [1966?]. 96 p.

White, Michael W. D. **Australian submarines: a history**. Canberra: AGPS Press, 1992. 284 p.

SUBMARINES -- AUSTRALIAN – WWI

Australia. Navy Office. Directorate of Public Information. **The discovery of AE2**. [Canberra]: [Directorate of Public Information, Navy], [1998].
Includes information and photographs about the discovery and documentation of the wreck of the Australian submarine AE2 scuttled off the Gallipoli Peninsula, Turkey, during World War 1. The site includes underwater photography by research team leader Dr Mark Spencer.
http://www.navy.gov.au/history/ae2/default.htm

Brenchley, Fred & Elizabeth Brenchley. **Stoker's Submarine [AE2]**. Pymble NSW: HarperCollins Australia, 2001. 304 p.

Çelik, Kenan and Ceyhan Koç, eds. **The Gallipoli Campaign International Perspectives 85 Years On, 24-25 April 2000, Çanakkale; conference papers**. Çanakkale: Çanakkale Onsekiz Mart Üniversity, 2001. 92 p.

Frame, Thomas R. **The shores of Gallipoli: naval dimensions of the Anzac Campaign**. Sydney: Hale & Iremonger, 2000. 256 p.

_____ and G.J. Swinden. **First in, last out, the Navy at Gallipoli**. Kenthurst [NSW.]: Kangaroo Press, 1990. 208 p.

Smyth, Dacre. **The submarine AE2 in World War I**. [Canberra]: A.C.T. Chapter, Naval Historical Society of Australia, 1974. 9 leaves.
Transcript of a lecture presented to the A.C.T. Chapter of the Naval Historical Society of Australia, Southern Cross Club, Woden, A.C.T. 30 April, 1974.

Stroker, Henry. **Straws in the wind**. London: Herbert Jenkins, 1925. 315 p.

Wester-Wemyss, Rosslyn Erskine Wemyss. **The navy in the Dardanelles campaign**. London, Hodder & Stoughton limited, [1924]. 288 p.

SUBMARINES -- AUSTRIAN -- GENERAL

Mayer, Horst Friedrich and Dieter Winkler. **Als die Schiffe tauchen lernten: die Geschichte der k.u.k. Unterseeboot-Waffe**. 2. Aufl. Wien: Verl. Österreich., Österr. Staatsdr., 1998. 188 p.

Trapp, Georg von. **Bis zum letzten Flaggenschuss; Erinnerungen eines osterreichischen U-Boots-Kommandanten**. Salzburg, Leipzig, A. Pustet, [c1935]. 247 p.

SUBMARINES -- BRAZILIAN -- GENERAL

Netto, Joao Palma. **CS-4 caca submarinos Gurupa: memorias de um marinheiro**. Salvador/BA: Jubiaba, c1984. 354 p.

SUBMARINES -- BRITISH -- GENERAL

Akermann, Paul. **Encyclopaedia of British submarines, 1901-1955**. [Liskeard, Eng.]: Maritime Books, c1989. 522 p.

Brown, David K., ed. **Nelson to Vanguard: warship development, 1923-1945**.
Annapolis, MD: Naval Institute Press, c2000. 224 p.
NPS/DKL Location: GENERAL V767 .B83 2000

Burgoyne, Alan Hughes. **Submarine navigation past and present**. London: Alexander Moring, Ltd., 1903. 2 v.
NPS/DKL Location: GENERAL VM365 .B9 V.1

Cocker, Maurice. **Observer's directory of Royal Naval submarines, 1901-1982**.
London: F. Warne; Annapolis, MD: Published and distributed in the U.S. by the U.S. Naval Institute, 1982. 128 p.
NPS/DKL Location: REFERENCE V859.G7 C63 1982

Cook, Graeme. **Silent marauders**. London: Hart-Davis, MacGibbon, 1976. 159 p.

Edmonds, Martin, ed. **100 years of "the trade": Royal Navy submarines, past, present & future**. Lancaster: CDISS, 2001. ([Bailrigg study; 6]). 262 p.
Based on the University of Lancaster Centre for Defence and International Security Studies conference entitled "A Centennial Celebration of the Royal Navy's Submarine Service" held at the Lancaster University Conference Centre, Sept. 27-29, 2000.

Evans, A. S. **Beneath the waves: a history of HM Submarine losses**. London: William Kimber, 1986. 439 p.

Gray, Edwyn. **Captains of war**. London: Leo Cooper, 1988. 275 p.

Hackforth-Jones, Gilbert. **The true book about submarines**. London: F. Muller, c1955. 141 p.

Jameson, William Scarlett, Sir. **Submariners [V.C.]**. London, P. Davies, [1962]. 207 p.

Kemp, Peter Kemp. **H.M. Submarines**. London: H. Jenkins, [1953]. 224 p.
NPS/DKL Location: BUCKLEY V210 .K3

Lenton, H. T. **Warships of the British & Commonwealth navies**. Shepperton (Sy.), Allan [1967]. 272 p.

_____. **Warships of the British & Commonwealth navies**. 2nd ed. Shepperton, Allan, 1969. 287 p.

_____. **Warships of the British & Commonwealth navies**. 3rd ed. Shepperton, Allan, 1971. 303 p.

Lipscomb, Frank Woodgate. **The British submarine**. London, A. & C. Black, [1954]. 269 p.

_____. **The British submarine**. 2nd, revised and expanded ed. Greenwich: Conway Maritime Press, 1975. 284 p.

Parker, John. **The silent service: the inside story of the Royal Navy's submarine heroes**. London: Headline, 2001. 294 p.

Preston, Antony. **The Royal Navy submarine service: a centennial history**. London: Conway Maritime, 2001. 192 p.

Ring, Jim. **We come unseen: the story of Britain's Cold War submariners**. London: John Murray, 2001. 270 p.

Sueter, Murray Fraser. **The Evolution of the submarine boat, mine and torpedo from the sixteenth century to the present time**. Portsmouth, J. Griffin, 1907. 384 p.

Tall, J. J. and Paul Kemp. **HM submarines in camera, 1901-1996**. Stroud, Gloucestershire: Sutton, 1996. 245 p.
NPS/DKL Location: GENERAL V859.G7 T35 1996

Winton, John. **The submariners: life in British submarines, 1901-1999; an anthology of personal experience**. London: Constable, 1999. 248 p.

SUBMARINES -- BRITISH -- WWI

Bennett, Mark H. J. **Under the periscope**. London, W. Collins sons & co. ltd., [1919]. 254 p.

Bower, John Graham. **The story of our submarines**. Edinburgh: William Blackwood & Sons, 1919. 297 p.
By "Klaxon."

Brodie, Charles Gordon. **Forlorn hope 1915: the submarine passage of the Dardanelles**. London: F. Books, 1956. 90 p.

Campbell, Gordon. **My mystery ships**. Garden City, NY, Doubleday, Doran & Company, inc., 1929. 318 p.
Originally published: London: Hodder & Stoughton, limited, [1928].

Carr, William Guy. **By guess and by God; the story of the British submarines in the war**. Garden City, NY, Doubleday, Doran & Company, Inc., 1930. 310 p.
Originally published: London: Hutchinson, [1930].
NPS/DKL Location: BUCKLEY D593 .C2

_____. **Hell's angels of the deep**. 5th impression. London, Hutchinson & co. ltd, [1932]. 288 p.

Carson, Edward. **The War on German Submarines**. London: Unwin, 1917. 8 p.

Chatterton, E. Keble. **Amazing adventure; a thrilling naval biography**. London, Hurst & Blackett, ltd., [1935]. 285 p.
"Narrating the naval career of Commander Godfrey Herbert."

_____. **Beating the U-boats**. London; New York: Hurst & Blackett, [1943]. 172 p.
Sequel to author's work: Fighting the U-boats.
NPS/DKL Location: BUCKLEY D593 .C38

_____. **Danger zone: the story of the Queenstown command**. Boston, Little, Brown, & company, 1934. 437 p.
Originally published: London: Rich & Cowan, ltd., 1934.
NPS/DKL Location: BUCKLEY D580 .C4

_____. **Fighting the U boats**. London: Hurst & Blackett, ltd., [1942]. 216 p.
NPS/DKL Location: BUCKLEY D593 .C4

_____. **Gallant gentlemen**. London: Hurst & Blackett, ltd., [1931]. 296 p.
NPS/DKL Location: BUCKLEY D581 .C4

Compton-Hall, Richard. **Submarines and the war at sea, 1914-18**. London: Macmillan, 1991. 345 p.

Currey, Edward Hamilton. **How we kept the sea**. London, New York [etc.]: T. Nelson & sons, [1917?]. 180 p.

DeLany, Walter S. **Bayly's Navy**. Washington, DC: Naval Historical Foundation, c1980. (Naval Historical Foundation publication; ser. 2, no. 25). 45 p.
NPS/DKL Location: GENERAL D594 .D44

Dommett, William Erskine. **Submarine vessels: including mines, torpedoes, guns, steering, propelling, and navigating apparatus, and with notes on submarine offensive and defensive tactics, and exploits in the present war**. London; New York: Whittaker & Co., 1915. 106 p.

Domville-Fife, Charles W. **Submarine warfare of to-day: how the submarine menace was met and vanquished, with descriptions of the inventions and devices used, fast boats, mystery ships, nets, aircraft, &c., &c., also describing the selection and training of the enormous personnel used in this new branch of the navy**.
Philadelphia: J.B. Lippincott co.; London: Seeley, Service & Co., 1920. 303 p.

Dorling, Taprell. **The secret submarine: A story of fighting by sea and land**. London: Blackie, 1916. 352 p.

Edwards, Kenneth. **We dive at dawn**. London: Rich & Cowan, ltd., [1939]. 388 p.
NPS/DKL Location: BUCKLEY V859.G7 E2

_____. **We dive at dawn**. Chicago, The Reilly & Lee co. [c1941]. 412 p.

Everitt, Don. **The K boats; a dramatic first report on the navy's most calamitous submarines**. New York, Holt, Rinehart & Winston, [1963]. 206 p.
Originally published: London, Harrap, [1963].

_____. **K boats: steam-powered submarines in World War** I. Annapolis, MD: Naval Institute Press, 1999. 144 p.
NPS/DKL Location: GENERAL V859.G7 E83 1999

Fayle, Charles Ernest. **Seaborne trade**. London, J. Murray, 1920-24. 3 v.
Contents: I. The cruiser period.--II. From the opening of the submarine campaign to the appointment of the shipping controller.--III. The period of unrestricted submarine warfare.
NPS/DKL Location: BUCKLEY D581 .F3 1920

Gibson, Richard Henry and Maurice Prendergast. **The German submarine war, 1914-1918**. New York: R.R. Smith inc., [1931]. 438 p.
Originally published: London: Constable, 1931.
NPS/DKL Location: GENERAL D591 .G4

_____. **The German submarine war, 1914-1918**. Annapolis, MD: Naval Institute Press, 2003. 438 p.

_____. **Germanskaia podvodnaia voina 1914-1918 gg. 1938**. 431 p.
Russian translation of The German submarine war, 1914-1918.

Gray, Edwyn. **A damned un-English weapon: the story of British submarine warfare, 1914-18**. London, Seeley, 1971. 259 p.

_____. **The underwater war; submarines, 1914-1918**. New York, Scribner [1972, c1971]. 259 p.
Originally published: London, Seeley, 1971 as A damned un-English weapon: the story of British submarine warfare, 1914-1918.

Higgins, A. Pearce. **Defensively-armed merchant ships and submarine warfare**. London, Stevens, 1917. 56 p.

Hurd, Archibald Spicer, Sir. **Submarines and zeppelins in warfare and outrage**. London, Sir J. Causton & sons, limited, 1916. 22 p.

Jellicoe, John Rushworth Jellicoe, Earl. **The crisis of the naval war**. London; New York: Cassell & company, ltd., 1920. 331 p.
NPS/DKL Location: BUCKLEY D581 .J4

_____. **The crisis of the naval war**. New York: George H. Doran company, [1921]. 331 p.
NPS/DKL Location: GENERAL D581 .J4

_____. **The submarine peril: the Admiralty policy in 1917**. Cassell, [1934]. 240 p.
NPS/DKL Location: BUCKLEY D593 .J4

_____. **Der U-Boot-Krieg, Englands schweste Stunde**. Berlin, Vorhut-Verlag C. Schlegel [19-]. 229 p.
German translation of The submarine peril.

Lambert, Nicholas A., ed. **The submarine service, 1900-1918**. Aldershot, Hants, England; Burlington, VT: Published by Ashgate for the Navy Records Society, 2001. (Publications of the Navy Records Society; vol. 142). 397 p.

Masters, David. **"I.D." New tales of the submarine war**. London, Eyre & Spottiswoode, 1935. 296 p.

_____. **The submarine war**. New York: Holt, [1935]. 296 p.
NPS/DKL Location: BUCKLEY D593 .M3

Newbolt, Henry John, Sir. **Submarine and anti-submarine**. London, New York [etc.] Longmans, Green & Co., 1918. 312 p.

Shankland, Peter and Anthony Hunter. **Dardanelles patrol**. New York: Scribner, [1964]. 191 p.
Originally published: London, Collins, 1964.
NPS/DKL Location: GENERAL D568.3 .S5

_____. **Dardanelles patrol**. London & Glasgow: Collins, 1965. 158 p.

_____. **Dardanelles patrol**. London: Mayflower Books, 1971. 159 p.

_____. **Dardanelles patrol**. London: Mayflower, 1978. 159 p.

_____. **Dardanelles patrol**. London: Granada, 1983. 159 p.

Smith, Gaddis. **Britain's clandestine submarines, 1914-1915**. New Haven, Yale University Press, 1964. (Yale historical publications. Wallace Notestein essays, 4). 155 p.
NPS/DKL Location: GENERAL D619 .S6

_____. **Britain's clandestine submarines, 1914-1915**. Hamden, CT: Archon Books, 1975, c1964. 155 p.
Originally published: New Haven: Yale University Press, 1964.

Wilson, Herbert Wrigley. **Hush; or The hydrophone service**. London, Mills & Boon, limited, [1920]. 188 p.

Wilson, Michael. **Baltic assignment: British submariners in Russia, 1914-1919**. London: L. Cooper in association with Secker & Warburg, 1985. 243 p.

_____. **Destination Dardenelles**. London: Cooper, 1988. 193 p.

SUBMARINES -- BRITISH -- WWII

Acworth, Bernard. **Life in a submarine**. [London], R. Tuck & Sons Ltd., [1941]. 46 p.

Banks, Arthur. **Wings of dawning: the battle for the Indian Ocean, 1939-1945**. Malvern Wells: Images Publishing, 1996. 416 p.

Baxter, Richard. **Stand by to surface**. Melbourne; London: Cassell & Co., 1944. 208 p.

Brown, David K., ed. **The design and construction of British warships, 1939-1945: the official record**. London: Conway Maritime Press, 1995-1996. 3 v.
Contents: [1]. Major surface vessels -- [2]. Submarines, escorts, and coastal forces -- [3]. Landing craft and auxiliary vessels.
NPS/DKL Location: GENERAL VA454 .D45 1995

Bruce, Henry J. **Twenty years under the sea**. S. Paul, [1939]. 228 p.
NPS/DKL Location: BUCKLEY VM981 .B8

Bryant, Benjamin. **Submarine command**. London: W. Kimber, 1975, c1958. 238 p.
Originally published: London, W. Kimber, [1958] as One man band.
NPS/DKL Location: GENERAL D784.G7 B94

Chalmers, William Scott. **Max Horton and the western approaches; a biography of Admiral Sir Max Kennedy Horton**. [London] Hodder & Stoughton, [1954]. 301 p.
NPS/DKL Location: BUCKLEY D780 .C4

Compton-Hall, Richard. **The underwater war, 1939-1945**. Poole, Dorset: Blandford Press; New York: Distributed by Sterling Pub., 1982. 160 p.
NPS/DKL Location: GENERAL D780 .C65 1982

Coote, John O. **Submariner**. 1st American ed. New York: Norton, 1992. 239 p.

Dickison, Arthur. **Crash dive: in action with HMS Safari, 1942-43**. Stroud, Gloucestershire: Sutton in association with the Royal Navy Submarine Museum, 1999. 212 p.
NPS/DKL Location: GENERAL D784.G7 D52 1999

Drummond, John Dorman. **H. M. U-boat**. London: W. H. Allen, 1958. 227 p.
Captured U-570 commissioned into the Royal Navy as HMS Graph.

_____. **H. M. U-boat**. London: Quality Book Club, 1958. 228 p.

Eldridge, A. W. C. **Just out of sight**. London: Minerva Press, 1998. 320 p.

Gibson, John Frederic. **Dark seas above**. Edinburgh, W. Blackwood, 1947. 286 p.
Appeared, in part, in Blackwood's magazine, "From a submariner's notebook."

Hart, Sydney. **Discharged dead; a true story of Britain's submarines at war**. London, Odhams Press, [1956]. 208 p.
NPS/DKL Location: GENERAL D784.G7 H2

_____. **Submarine Upholder**. London: Oldbourne, [1960]. 207 p.

Hezlet, Arthur Richard, Sir. **HMS Trenchant: from Chatham to the Banka Strait**. London: Leo Cooper, 2000. 184 p.

Jackson, Robert. **The Royal Navy in World War II**. Annapolis, MD: Naval Institute Press, c1997. 176 p.
NPS/DKL Location: GENERAL D771 .J24 1997

Jewell, Norman Limbury Auchinleck. **Secret mission submarine**. Chicago: Ziff-Davis Pub. Co., [1945]. 159 p.
NPS/DKL Location: GENERAL D784.G7 J5

Keeble, Peter (L. A. J.). **Ordeal by water**. London; New York: Longmans, Green, [1957]. 216 p.
NPS/DKL Location: BUCKLEY D770 .K2

_____. **Ordeal by water**. London: Pan Books, 1959. 205 p.

Kemp, Paul. **British submarines in World War Two**. Poole, Dorset: Arms & Armour Press; [New York, NY: Distributed in the USA by Sterling Pub. Co., c1987]. (Warships illustrated; no. 11). 64 p.

_____. **The T-class submarine: the classic British design**. London: Arms & Armour. 1990. 160 p.

King, William Donald. **Dive and attack: a submariner's story**. Rev. ed. London: W. Kimber, 1983. 236 p.
Revised edition of Stick and the stars.

_____. **The stick and the stars**. London, Hutchinson [1958]. 192 p.

Lambert, John and David Hill. **The submarine Alliance**. Annapolis, MD: Naval Institute Press, 1986. (Anatomy of the ship). 120 p.
NPS/DKL Location: GENERAL VA458.A48 L36 1986

Leasor, James. **Boarding party**. 1st American ed. Boston: Houghton Mifflin, 1979, c1978. 203 p.
Originally published: London: Heinemann, 1978

Lenton, H. T. **British submarines**. Garden City, NY, Doubleday, [1972]. (Navies of the Second World War). 160 p.
Originally published: London: MacDonald, [1972].

Low, A. M. **The submarine at war**. New York, Sheridan House [1942]. 305 p.
Originally published: London, Melbourne, Hutchinson & co., ltd. 1941.
NPS/DKL Location: BUCKLEY V210 .L9

Macintyre, Donald G. F. W. and Robert B. Carney. **U-boat killer**. London, Weidenfeld & Nicolson, [1956]. 179 p.
NPS/DKL Location: BUCKLEY D780 .M2

_____ and _____. **U-boat killer**. New York, Norton [1957, c1956]. 239 p.

_____ and _____ **U-Boat killer**. Annapolis, MD: Naval Institute Press, c1976. 175 p.

_____ and _____. **U-Boat killer: fighting the U-Boats in the Battle of the Atlantic**. London: Cassell, 1999. (Cassell military paperbacks). 179 p.

Mackenzie, Hugh. **The sword of Damocles: some memories of Vice Admiral Sir Hugh Mackenzie, KCB, DSO +, DSC**. [Gosport]: Royal Navy Submarine Museum, c1995. 276 p.

Mars, Alastair. **British submarines at war, 1939-1945**. [Annapolis? MD], Naval Institute Press, [c1971]. 256 p.

_____. **British submarines at war, 1939-1945**. London, Kimber, 1971. 256 p.

_____. **HMS Thule intercepts**. London: Elek Books, 1956. 241 p.
Sequel to Unbroken, the story of a submarine.

_____. **Submarines at war, 1939-1945**. [London]: Corgi Books, [1974]. 272 p.
Originally published: London, Kimber, 1971, as British submarines at war, 1939-1945.

_____. **Unbroken, the story of a submarine**. London, F. Muller, [1953]. 223 p.
NPS/DKL Location: GENERAL D784.G7 M2

Masters, David. **Up periscope**. New York, Dial press, [1943]. 275 p.
Originally published: London, Eyre & Spottiswoode, 1942.
NPS/DKL Location: BUCKLEY D780 .M3

Nethercoate-Bryant, Keith T. et al, ed. **Submarine memories: our time in boats; some of the lesser known facts from the Gatwick Submarine Archive**. Upper Beeding, W. Sussex: Gatwick Submarine Archives, [1994?]. 141 p.

Oram, Harry Percy Kendall. **Ready for sea**. London, Seeley, 1974. 250 p.

Ostric, Zvonko and Stjepan Vekaric. **Agonija na dnu mora**. Beograd, Rad, 1958. (Iza kulisa). 320 p.

Roskill, Stephen Wentworth. **Das Geheimnis um U 110** [Übers. aus d. Engl.: Dietrich Niebuhr]. Frankfurt a.M.: Verl. f. Wehrwesen Bernard & Graefe, 1960. 147 p.
German translation of The secret Capture.

_____. **The secret capture**. London: Collins, 1959. 156 p.
Capture of the U-110, recommissioned as HMS Graph.

Simpson, George Walter Gillow. **Periscope view: a professional autobiography**. London, Macmillan, 1972. 315 p.

Soltikow, Michael Alexander, Graf. SOS--**"Thetis," 36 stunden zwischen börse und tod**. Dresden, F. Müller, 1944. 167 p.

_____. **S. O. S. Thétis! Trente-six heures entre la bourse et la mort**. Paris, Aux Armes de France, 1943. 116 p.

Tayenthal, Wilhelm. **Britisches Unterseeboot "Thetis": Stationen einer Katastrophe in Friedenszeiten**. Hamburg: Maximilian-Verl. Schober, [1998]. (Schiffe, Menschen, Schicksale; Nr. 59). 46 p.

Treadwell, Terry C. **Strike from beneath the sea. a history of aircraft-carrying submarines**. [New ed.]. Brimscombe Port. Tempus, 1999. 191 p.
Originally published: London: Conway Maritime Press, 1985, as Submarines with wings: the past, present and future of aircraft-carrying submarines

Trenowden, Ian. **The hunting submarine: the fighting life of HMS Tally-Ho**. [Bristol]: Crecy, 1994. 224 p.

Turner, John Frayn. **Periscope patrol: the saga of Malta submarines**. London: G. G. Harrap, 1957. 218 p.

Warren, Charles Esme Thornton and James Benson. **"The Admiralty regrets ..." the story of His Majesty's submarine Thetis and Thunderbolt.** London, G.G. Harrap & Co., [1958]. 223 p.
NPS/DKL Location: BUCKLEY VA458.T46 W37 1958

_____ and _____. **Only four escaped: the sinking of the submarine Thetis.** William Sloane Associates, 1959. 219 p.
Originally published: London: London, G.G. Harrap & Co., 1958 as The Admiralty regrets.
NPS/DKL Location: GENERAL V859.G7 W2

_____ and _____. **Will not we fear: the story of His Majesty's submarine "Seal" and of Lieutenant-Commander Rupert Lonsdale.** W. Sloane Associates, [1961]. 228 p.
NPS/DKL Location: GENERAL D784.G7 W2

Wemyss, David Edward Gillespie. **Walker's Groups in the Western Approaches.** Liverpool: Liverpool Daily Post & Echo, 1948. 172 p.

Wingate, John. **The fighting Tenth: the Tenth Submarine Flotilla and the siege of Malta.** London: L. Cooper, 1991. 384 p.

Young, Edward Preston. **One of our submarines.** London, R. Hart-Davis, 1952. 316 p.
NPS/DKL Location: GENERAL D784.G7 Y7

_____. **One of our submarines.** Revised ed. London, Pan, 1968. 272 p.

_____. **One of our submarines.** Ware, Hertfordshire: Wordsworth Editions, 1997. 316 p.
Originally published: London: R. Hart-Davis, 1952.

_____. **Undersea patrol.** New York, McGraw-Hill [1953, c1952]. 298 p.
Originally published: London, R. Hart-Davis, 1952 as One of our submarines.

SUBMARINES – CANADIAN -- GENERAL

Ferguson, Julie H. **Through a Canadian periscope: the story of the Canadian Submarine Service.** Toronto; Niagara Falls, NY: Dundurn Press, c1995. 364 p.
NPS/DKL Location: GENERAL V859.C3 F47 1995

Perkins, J. David. **The Canadian submarine: service in review.** St. Catharines, Ont.: Vanwell Pub., c2000. 208 p.

_____. **Canadian wartime submariners.** Boutiliers Point, N.S.: Seaboot Productions, 1994. 40 p.

_____. **Submarine sailor: the First World War adventures of a Canadian submarine captain**. Boutiliers Point, N.S.: Seaboot Productions, c1994. 44 p.

SUBMARINES -- CHINESE -- GENERAL

Lewis, John Wilson and Xue Litai. **China's strategic seapower: the politics of force modernization in the nuclear age**. Stanford, CA: Stanford University Press, c1994. 393 p.
NPS/DKL Location: GENERAL VA633 .L48 1994

SUBMARINES -- DANISH -- GENERAL

Berg, Hans Chr, Dahlerup Koch, P.B. Nielsen. **U-både gennem 75 år: det danske ubådsvåben 1909-84**. København: Forum, 1984. (Marinehistorisk selskabs skrift; 20). 153 p.

Steensen, Robert Steen. **Vore undervandsbsde gennem 50 sr, 1909-1959**. [K(benhavn] Marinehistorisk selskab; i kommission hos Munksgaard, 1960. 309 p.

SUBMARINES -- DUTCH -- GENERAL

Edwards, Hugh. **Islands of angry ghosts**. London, Hodder & Stoughton, [1966]. 208 p.

_____. **Islands of angry ghosts**. Sydney, etc.: Angus & Robertson, 1969. 207 p.

Froma, Tonny. **Ja, het moest: de personele verliezen van de Onderzeedienst tijdens de Tweede Wereldoorlog**. Bergen: Bonneville, c1997. 263 p.

SUBMARINES -- FRENCH -- GENERAL

Antier, Jean Jacques. **Histoire mondiale du sous-marin**. Paris, R. Laffont, 1968. 391 p.

_____. **L'aventure heroique des sous-marins francais: 1939-1945**. [Paris]: Editions maritimes & d'outre-mer, c1984. (Collection "Embruns de l'histoire"). 347 p.

_____. **La flotte se saborde, Toulon 1942**. Paris: Presses de la Cité, c1992. 368 p.:

_____. **Les sous-mariniers**. Paris: J. Grancher, c1976. (C'Θtaient des hommes). 252 p.

_____. **Les sous-mariniers**. Nouv. Ɵd. rev. et augm. Rennes: Editions Ouest-France, c1994. 275 p.

Chourgnoz, Jean-Marie. **Les sous-marins francais: les bateaux noirs**. Boulogne: ETAI, c1998. 167 p.

Chatelle, Albert. **La base navale du Havre et la guerre sous-marine secrète en Manche, 1914-1918**. Paris: Éditions Médicis, 1949. 261 p.

France. Marine. Groupe d'action sous-marine. **Chasseurs de sous-marins: le Groupe d'action sous-marine / ouvrage realise par les equipages du Groupe d'action sous-marine**. Paris: Addim, [1996]. (Collection Les armes et les hommes). 93 p.

Garier, Gerard. **L'odyssee technique et humaine du sous-marin en France**. Bourg-en-Bresse: Marines, [1995-]. v.
V. 2 -- Des Emeraudes (1905-1906) au Charles Brun (1908-1913).

Griffi, Toussaint and Laurent Preziosi. **Premiere mission en Corse occupee: avec le sous-marin Casabianca, decembre 1942-mars 1943**. Paris: Editions L'Harmattan, c1988. (Collection Chemins de la memoire). 191 p.

Guierre, Maurice Casimir Lucien. **[L']épopée de Surcouf et le commandant Louis Blaison**. Paris, Ed. Bellenand, [1952]. 250 p.

L'Herminier, Jean. **Casabianca, 27 novembre 1942-13 septembre 1943**. Paris, Editions France-Empire, [1949]. 315 p.

_____. **Casabianca**. Paris, Presses Pocket, [1963]. 250 p.

Huan, Claude. **Les sous-marins francais, 1918-1945**. Bourg en Bresse, France: Marines edition, [1995?]. 247 p.

Laurens, Adolphe. **Introduction à l'étude de la guerre sous-marine**. Paris, A. Challamel, 1921. 233 p.

Pasquelot, Maurice. **Les sous-marins de la France libre: 1939-1945**. Paris: Presses de la Cite, c1981. (Troupes de choc). 285 p.

Sacaze, Rene. **Le Casabianca avant L'Herminier**. Paris, editions France-Empire, [1962]. 316 p.

Thomazi, Auguste Antoine. **Sous-marins et croiseurs français**. Paris, Plon, [1947]. 120 p.

Valentiner, Max. **La terreur des mers; mes aventures en sous-marin, 1914-1918**. Paris, Payot, 1931. 252 p.

SUBMARINES -- GERMAN -- GENERAL

Botting, Douglas and the editors of Time-Life Books. **The U-boats**. Alexandria, VA: Time-Life Books, c1979. 176 p.
NPS/DKL Location: GENERAL V210 .B67 MISSING

Gabler, Ulrich. **Unterseebootbau**. 4., überarb. u. erw. Aufl. Bonn: Bernard u. Gr., 1996. 167 p.

Gröner, Erich. **Die deutschen Kriegsschiffe, 1815-1945**. München: Bernard & Graefe, c1982-c1994. 8 v. in 10.
Originally published: München-Berlin. J. F. Lehmanns verlag, 1937.
Incomplete Contents: Bd. 3. U-Boote, Hilfskreuzer, Minenschiffe, Netzleger, Sperrbrecher.

_____. **German warships, 1815-1945**. Annapolis, MD: Naval Institute Press, 1990-1991. 2 v.
English translation of: Die deutschen Kriegsschiffe, 1815-1945.
NPS/DKL Location: GENERAL VA513 .G6813 1990

_____, Dieter Jung and Martin Maass. **Die deutschen Kriegsschiffe, 1815-1945**. München: Bernard & Graefe, c1982-c1994. 8 v. in 10

Guitard, Pierre. **U-boat warfare**. London, Pallas publishing company limited, [1939]. 80 p.

Hadley, Michael L. **Count not the dead: the popular image of the German submarine**. Annapolis, MD: Naval Institute Press, c1995. 253 p.
NPS/DKL Location: GENERAL V859.G3 H33 1995

_____. **Der Mythos der deutschen U-Bootwaffe**. Hamburg: Mittler, 2001. 207 p.

Jeschke, Hubert. **U-Boottaktik. Zur dt. U-Boottaktik 1900-1945**. Freiburg, Rombach, (1972). (Einzelschriften zur militärischen Geschichte des Zweiten Weltkrieges, 9). 120 p.

Kemp, Paul. **U-boats destroyed: German submarine losses in the World Wars**. Annapolis, MD: Naval Institute Press, 1997. 288 p.
NPS/DKL Location: GENERAL D781 .K46 1997

_____. **U-boats destroyed: German submarine losses in the World Wars**. London: Arms & Armour, 1999. 288 p.

Kohl, Fritz and Eberhard Rössler. **The Type XXI U-boat**. Annapolis, MD: Naval Institute Press, c1991. 127 p.
English translation originally published as "Ubootyp XXI" 1988.

Krause, Günter. **U-Boot und U-Jagd**. 1. Aufl. Berlin: Militärverlag der Deutschen Demokratischen Republik, c1984. 263 p.

Kruska, Emil und Eberhard Rössler. **Walter-U-Boote**. München, J. F. Lehmann, [1969]. (Wehrwissenschaftliche Berichte, Bd. 8). 216 p.

Kurzak, Karl Heinz. **Unterseeboote und Torpedos mit Kreislaufantrieb**. [Berlin, Im Waldwinkel 5a, Selbestverl. Rössler], 1969. 162 p.

Luntinen, Pertti. **Saksan keisarillinen laivasto Itamerella: aikeet, suunnitelmat ja toimet**. Helsinki: SHS, 1987 [i.e. 1988]. (Historiallisia tutkimuksia; 143). 262 p.

Miller, David. **Deutsche U-Boote 1914-1945: Geschichte, Entwicklung, Ausrüstung**. 1. Aufl. Königswinter: Heel Verlag GmbH, 2003. 208 p.

_____. **U-boats: the illustrated history of the raiders of the deep**. 1st ed. Washington, DC: Brassey's, c2000. 208 p.
NPS/DKL Location: GENERAL V859.G3 M555 2000

Moller, Eberhard. **Kurs Atlantik: die deutsche U-Boot-Entwicklung bis 1945**. 1. Aufl. Stuttgart: Motorbuch, 1995. 286 p.

_____ and Werner Brack. **Enzyklopädie deutscher U-Boote: von 1904 bis zur Gegenwart**. 1. Aufl. Stuttgart: Motorbuch-Verlag, 2002. 304 p.

Nohse, Lutz und Eberhard Rössler. **Konstruktionen für die Welt: Geschichte der Gabler-Unternehmen IKL und MG**. Herford: Koehler, c1992. 184 p.

_____ and _____. **Moderne Küsten-Uboote**. München, J.F. Lehmann, [c1972]. (Wehrwissenschaftliche Berichte, Bd. 12). 139 p.

Nöldeke, Hartmut and Volker Hartmann. **Der Sanitätsdienst in der deutschen U-Boot-Waffe und bei den einkampfverbänden: Geschichte der deutschen U-Boot-Medizin**. Hamburg; Berlin; Bonn: Mittler, 1996. 280 p.

Nowarra, Heinz J. **Unterseeboot Typ VII**. Friedberg: Podzun-Pallas, 1977. (Das Waffen-Arsenal; Bd. 37). 46 p.

Preston, Antony. **U-boats**. London: Arms & Armour Press, 1978. 192 p.

Prim, Joseph and Mike McCarthy. **Those in peril: the U-boat menace to Allied shipping in Newfoundland and Labrador waters, World War I and World War II.** St. John's, Nfld.: Jesperson Pub., 1995. 176 p.

Ripley, Tim. **Die deutschen Spezialeinheiten und ihre Waffensystem 1939 - 1945: Panzer - Kampfflugzeuge - U-Boote - V1 - V2.** 1. Aufl. Klagenfurt: Neuer Kaiser Verlag GmbH, 2003. 192 p.

Rössler, Eberhard. **Die deutschen U-Boote und ihre Werften: eine Bilddokumentation über den dt. U-Bootbau in zwei Bänden.** München: Bernard & Graefe, c1979-1980. 2 v.

_____. **Die deutschen Uboote und ihre Werften: eine Bilddokumentation über den deutschen Ubootbau von 1935 bis heute.** Koblenz: Bernard & Graefe, c1990. 335 p.

_____. **Geschichte des deutschen Ubootbaus.** 2., überarbeitete und erw. Aufl. Koblenz: Bernard & Graefe, c1986-c1987. 2 v.
Originally published: München: J. F. Lehmanns Verlag, c1975.

_____. **Die grossen Walter-Uboote: Typ XVIII und Typ XXVI.** Bonn: Bernard & Graefe, c1998. 88 p.

_____. **Die Sonaranlagen der deutschen U-Boote: Entwicklung, Erprobung und Einsatz akustischer Ortungs- und Täuschungseinrichtungen für Unterseeboote in Deutschland.** Herford: Koehler, c1991. (Warengruppe; Nr. 36). 128 p.

_____. **Die Torpedos der deutschen U-Boote: Entwicklung, Herstellung und Eigenschaften der deutschen Marine-Torpedos.** Herford: Koehler, c1984. 271 p.

_____. **The U-boat: the evolution and technical history of German submarines** [translated by Harold Erenberg]. 1st English-language ed., rev. and expanded. Annapolis, MD: Naval Institute Press, 1981. 384 p.
English translation of: Geschichte des deutschen Ubootbaus.
NPS/DKL Location: GENERAL VM365 .R6413

_____. **The U-boat: the evolution and technical history of German submarines** [translated by Harold Erenberg]. London: Arms & Armour Press, 1981. 384 p.
English translation of: Geschichte des deutschen Ubootbaus.

_____. **U-Boottyp XXI.** 2., verb. Aufl. München, J. F. Lehmann, 1967. (Wehrwissenschaftliche Berichte, Bd. 1). 159 p.

_____. **U-Boottyp XXI.** 3., verb. Aufl. München: Bernard & Graefe, 1980. (Wehrtechnische Handbücher). 160 p.

_____. **U-Boottyp XXIII**. München, J.F. Lehmann, 1967. (Wehrwissenschaftliche Berichte, Bd. 4). 116 p.

_____. **Die Unterseeboote der Kaiserlichen Marine**. Bonn: Bernard & Graefe, c1997. 232 p.

Tarrant, V. E. **U-Boat Offensive, 1914-1945**. Annapolis, MD: Naval Institute Press, c1989. 190 p.
NPS/DKL Location: GENERAL V859.G3 T37 1989

Terraine, John. **Business in great waters: the U-boat wars, 1916-1945**. London: L. Cooper, 1989. 841 p.
Simultaneously published: New York: Putman, 1989, as The U-boat wars, 1916-1945.

_____. **Business in great waters: the U-boat wars, 1916-1945**. London: Mandarin. 1990, c1989. 840 p.
Originally published: London: L. Cooper, 1989

_____. **Business in great waters: the U-boat wars, 1916-1945**. London: Wordsworth Editions, 1999. (Wordsworth military library). 841 p.
Originally published: London: L. Cooper, 1989

_____. **The U-boat wars, 1916-1945**. New York: Putnam, c1989. 841 p.
NPS/DKL Location: GENERAL D591 .T47 1989

_____. **The U-boat wars, 1916-1945**. 1st Owl book ed. New York: Holt, 1990. 841 p.
Reprint, originally published: New York: Putnam, c1989.

Williamson, Gordon. **German U-boat crews 1914-45**. London: Osprey, c1995. 64 p.

_____. **U-boat crews, 1914-45**. London: Osprey, 1998,c1995. (Elite series; 60). 64 p.
NPS/DKL Location: GENERAL VC305.G3 W55 1995

SUBMARINES -- GERMAN -- WWI

Ajax, pseud. **The German pirate; his methods and record**. New York, George H. Doran company [c1918]. 124 p.

Anon. **U-Boote im Eismeer**. Berlin, A. Scherl G.m.b.H., [1916]. 106 p.

Archer, William. **The pirate's progress; a short history of the U-boat**. New York, London, Harper & brothers, 1918. 106 p.

NPS/DKL Location: BUCKLEY D591 .A7

_____. **The pirate's progress; a short history of the U-boat**. London, Chatto & Windus, 1918. 96 p.

_____. **Sørøverens færd** [tr. by Johan Warburg]. København, V. Pio, 1918. 84 p.
Danish translation of The pirate's progress.

Art, Robert J. **The influence of foreign policy on seapower: new weapons and Weltpolitik in Wilhelminian Germany**. Beverley Hills, Sage Publications [c1973]. (A Sage professional paper. International studies series ser. no. 02-019). 49 p.

Bailey, Thomas Andrew and Paul B. Ryan. **The Lusitania disaster: an episode in modern warfare and diplomacy**. 383 p.
NPS/DKL Location: GENERAL D592.L8 B23

Bateman, Charles T. **U-boat devilry, illustrating the heroism and endurance of merchant seamen**. London, New York, Hodder & Stoughton, 1918. 175 p.

Bauer, Hermann. **Als Fuhrer der U-Boote im Weltkriege; der Eintritt der U-Boot-Waffe en die Seekriegfuhrung**. Leipzig, Koehler & Amelang, [1942]. 470 p.

_____. **Das Unterseeboot: seine Bedeutung als Teil einer Flotte, seine Stellung im Volkerrecht, seine Kriegsverwendung, seine Zukunft**. Berlin: Mittler & Sohn, 1931. 140 p.

Birnbaum, Karl E. **Peace moves and U-boat warfare: a study of Imperial Germany's policy towards the United States, April 18, 1916-Jan. 9, 1917**. Stockholm: Almquist & Wiksell, [1958]. (Acta Universitatis Stockholmiensis. Stockholm studies in history. 2). 388 p.
NPS/DKL Location: GENERAL D619 .B6

_____. **Peace moves and U-boat warfare: a study of Imperial Germany's policy towards the United States, April 18, 1916-Jan. 9, 1917**. [Hamden, CT]: Archon Books, 1970. 388 p.
Originally published: Stockholm: Almquist & Wiksell, [1958].

Biskupski, Stanislaw. **Uwaga, peryskop!** [Wyd. 1]. Warszawa, Wiedza Powszechna, 1962. 283 p.

Bongard, Willy. **Die Zentrumsresolution vom 7. Oktober 1916**. Koln: Orthen, 1937. 63 p.
Thesis (Ph.D.), Universitat Koln, 1937.

Busch, Fritz Otto and George Günther Forstner, eds. **Unsere marine im weltkrieg**. Berlin, Brumen-verlag, W. Bischoff, [c1934]. 514 p.

Contents includes Der U-books-krieg.

Clark, William Bell. **When the U-boats came to America**. Boston, Little, Brown & company, 1929. 359 p.
NPS/DKL Location: BUCKLEY D591 .C5

Clephane, Lewis P. and Naval History Division. **History of the Naval Overseas Transportation Service in World War I**. Washington, Naval History Division; [for sale by the Supt. of Docs., U.S. Govt. Print. Off.] 1969. 283 p.
NPS/DKL Location: GENERAL D589.U6 C6

Coles, Alan. **Three before breakfast: a true and dramatic account of how a German U-boat sank three British cruisers in one desperate hour**. Havant [Eng.]: K. Mason; White Plains, NY: distributed by Sheridan House, c1979. 192 p.

Crompton, Iwan and Werner von Langsdorff. **Englands Verbrechen an U 41, der zweite "Baralong"-Fall im Weltkreig**. Gutersloh, C. Bertelsmann, [1940]. 224 p.

Dinklage, Ludwig. **U-boot-fahrer und kamelsreiter: kriegsfahrten eines deutschen unterseebootes**. Stuttgart: Franckh'sche verlagshandlung, 1939. 212 p.

Ellis, Frederick D. **The tragedy of the Lusitania: embracing authentic stories by the survivors and eye-witnesses of the disaster, including atrocities on land and sea, in the air, etc.** [Philadelphia, PA?]: National publishing co., 1915. 320 p.

Ernst, Georg. **Bis zur letzten Stunde: Illusion und Wirklichkeit**. Hamburg: E.S. Mittler, c1995. 223 p.

Fechter, Hans. **In der alarmkoje von U35**. Berlin-Wien, Ullstein & co., 1918. 151 p.

Forstner, George Gunther, freiherr von. **The journal of submarine commander von Forstner** [tr. by Mrs. Russell Codman]. Boston & New York, Houghton Mifflin company, [c1917]. 135 p.

Freiwald, Ludwig. **Last days of the German fleet** [translated by Martin Moore]. London, Constable & Company Ltd., 1932. 318 p.
English translation of Die verratene flotte. Aus den letzten Tagen der deutschen Kriegsmarine.

_____. **Schlachtkreuzer im Nebel**. Munchen, J.F. Lehmann, 1934. 70 p.

_____. **U-Boots-Maschinist Fritz Kasten: Das Frontb. d. dt. Kriegsmarine**. München: J. F. Lehmanns Verl., 1933. 327 p.

_____. **Die verratene Flotte: Aus den letzten Tagen d. deutschen Kriegsmarine**. München: J. F. Lehmanns Verl., 1931. 298 p.

_____. **Die verratene Flotte: Aus d. letzten Tagen d. dt. Kriegsmarine**. 2. Aufl. München: Lehmanns Verl., 1937. 297 p.

Frost, Wesley. **German submarine warfare; a study of its methods and spirit, including the crime of the "Lusitania," a record of observations and evidence**. New York, London, D. Appleton & company, [c1918]. 243 p.

Fürbringer, Werner. **Alarm! Tauchen!! U-boot in kampf und sturm**. Berlin, Im Deutschen verlag, [c1933]. 257 p.

_____. **Fips: legendary U-boat commander, 1915-1918** [translated from the German by Geoffrey Brooks]. Annapolis, MD: Naval Institute Press, c1999. 146 p.
English translation of Alarm! Tauchen!!

Fuss, Richard. **Der U-Boot-Krieg des Jahres 1915; ein Kapitel auswartiger Politik im Weltkriege**. Stuttgart, Kolhammer, 1936. (In Beitrage zur Geschichte der nachbismarckischen Zeit und des Weltkriegs, Hft. 33). 96 p.

George, S. C. **Jutland to junkyard; the raising of the scuttled German High Seas Fleet from Scapa Flow; the greatest salvage operation of all time**. Cambridge, Stephens, 1973. 176 p.

Germany. Marine-Archiv. **Der Krieg zur See 1914-1918**. Berlin: Mittler, 1920-1966. 22 v.
Contents: Der Krieg in der Nordsee. (7 v.) -- Der Krieg in der Ostsee. (3 v.) -- Der Kreuzerkrieg in den auslandischen Gewassern. (3 v.) -- Der Krieg in den Turkischen Gewassern. (2 v.) -- Die Kampfe der kaiserlichen Marine in den deutschen Kolonien. (1 v.) -- Der Handelskrieg mit U-Booten. (5 v.) -- Die Uberwasserstreitkrafte und ihre Technik. (1 v.)

Gibson, Richard Henry and Maurice Prendergast. **The German submarine war, 1914-1918**. New York: R.R. Smith inc., [1931]. 438 p.
Originally published: London: Constable, 1931.
NPS/DKL Location: GENERAL D591 .G4

_____ and _____. **The German submarine war, 1914-1918**. 2d ed. London, Constable, [1931]. 438 p.

Grant, Robert M. **U-boat intelligence, 1914-1918**. [Hamden, CT]: Archon Books, 1969. 192 p.
Originally published: London: Putnam, 1969.
NPS/DKL Location: GENERAL VB230 .G7

Gray, Edwyn. **The U-boat war, 1914-18**. New York, Scribner, [1972]. 280 p.
Originally published: London: Seeley, 1972 as The killing time.

Gross, Gerhard Paul. **Die Seekriegfuhrung der Kaiserlichen Marine im Jahre 1918.** Frankfurt am Main; New York: P. Lang, c1989. (Europaische Hochschulschriften. Reihe III, Geschichte und ihre Hilfswissenschaften; Bd. 387). 574 p.

Grosse, Karl Friedrich. **Politische und militarische Bedeutung des Unterseebootskrieges 1914/18.** Rosenheim, Druck Rosenheimer anzeiger, 1937. 163 p.

Hashagen, Ernst. **U-boats westward!** [translated by Lieutenant-Commander Vesey Ross]. New York, London, G. P. Putnam's sons, 1931. 247 p.
Previous edition entitled: The log of a U-boat commander; or, U-boats westward -- 1914-1918.
NPS/DKL Location: BUCKLEY D591 .H3

_____. **U-Boote westwarts!: meine Fahrten um England, 1914-1918.** 3. Aufl. Berlin: E. S. Mittler, 1931. 219 p.

Heitmann, Jan. **Unter Wasser in die Neue Welt: Handelsunterseeboote und kaiserliche Unterseekreuzer im Spannungsfeld von Politik und Kriegfuhrung.** Berlin: Berlin Verlag Arno Spitz, c1999. 365 p.

Hirose, Hikota. **Emono o motomete: Doitsu sensuikan no daikatsuyaku.** Tokyo: Kaigun Kenkyusha, Showa 3 [1928]. 238 p.

Hoehling, A. A. and Mary Hoehling. **The last voyage of the Lusitania.** [1st ed.]. New York: Holt, [1956]. 255 p.
NPS/DKL Location: BUCKLEY D592.L8 H6

Hubernagel, Wilhelm. **Die Bekanntmachungen des U-Boot-Handelskrieges: vom 4. Februar 1915 und 1. Februar 1917.** Greifswald: E. Hartmann, 1927. Dissertation, Marburg. 89 p.

James, Henry Johnson. **German subs in Yankee waters: first world war.** New York: Gotham house, [c1940]. 208 p.
NPS/DKL Location: BUCKLEY D591 .J2

Johnson, Rody. **Different battles: the search for a World War II hero.** Manhattan, KS: Sunflower University Press, c1999. 172 p.

Jung, Hermann Albert Karl. **Krieg unter Wasser; der Opfertod der Funftausend.** Oldenburg i.O., Berlin, G. Stalling, [1940, c1939]. 222 p.

Kaulisch, Baldur. **Die Auseinandersetzung um den uneingeschränkten U-Boot-Krieg innerhalb der herrschenden Klasse Deutschlands während des ersten Weltkrieges <Herbst 1914 bis Frühjahr 1917>.** Berlin,, 1970. 232 p.

_____. **U-Bootkrieg 1914/1918**. Berlin: Deutscher Verlag der Wissenschaften, 1976. (Illustrierte historische Hefte; 4). 39 p.

König, Paul. **Atlanterhavsfart**. Kjøbenhavn, H. Aschehoug, 1916. 108 p.

_____. **El 1.er viaje del submarino mercante "Deutschland**. Buenos Aires, M. Schneider, 1918. 142 p.

_____. **Die fahrt der Deutschland**. Berlin, Ullstein & co., 1916. 152 p.

_____. **Die fahrt der Deutschland**. Berlin, Ullstein & co., 1917. 157 p.

_____. **Die fahrt der Deutschland, das erste untersee-frachtschiff**. New York, Hearst's International library co. [1916]. 254 p.

_____. **Fahrten der U Deutschland im weltkrieg**. Berlin, Ullstein, [c1937]. 188 p. Originally published as Die fahrt der Deutschland in 1916; new enlarged edition.

_____. **Voyage of the Deutschland** [translated by Vivien Ellis]. London, C. Arthur Pearson, 1917. 126 p.

_____. **Voyage of the Deutschland, the first merchant submarine**. New York, Hearst's International Library Co., 1916. 247 p.
NPS/DKL Location: BUCKLEY D592.D4 K7

_____. **Voyage of the Deutschland: the first merchant submarine**. Annapolis, MD: Naval Institute Press, 2001. (Classics of naval literature). 264 p.
Written by Ernst Bischof and published under the name of Capt. Paul König.

Krause, Andreas. **Scapa Flow: die Selbstversenkung der wilhelminischen Flotte**. Berlin: Ullstein, c1999. 431 p.

Kruger, Wolfgang. **Der Entschluss zum uneingeschrankten U-Bootkrieg im Jahre 1917 und seine volkerrechtliche Rechtfertigung**. Berlin, Mittler, 1959. (Marine Rundschau, Zeitschrift fur Seewesen Beiheft 5). 71 p.

Langsdorff, Werner von. **U-Boote am Feind: 45 deutsche U-boot-Fahrer erzahlen**. Gutersloh, C. Bertelsmann [c1937]. 367 p.

Lauriat, Charles Emelius. **The Lusitania's last voyage: being a narrative of the torpedoing and sinking of the R.M.S. Lusitania by a German submarine off the Irish coast May 7, 1915**. Boston; New York: Houghton Mifflin company, 1915. 158 p.
NPS/DKL Location: BUCKLEY D592.L8 L3

Lehmann, Ernst. **Deutschlands Unterseeboot-Sorge: Predigt, geh. am Sonntag nach d. 1. Febr. 1917**. [s. l.]: [s. n.], 1917. 8 p.

Lorey, Hermann. **Der Krieg in den türkischen Gewässern**. Berlin, E.S. Mittler, 1928-38. (Der Krieg zur See, 1914-1918). 2 v.

Mattes, Klaus. **Die Seehunde: Klein-U-Boote: letzte deutsche Initiative im Seekrieg, 1939-1945**. Hamburg: Mittler, [1995]. 224 p.

Messimer, Dwight R. **Escape**. Annapolis, MD: Naval Institute Press, c1994. 266 p.

_____. **The merchant U-boat: adventures of the Deutschland, 1916-1918**. Annapolis, MD: Naval Institute Press, c1988. 234 p.
NPS/DKL Location: GENERAL D592.D4 .M47 1988

Michelsen, Andreas. **Der U-bootskrieg, 1914-1918**. Leipzig, Hase & Koehler, c1925. 207 p.

Mielke, Otto. **Unterseeboot "U-38": Kapitänleutnant Valentiers besondere Einsätze im 1. Weltkrieg**. Kiel: Stade, [2001]. (Schiffe, Menschen, Schicksale; Nr. 91). 46 p.

_____. **Unterseeboot "U 21": Dardanellenkämpfe 1915**. Hamburg: Maximilian-Verl. Schober, 1999. (Schiffe, Menschen, Schicksale; Nr. 66). 46 p.

Neureuther, Karl and Claus Bergen, eds. **U-boat stories; narratives of German U-boat sailors** [translated by Eric Sutton]. London: Constable, 1931. 207 p.
English translation of Wir leben noch!

_____ and _____. **Wir leben noch! Deutsche seehelden im U-bootkampf**. Stuttgart: Union deutsche verlagsgesellschaft, [c1930]. 206 p.

Niemoller, Martin. **From U-boat to concentration camp: the autobiography of Martin Niemoller**. London: William Hodge & Co., 1939. 281 p.
Originally published as From U-boat to pulpit, 1936.

_____. **Vom U-Boot zur Kanzel**. Berlin: M. Warneck, 1934. 210 p.

_____and Henry Smith Leiper. **From U-boat to pulpit, including an appendix, From pulpit to prison**, by Henry Smith Leiper. [Translated by Commander D. Hastie Smith]. Chicago; New York: Willett, Clark & company, 1937. 223 p.
NPS/DKL Location: BUCKLEY D591 .N6

O'Sullivan, Patrick. **The Lusitania: unravelling the mysteries**. Dobbs Ferry, NY: Sheridan House, 2000. 194 p.
Originally published: Westlink Park, Doughcloyne, Cork: Collins Press, c1998.

Preston, Diana. **Lusitania: an epic tragedy**. New York: Walker & Co., 2002. 528 p.

Ramlow, Gerhard. **Ausgelaufen westwarts**. Potsdam, L. Voggenreiter, [c1937]. 143 p.

Ramsay, David. **Lusitania: saga and myth**. Rochester, Kent: Chatham, 2001. 308 p.

Ringelnatz, Joachim. **Unterseeboot "U 9": U-Boot-Krieg 1914/15**. Hamburg: Maximilian-Verl. Schober, [1998]. (Schiffe, Menschen, Schicksale; Nr. 61). 46 p.

Ritter, Paul. **Ubootsgeist; abenteuer und fahrten im Mittelmeer**. Leipzig, K. F. Koehler, [c1935]. 246 p.

Rose, H. Wickliffe. **Brittany patrol; the story of the suicide fleet**. New York, W. W. Norton & Company, inc. [c1937]. 367 p.
NPS/DKL Location: BUCKLEY D589.U6 R7

Rose, Hans. **Auftauchen! Kriegsfahrten von "U 53"**. 6. Aufl. [Essen], Essener Verlaganstalt, 1941. 315 p.

Salvadori, Alberto. **La fine di due imperi: Scapa Flow e la resa di Singapore**. Roma: Rivista Marittima, 1998. [Rivista marittima. 1998, no. 6 (Supplement)]. 132 p.

Schröder, Joachim. **Die U-Boote des Kaisers: die Geschichte des deutschen U-Boot-Krieges gegen Großbritannien im Ersten Weltkrieg**. Lauf a.d. Pegnitz: Europaforum-Verl., 2001. (Subsidia academica: Reihe A, Neuere und neueste Geschichte; Bd. 3). 504 p.
Dissertation, Dortmund, Univ., 1999.

Schultze-Bahlke, Georg. **U-Boote**. Berlin: Richard Carl Schmidt, 1918. (Motorschiff-Bibliothek; Bd. 3). 203 p.

Selow-Serman, K. E. **U-Boot-Abenteuer im Sperrgebiet**. Berlin, A. Scherl, [c1917]. 112 p.

Simpson, Colin. **The Lusitania**. Boston: Little, Brown, 1972. 303 p.
NPS/DKL Location: GENERAL D592.L8 S5

_____. **The Lusitania**. London, England; New York, N.Y., USA: Penguin Books, c1983. 325 p.
Originally published: London: Longman, 1972.

Spencer, Samuel R. **Decision for war, 1917; the Laconia sinking and the Zimmermann telegram as key factors in the public reaction against Germany**. Rindge, NH, R.R. Smith, 1953. 109 p.

Spiegel von und zu Peckelsheim, Edgar. **The adventures of the U-202: an actual narrative**. New York: The Century co., 1917. 202 p.
NPS/DKL Location: BUCKLEY D591 .S7

_____. **Kriegstagebuch "U 202": angefangen den 12th. April 19.., abgeschlossen den 30th. April 19...** Berlin: Scherl, 1916. 138 p.

_____. **Kriegstagebuch U 202 und Oberheizer Zenne, der letzte Mann der Wiesbaden**. Berlin: Scherl, 1917. 174 p.

_____. **U. boat 202, the war diary of a German submarine** [tr. by Captain Barry Domvile]. London, A. Melrose Ltd., 1919. 170 p.

_____. **U-Boot im Fegefeuer**. Berlin: Scherl, 1930. 211 p.

_____. **U-Boot im Fegefeuer**. Berlin: Scherl, 1940. 211 p.

_____. **U-Boot im Fegefeuer: ein Buch über den U-Bootkrieg 1914 - 1918**. [2. Aufl.]. Preetz/Holstein: Gerdes, 1963. 304 p.

Spindler, Arno. **La guerra al commercio con i sommergibili** [tr. Raffaele de Courten and Wladimiro Pini]. Roma, Instituto poligrafico dello stato, Libreria, 1934-. 3 v.
Italian translation of Der Handelskrieg mit U-Booten.

_____. **Der Handelskrieg mit U-Booten**. Berlin, E.S. Mittler & Sohn, 1932-. (Der Krieg zur See, 1914-1918). v.

Steinwager, Leonhard. **U-boot, Englands tod!** München, J. F. Lehmann, 1918. 48 p.

Stokes, Roy. **Death in the Irish Sea: the sinking of the RMS Leinster**. Wilton, Cork: Collins Press, 1998. 153 p.

Termote, Tomas. **Verdwenen in de Noordzee: geschiedenis van de Duitse U-botenaan de Belgische kust in de Eerste Wereldoorlog en opheldering over het lot van vijftien verdwenen onderzeeers**. Erpe-Mere: De Krijger, [1999] . (Belgie in oorlog. Speciaal; nr. 2). 164 p.

Thomas, Lowell. **Raiders of the deep**. Doubleday, Doran, 1928. 363 p.
NPS/DKL Location: GENERAL D591 .T4

_____. **Raiders of the deep**. Garden City, NY: Garden City Pub. Co., [c1928]. 363 p.
NPS/DKL Location: BUCKLEY D591 .T4

_____. **Ritter der Tiefe** [ubersetzt und bearbeitet von C. Freiherr v. Spiegel]. Gutersloh, C. Bertelsmann [c1930]. 360 p.

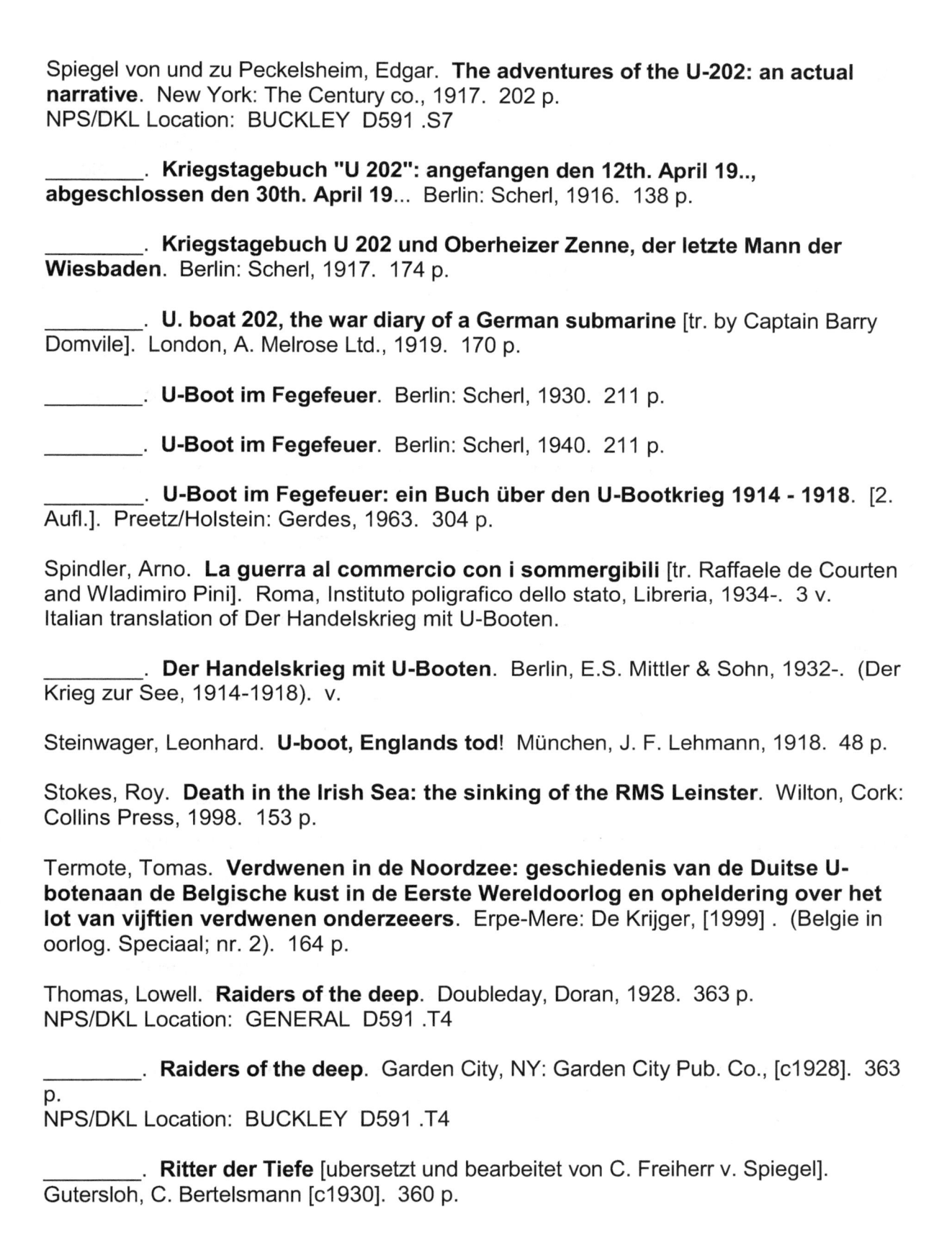

English translation of Raiders of the deep.

United States. National Archives and Records Service. **U-boats and T-boats, 1914-1918**. Washington: National Archives and Records Service, U.S. General Services Administration, 1984 [i.e. 1985]. (Guides to the microfilmed records of the German Navy, 1850-1945; no. 1). 355 p.
NPS/DKL Location: FEDDOCS AE 1.112/2:1

United States. Office of Naval Records and Library. **German submarine activities on the Atlantic coast of the United States and Canada**. Washington, Govt. Print. Off., 1920. (Publication number 1). 163 p.
NPS/DKL Location: GENERAL D589.U5 A4 NO.1 1920

Van der Vat, Dan. **The grand scuttle: the sinking of the German Fleet at Scapa Flow in 1919**. London: Hodder & Stoughton, 1982. 240 p.

_____. **The grand scuttle: the sinking of the German Fleet at Scapa Flow in 1919**. Annapolis, MD: Naval Institute Press, 1986. 240 p.

Van der Vat, Dan. **The grand scuttle: the sinking of the German Fleet at Scapa Flow in 1919**. Edinburgh: Birlinn, 1997. 240 p.

Wiebicke, Karl. **Die manner von U96: erinnerungen an fahrten unseres U-Bootes**. Leipzig: Koehler, 1934. 208 p.

Wood, Lawson. **The bull & the barriers: the wrecks of Scapa Flow**. Stroud, Gloucestershire: Tempus, 2000. 127 p.

SUBMARINES -- GERMAN -- WWII

Alaluquetas, Jacques. **Un loup gris dans l'Atlantique: l'U68 au combat**. Paris: Grancher, c1999. (Témoignages pour l'histoire). 260 p.

Alman, Karl [pseudo.] -- see Kurowski, Franz.

Alvensleben, Oda von. **Unterseebootskrieg und Volkerrecht**. Stuttgart und Berlin, Deutsche Verlags-Anstalt, 1916. (Der Deutsche Krieg; politische Flugschriften 81./82. Hft). 82 p.

Anbrosius, Hans Heinrich. **Die Verluste der neutralen Handelsschiffahrt im Gegenwärtigen Seekrieg**. Berlin, Junker und Dünnhaupt, 1940. (Schriften des Deutschen Instituts für aussenpolitische Forschung und des Hamburger Instituts für auswärtige politik; 87). 78 p.

Andrade, Allan. **Leopoldville troopship disaster: in memoriam**. Limited ed. Flushing, NY: A. Andrade, c1999. 72 p.
Ft. Benning, Georgia, ceremony, October 22, 1999.

_____. **S.S. Leopoldville disaster, December 24, 1944**. 1st ed. New York: Tern Book Co., 1997. 268 p.

Antonov, Aleksandr Mikhailovich. **Germanskie elektrolodki XXI i XXIII serii**. Sankt-Peterburg: Izd-vo "Gangut", 1997. (Biblioteka "Gangut"--Korabli mira; 1). 48 p.

Arendt, Rudolf. **Letzter Befehl, Versenken: deutsche U-Boote im Schwarzen Meer, 1942-1944: Erinnerungen eines U-Boot-Kommandanten**. Hamburg: Verlag E.S. Mittler, 1998. 236 p.

_____. **Letzter Befehl: versenken: deutsche U-Boote im Schwarzen Meer 1942 - 1944; Erinnerungen eines U-Boot-Kommandanten**. 1. Aufl. [München]: Ullstein, 2003. 366 p.

Bailey, Thomas Andrew and Paul B. Ryan. **Hitler vs. Roosevelt: the undeclared naval war**. New York: Free Press, c1979. 303 p.
NPS/DKL Location: GENERAL D773 .B17

Barker, Ralph. **Goodnight, sorry for sinking you: the story of the S.S. City of Cairo**. London: Collins, 1984. 251 p.

Baroth, Hans Dieter and Fabian Becker. **U-Boot-Bunker "Valentin": Kriegswirtschaft und Zwangsarbeit Bremen-Farge 1943-45**. Bremen: Ed. Temmen, [1996]. 111 p.

Bartsch, Max. **Was jeder vom deutschen U-boot wissen muss**. Berlin, W. Limpert, [c1940]. 40 p.

Beasant, John. **Stalin's silver: the sinking of the USS John Barry**. 1st U.S. ed. New York: St. Martin's Press, 1999. 216 p.
Originally published: London: Bloomsbury, 1995.

Bendert, Harald. **U-Boote im Duell**. Berlin; Hamburg: E.S. Mittler, 1996. 190 p.

Bengtsson, Roger and Jurgen von Zweigberg. **Den torpederade gotlandsbåten Hansa: människor, minnen, mysterier**. [Höganäs]: Wiken, c1992. 164 p.

Bercuson, David Jay and Holger H. Herwig. **Deadly seas: the story of the St. Croix, the U305 and the Battle of the Atlantic**. Toronto: Random House of Canada, c1997. 346 p.

Berenbrok, Hans Dieter [Cajus Bekker, pseud]. **Hitler's naval war** [translated and edited by Frank Ziegler]. Garden City, NY: Doubleday, 1974. 400 p.

English translation of **Verdammte See**.
NPS/DKL Location: GENERAL D771 .B44 1974

_____. **Hitler's naval war** [translated and edited by Frank Ziegler]. London: Macdonald, 1974. 400 p.
English translation of Verdammte See.

_____. **Verdammte See: ein Kriegstagebuch der deutschen Marine**. Berlin, Darmstadt, Wien: Dt. Buchgemeinschaft, 1973. 392 p.

_____. **Verdammte See: ein Kriegstagebuch der deutschen Marine**. 5. Aufl. Oldenburg: Stalling, 1974. 392 p.
Originally published: [Oldenburg], G. Stalling, [c1971].

_____. **Verdammte See: ein Kriegstagebuch der deutschen Marine**. Ungekürzte Ausg. Frankfurt (M.), Berlin, Wien: Ullstein, 1974, c1971. 383 p.

_____. **Verdammte See: ein Kriegstagebuch der deutschen Marine**. Stuttgart, Hamburg, München: Dt. B_cherbund, [1975]. 392 p.

_____. **Verdammte See: ein Kriegstagebuch der deutschen Marine**. Berecht. Ausg. Klagenfurt: Verlag Buch u. Welt, c1978. 392 p.

_____. **Verdammte See: ein Kriegstagebuch der deutschen Marine**. Herford: Koehler, 1978. 392 p.

_____. **Verdammte See: ein Kriegstagebuch der deutschen Marine**. Sonderausg. Köln: Naumann und Göbel, [1985 ?]. 384 p.

_____. **Verdammte See: ein Kriegstagebuch der deutschen Marine**. Ungekürzte Ausg. Frankfurt/M; Berlin: Ullstein, 1996. 397 p.

_____. **Verdammte See: ein Kriegstagebuch der deutschen Marine**. Augsburg: Bechtermünz-Verlag, 1998. 368 p.

_____. **Verdammte See: ein Kriegstagebuch der deutschen Marine 1939 - 1945**. Gräfelfing vor München: Urbes-Verl., [1991]. 368 p.

_____. **Verdammte See: ein Kriegstagebuch der deutschen Marine 1939 - 1945**. Gräfelfing vor München: Urbes-Verl., [1993 ?]. 369 p.

Bernig, Heinrich [pseud.] -- see Kurowski, Franz.

Blair, Clay. **Hitler's U-boat war**. 1st. ed. New York: Random House, c1996-c1998. 2 v.
Contents: v. 1. The hunters, 1939-1942. -- v. 2. The hunted, 1942-1945.
NPS/DKL Location: GENERAL D781 .B53 1996

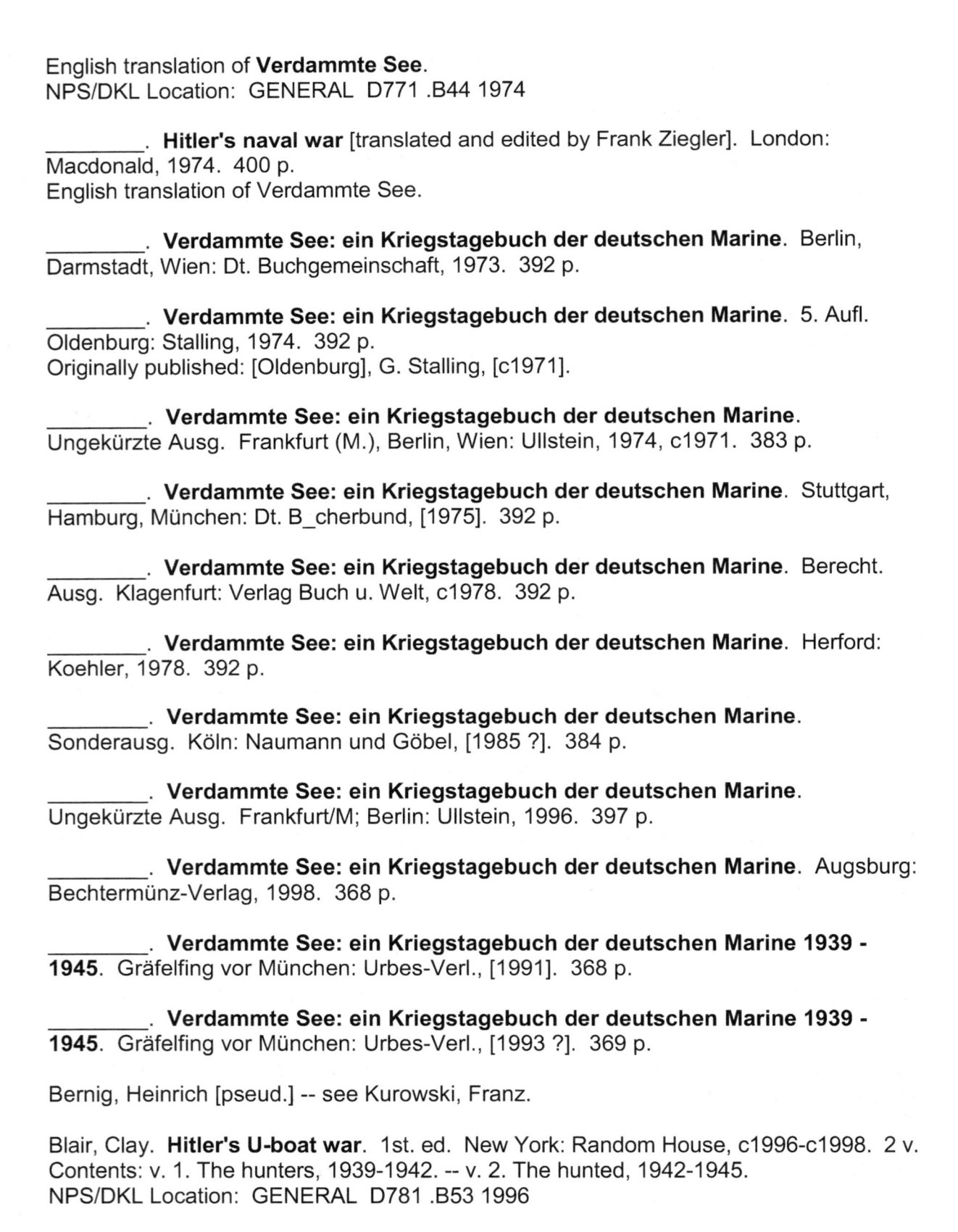

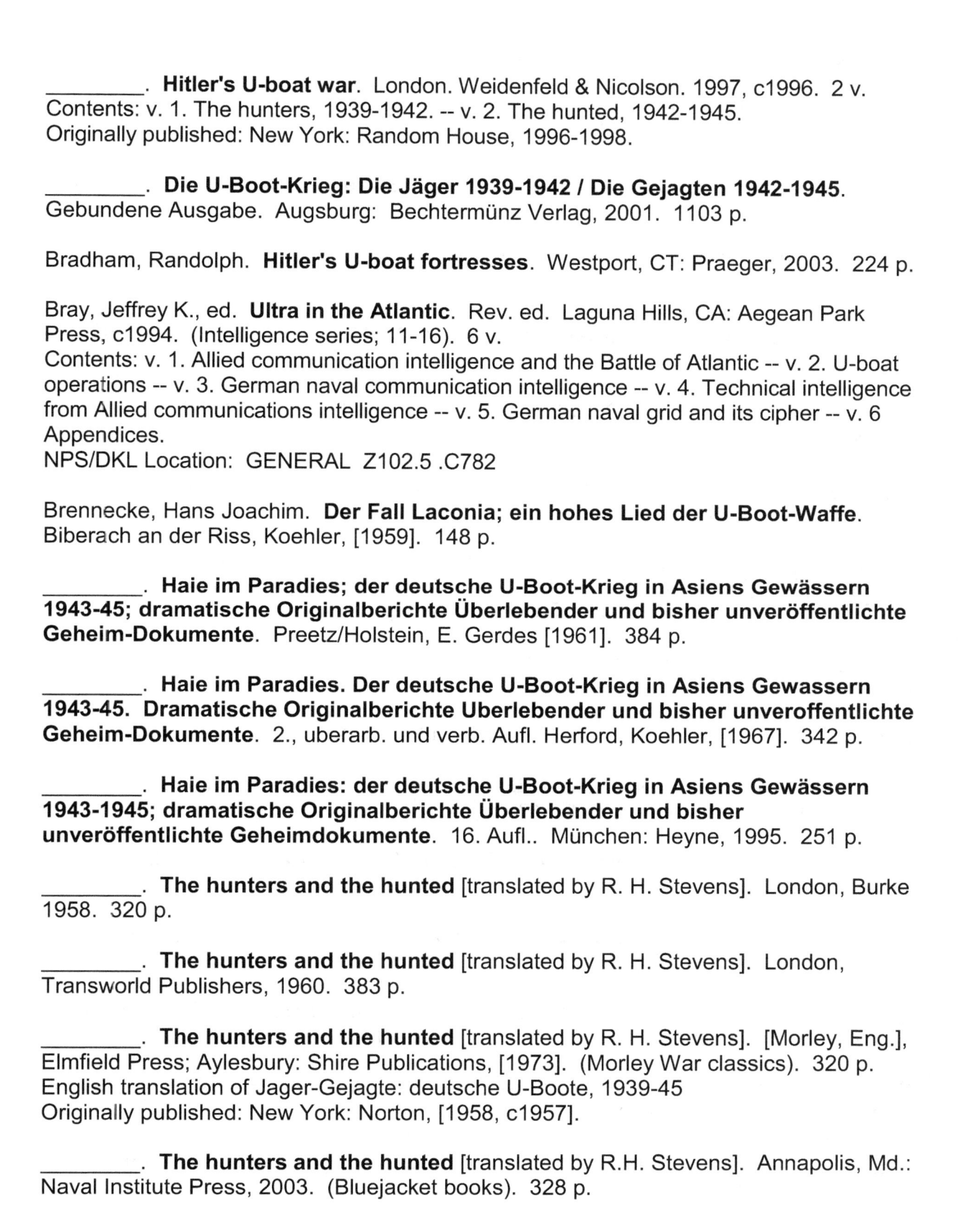

_____. **Hitler's U-boat war**. London. Weidenfeld & Nicolson. 1997, c1996. 2 v.
Contents: v. 1. The hunters, 1939-1942. -- v. 2. The hunted, 1942-1945.
Originally published: New York: Random House, 1996-1998.

_____. **Die U-Boot-Krieg: Die Jäger 1939-1942 / Die Gejagten 1942-1945**.
Gebundene Ausgabe. Augsburg: Bechtermünz Verlag, 2001. 1103 p.

Bradham, Randolph. **Hitler's U-boat fortresses**. Westport, CT: Praeger, 2003. 224 p.

Bray, Jeffrey K., ed. **Ultra in the Atlantic**. Rev. ed. Laguna Hills, CA: Aegean Park
Press, c1994. (Intelligence series; 11-16). 6 v.
Contents: v. 1. Allied communication intelligence and the Battle of Atlantic -- v. 2. U-boat
operations -- v. 3. German naval communication intelligence -- v. 4. Technical intelligence
from Allied communications intelligence -- v. 5. German naval grid and its cipher -- v. 6
Appendices.
NPS/DKL Location: GENERAL Z102.5 .C782

Brennecke, Hans Joachim. **Der Fall Laconia; ein hohes Lied der U-Boot-Waffe**.
Biberach an der Riss, Koehler, [1959]. 148 p.

_____. **Haie im Paradies; der deutsche U-Boot-Krieg in Asiens Gewässern
1943-45; dramatische Originalberichte Überlebender und bisher unveröffentlichte
Geheim-Dokumente**. Preetz/Holstein, E. Gerdes [1961]. 384 p.

_____. **Haie im Paradies. Der deutsche U-Boot-Krieg in Asiens Gewassern
1943-45. Dramatische Originalberichte Uberlebender und bisher unveroffentlichte
Geheim-Dokumente**. 2., uberarb. und verb. Aufl. Herford, Koehler, [1967]. 342 p.

_____. **Haie im Paradies: der deutsche U-Boot-Krieg in Asiens Gewässern
1943-1945; dramatische Originalberichte Überlebender und bisher
unveröffentlichte Geheimdokumente**. 16. Aufl.. München: Heyne, 1995. 251 p.

_____. **The hunters and the hunted** [translated by R. H. Stevens]. London, Burke
1958. 320 p.

_____. **The hunters and the hunted** [translated by R. H. Stevens]. London,
Transworld Publishers, 1960. 383 p.

_____. **The hunters and the hunted** [translated by R. H. Stevens]. [Morley, Eng.],
Elmfield Press; Aylesbury: Shire Publications, [1973]. (Morley War classics). 320 p.
English translation of Jager-Gejagte: deutsche U-Boote, 1939-45
Originally published: New York: Norton, [1958, c1957].

_____. **The hunters and the hunted** [translated by R.H. Stevens]. Annapolis, Md.:
Naval Institute Press, 2003. (Bluejacket books). 328 p.

NPS/DKL Location: GENERAL D781 .B713 2003

_____. **Jäger-Gejagte: deutsche U-Boote, 1939-45**. 3. Aufl. Biberach an der Riss: Koehlers, c1956. 382 p.

_____. **Jäger - Gejagte: deutsche U-Boote 1939-1945**. 6. Aufl. Herford: Koehler, 1982. 432 p.

_____. **Jäger-Gejagte: deutsche U-Boote, 1939-45**. 4. Aufl. München: Heyne, 1994. 437 p.

_____. **Jäger-Gejagte: deutsche U-Boote, 1939-45**. 8. Aufl. Biberach an der Riss: Koehlers, 2000. 494 p.

_____. **Die Wende im U-Boot-Krieg: Ursachen und Folgen, 1939-1943**. Herford: Koehler, c1984. 361 p.

_____. **Die Wende im U-Boot-Krieg: Ursachen und Folgen 1939-1943**. München: Heyne, 1991. 573 p.

Breyer, Siegfried. **German U-boat Type XXI** [translated from the German by Ed Force]. Atglen, PA: Schiffer Pub. Ltd., c1999. (Schiffer military history book). 49 p. English translation of Wunderwaffe Elektro-Uboot Typ XXI.

_____. **"Wunderwaffe" Elektro-Uboot Typ XXI**. Wölfersheim-Berstadt: Podzun-Pallas, c1996. 48 p.

Brustat-Naval, Fritz. **Ali Cremer: U 333**. Frankfurt/M; Berlin: Ullstein, 1994. 400 p.

_____ and Teddy Suhren. **Nasses Eichenlaub: als Kommandant u.F.d.U. im U-Boot-Krieg**. Herford: Koehler, 1983. 175 p.

_____ and _____. **Nasses Eichenlaub: als Kommandant u. F.d.U. im U-Boot-Krieg**. 3. Aufl.. Herford: Koehler, 1987. 175 p.

_____ and _____ **Nasses Eichenlaub: Als Kommandant und F.d.U. im U-Boot-Krieg**. Berlin: Ullstein, TB-Vlg., 1995. 175 p.

_____ and _____. **Nasses Eichenlaub: als Kommandant und F.d.U. im U-Boot-Krieg**. Sonderausg. Hamburg: Koehler, 1998. 153 p.

Buchheim, Lothar Günther. **Die U-Boot-Fahrer: die Boote, die Besatzungen und ihr Admiral**. München: C. Bertelsmann, c1985. 307 p.

_____. **Die U-Boot-Fahrer: die Boote, die Besatzungen und ihr Admiral**. Munchen: Piper, c1998. 307 p.

_____. **U-Boot-Krieg**. München; Zürich: Piper, 1976. 308 p.

_____. **Zu Tode gesiegt: der Untergang der U-Boote**. 1. Aufl. München: C. Bertelsmann, c1988. 307 p.

_____ and Alexander Rost. **Jäger im Weltmeer**. 1. Aufl. Hamburg: Hoffmann und Campe, 1996. 63 p.
Originally published: Berlin Suhrkamp, 1943.

_____ and Michael Salewski. **U-boat war** [translated by Gudie Lawaetz]. New York: Bonanza Books; Distributed by Crown Publishers, 1986. 300 p.
English translation of: U-Boot-Krieg.
Originally published: New York: Knopf, 1978.

Busch, Fritz-Otto. **Krieg der "Grauen Wölfe": die Feindfahrten des Unteseebootes U 110**. Rastatt: E. Pabel, c1986. (Die deutsche Wehrmacht im II. Weltkrieg). 108 p.

_____. **U-Bootsfahrten**. Leipzig: F. Schneider, c1934. 95 p.

Busch, Harald. **So war der U-Boot-Krieg**. Bielefeld: Dt. Heimat-Verl., 1952. 391 p.

_____. **So war der U-Boot-Krieg**. 2. Aufl.. Bielefeld: Dt. Heimat-Verl., 1954. 472 p.

_____. **So war der U-Boot-Krieg: Totentanz der Sieben Meere**. Rastatt i. Baden: Pabel, 1960. 381 p.

_____. **U-boats at war** [translated from the German by L.P.R. Wilson]. New York, Ballantine Books, [1955]. 176 p.
English translation of So war der U-Boot-Krieg.

Busch, Rainer and Hans-Joachim Roll. **German U-boat commanders of World War II: a biographical dictionary** [translated by Geoffrey Brooks]. London: Greenhill Books; Annapolis, MD: Naval Institute Press, c1999. 301 p.
English translation of Deutschen U-Boot-Kommandanten.
NPS/DKL Location: GENERAL CT32 .B87 1999

_____ and _____. **Der U-Boot-Krieg, 1939-1945**. Hamburg: Mittler & Sohn, [1996-1998]. 4 v.
Incomplete Contents: 1. Die deutschen U-Boot-Kommandanten -- 2. Der U-Boot Bau auf deutschen Werften -- 4. Deutsche U-Boot-Verluste von September 1939 bis Mai 1945.

Caram, Ed, comp. **Records and reports of the U-352: WW II German U-boat in the N.C. graveyard of the Atlantic**. 1st ed. [S.l.: s.n.], c1983. 1 v.

_____. **U-352, the sunken German Uboat in the graveyard of the Atlantic.** 1st ed. [s.l.: E. Caram], c1987. 105 p.

Caulfield, Malachy Francis. **A night of terror; the story of the Athenia affair.** London, Muller, [1958]. 222 p.

Cernigoi, Enrico. **U-Boote: battaglia nell'Atlantico.** Firenze: Giunti, c1999. 63 p.

Cheatham, James T. **The Atlantic turkey shoot: U-Boats off the Outer Banks in World War II.** Greenville, NC: Williams & Simpson, [c1990]. 61 p.

Chewning, Alpheus J. **The approaching storm: U-boats off the Virginia coast during World War II.** Lively, VA: Brandylane, c1994. 171 p.

Cosgrove, Brian A. **From the Sea Versus The U-Boat.** Newport, RI: Naval War College Dept of Operations. (ADA279488). 1994. 36 p.
Abstract: This paper will analyze World War II U-boat operations against Allied sealift with focus on the period from May 1943 to the end of the war. It will show the relevance of the operational and strategic decisions of this historical campaign to the challenges of today's potential regional conflicts. In 1943, Allied technological innovations and convoy employment precipitated a decline in U-boat successes and changes to the final portion of the U-boat campaign produced fewer U-boat victories, yet remained an effective operational scheme. It is relevant that the inability of Allied forces to consistently thwart successful U-boat attacks, along their own coastlines, emphasizes a weakness in our Naval Strategy Today, insufficient and usually lightly protected sealift. The Navy and Marine Corps joint White Paper, From the Sea, articulates Navy support of the National Security and National Military Strategies of the United States with a commitment to concentrate more on capabilities required in the complex operating environment of the 'littoral' or coastlines of the earth.
NPS/DKL Location: MICROFORM: ADA279488

Cremer, Peter Erich and Fritz Brustat-Naval. **U 333: the story of a U-boat ace** [translated by Lawrence Wilson]. London: The Bodley Head, 1984. 303 p.
English translation of Ali Cremer, U 333.

_____ and _____. **U-boat commander: a periscope view of the battle of the Atlantic.** [translated by Lawrence Wilson]. Annapolis, MD: Naval Institute Press, c1984. 244 p.
NPS/DKL Location: GENERAL D781 .C7413 1984A

Davis, Donald L. **A New Look at the Battle of the Atlantic.** Newport, RI: Naval War College; Available from National Technical Information Service, Springfield, VA, 1993. (ADA266796). 37 p.
Abstract: *The paper examines the Battle of the Atlantic from an operational rather than the usual strategic perspective. The impressive achievements of the small force of German submarines against such overwhelming odds was a direct result of Admiral Karl Doenitz's skillful practice of the Operational Art. An examination of his attributes and methods may provide useful guidance for the commander of the small, austere force of the future. Superior numbers or technology does not guarantee for military success. Sound doctrine, vision, operational excellence, initiative and audacity, on the other hand, can produce substantial advantages. The paper also cautions that the dramatic allied reconstitution which did so much to turn the tide in the Battle of the Atlantic, is unlikely to reoccur and that the large, costly multi-purpose*

weapons platforms of today may be as ineffective in fighting the low intensity naval battle on the littoral as the large capital ships were in the Battle of the Atlantic.
NPS/DKL Location: MICROFORM ADA266796

Donitz, Karl. **The conduct of the war at sea: an essay**. Washington: Division of Naval Intelligence, 1946. 34 p.

_____. **40 Fragen**. 3., neugestaltete Aufl. München: Bernard & Graefe, c1979. 230 p.
English translation of La guerre en 40 questions.
Earlier editions published as Deutsche Strategie zur See im Zweiten Weltkrieg.

_____. **40 [i.e. Vierzig] Fragen**. Vierte Auflage. München: Bernard & Graefe, c1980. 230 p.

_____. **Deutsche Strategie zur See im Zweiten Weltkrieg; die Antworten des Grossadmirals auf 40 Fragen**. Frankfurt am Main, Bernard & Graefe, 1970. 230 p.

_____. **Memoirs: ten years and twenty days** [translated by R. H. Stevens, in collaboration with David Woodward]. [1st ed.] Cleveland, World Pub. Co., [1959]. 500 p.
English translation of Zehn Jahre und zwanzig Tage.
NPS/DKL Location: GENERAL D781 .D6

_____. **Memoirs: ten years and twenty days** [translated by R. H. Stevens, in collaboration with David Woodward]. Westport, CT: Greenwood Press, 1976, c1959. 500 p.

_____. **Die U-Bootswaffe**. Berlin: E. S. Mittler, c1939. 65 p.

_____. **Zehn Jahre und zwanzig Tage**. Bonn, Athenaum, 1958. 512 p.

_____. **Zehn Jahre und Zwanzig Tage**. [2. Aufl.]. Frankfurt am Main, Athenaum Verlag, 1963. 490 p.

_____. **Zehn Jahre und zwanzig Tage**. [4. Aufl.]. Frankfurt am Main, Bernard & Graefe fur Wehrwesen, 1967. 491 p.

Dupuy, Trevor N. **The naval war in the west: the wolf packs**. London: Ward, 1965. (The illustrated history of World War II; v.5). 68 p.

Edwards, Bernard. **Donitz and the wolf packs**. London: Brockhampton Press, 1999, c1996. 240 p.

Enders, Gerd. **Deutsche U-Boote zum Schwarzen Meer: 1942 - 1944; eine Reise ohne Wiederkehr**. 2. Aufl.. Hamburg; Berlin; Bonn: Mittler, 2001. 130 p.

Ernst, Georg. **Bis zur letzten Stunde: der U-Boot-Krieg 1939 – 1945**. Augsburg: Bechtermünz-Verl., 1999. 223 p.

_____. **Bis zur letzten Stunde: Illusion und Wirklichkeit**. Hamburg; Berlin; Bonn: Mittler, 1995. 223 p.

_____. **Bis zur letzten Stunde: Illusion und Wirklichkeit**. 1. Aufl. [München]: Ullstein, 2003. 377 p.

Essex, James W. **Victory in the St. Lawrence: Canada's unknown war**. Erin, Ontario: Boston Mills Press, c1984. 159 p.

Filipowski, Sean R. **Operation PAUKENSCHLAG: An Operational Analysis**. Newport, RI: Naval War College; Available from National Technical Information Service, Springfield, VA, 1994. (ADA279625). 44 p.
Abstract: *Operation Paukenschlaq, a German U-boat operation against Allied shipping along the East Coast of the United States and Canada in early 1942, is analyzed from the perspective of the operational level of war. The plan and its execution are examined to provide conclusions and lessons learned for future operational planning considerations. Chapter One provides a short historical summary of the German U-boat Force and the Battle of the Atlantic. Chapter Two analyzes the operational design of Paukenschlaq. Chapter Three discusses the execution of the operation. Finally, Chapter Four offers information from the operation which could be useful for future commanders. This analysis of Operation Paukenschlag shows that an operation conceived, planned, and executed in as short as time as Paukenschlaq was, can be successful, provided several critical factors prevail.*
NPS/DKL Location: MICROFORM ADA279625

Fock, Harald. **Marine-Kleinkampfmittel: bemannte Torpedos, Klein-U-Boote, Klein-Schnellboote, Sprengboote; gestern - heute - morgen**. 1. Aufl. Herford: Koehler, 1982. 200 p.

Frank, Wolfgang. **Enemy submarine: the story of Gunther Prien, captain of U47**. London: W. Kimber, 1954. 200 p.
NPS/DKL Location: BUCKLEY D780 .F7

_____. **Prien greift an; nach aufzeichnungen des verfassers an bord undden beim befehlshaber der unterseeboote vorliegenden dienstlichen kriegstagebuchern des korvettenkapitans Gunther Prien**. Hamburg, H. Kohler, [c1942]. 281 p.

_____. **The Sea Wolves; the story of German U-boats at war** [translated by R. O. B. Long]. New York, Rinehart, [1955]. 340 p.
English translation of Die Wölfe und der Admiral.
NPS/DKL Location: BUCKLEY D781 .F7

_____. **The Sea Wolves: the story of German U-boats at war** [translated by R.O.B. Long]. Weidenfeld & Nicolson, 1955. 251 p.
Abridged English translation of Die Wölfe und der Admiral.

_____. **Der Stier von Scapa Flow; Leben und Taten des U-Boot-Kommandanten Günther Prien. Nach Aufzeichnungen des Verfassers an Bord U 47 und den seinerzeit beim Befehlshaber der Unterseeboote vorliegenden dienstlichen Kriegstagebüchern des Korvettenkapitäns Günther Prien, sowie deutschen und englischen Dokumenten der Kriegs- und Nachkriegszeit.** [Oldenburg (Oldb.)] G. Stalling, [1958]. 292 p.

_____. **U-Boote contre les marines alliées.** [Grenoble], Arthaud, [1956]. 429 p.
French translation of Die Wölfe und der Admiral.

_____. **Die Wölfe und der Admiral; der Roman der U-Boote.** Oldenburg (Oldb) G. Stalling, [1953]. 552 p.

_____. **Die Wölfe und der Admiral; Triumph und Tragik der U-Boote.** 3. Aufl. Oldenburg (Oldb), G. Stalling, [1957]. 580 p.

_____. **Die Wölfe und der Admiral: [U-Boote im Kampfeinsatz - Triumpf und Tragik].** 6. Aufl. Bergisch Gladbach: Lübbe, 1995. 752 p.

_____. **Die Wölfe und der Admiral: U-Boote im Kampfeinsatz.** [Augsburg]: Bechtermünz, 2001. 751 p.

Franks, Norman L. R. **Conflict over the bay.** London: William Kimber, 1986. 284 p.

Freyer, Paul Herbert. **Der Tod auf allen Meeren.** (2., durchges. Aufl.). [Berlin], Deutscher Militärverl, [1971]. 390 p.
Originally published: Berlin, Deutscher Militärverl., 1970.

_____. **Der Tod auf allen Meeren.** (4., durchges. Aufl.). [Berlin] Militärverl. d. DDR, [1973]. 390 p.

_____. **Die Versenkung der Athenia.** [Berlin], Deutscher Militärverl., [1970]. (Tatsachen, 104). 61 p.

Fuhren, Franz. **Kapitanleutnant Schepke erzahlt.** Minden (Westf.), W. Kohler, 1943. 26 p.
Biography of the Commander, U-100, rammed and depth charged by the British destroyers HMS Walker and HMS Vanoc.

Gallery, Daniel V. **Twenty million tons under the sea.** Chicago, H. Regnery Co., 1956. 344 p.
History of the U 505.
NPS/DKL Location: BUCKLEY D782.U18 G2

_____. **Twenty million tons under the sea.** Annapolis, MD: Naval Institute Press, [2001]. (Bluejacket books). 344 p.

Originally published: Chicago: Regnery, 1956.

Gamelin, Paul. **Les bases sous-marines allemandes de l'Atlantique et leurs defenses: 1940-1945**. La Baule: Editions des Paludiers, c1981. 99 p.

Gannon, Michael. **Black May**. 1st ed. New York: HarperCollins Publishers, c1998. 492 p.
NPS/DKL Location: GENERAL D781 .G356 1998

_____. **Operation Drumbeat: the dramatic true story of Germany's first U-boat attacks along the American coast in World War II**. 1st ed. New York: Harper & Row, c1990. 490 p.
NPS/DKL Location: GENERAL D781 .G36 1990

_____. **Operation Paukenschlag: der deutsche U-Boot-Krieg gegen die USA** [Dt. übertr. von Klaus-Dieter Schmidt]. Frankfurt/M; Berlin: Ullstein, 1992. 510 p.
German translation of Operation Drumbeat.

_____. **Operation Paukenschlag: der deutsche U-Boot-Krieg gegen die USA**. [Übers.: Klaus-Dieter Schmidt]. Augsburg: Bechtermünz-Verl., 1997. 510 p.
German translation of Operation Drumbeat.

_____. **Operation Paukenschlag: Der deutsche U-Boot-Krieg gegen die USA** [Aus dem Amerikan von Klaus-Dieter Schmidt]. Berlin: Ullstein TB-Vlg., 1998. 550 p.
German translation of Operation Drumbeat.

_____. **Schwarzer Mai: die Entscheidung im U-Boot-Krieg** [aus dem Amerikan von Horst Rehse]. 1. Aufl.. München: Ullstein, 2001. 699 p.
German translation of Black May.

Germany. Wehrmacht. Oberkommando. **1939 gegen England, Berichte und Bilder, herausgegeben vom Oberkommando der Wehrmacht**. Berlin, Zeitgeschichte-Verlag W. Andermann, 1940. 156 p.

Gasaway, E. B. **Grey wolf, grey sea**. New York, Ballantine Books, [1970]. 245 p.
Jochen Mohr and the U-124.

Giese, Otto and James E. Wise, Jr. **Shooting the war: the memoir and photographs of a U-boat officer in World War II**. Annapolis, MD: Naval Institute Press, c1994. 289 p.

Gmeline, Patrick de. **Sous-marins allemands au combat**. Paris: Presses de la Cité, c1997. 567 p.

Great Britain. Admiralty. **His Majesty's submarines** [prepared for the Admiralty by the Ministry of Information]. London, H. M. Stationery Off. [c1945]. 64 p.

Great Britain. Central Office of Information. **The battle of the Atlantic. The official account of the fight against the U-boats, 1939-1945.** [Prepared for the Admiralty and the Air Ministry]. London, H.M. Stationery Off., [1946]. 104 p.

Grooms, Bruce E. **Critical German Submarine Operations Versus Allied Convoys During March 1943: An Operational Analysis.** Newport, RI: Naval War College; Available from National Technical Information Service, Springfield, VA, 1993. (ADA264185). 46 p.
Abstract: *German submarine operations against allied convoys, during March 1943 is critically analyzed from an operational perspective. The theater commander's operational scheme is dissected for the purpose of identifying lessons which can be applied to the planning and execution of today's theater operations. A brief historical account of the early phases of the war and the events and decisions which preceded the critical convoy battles will be followed by an analysis of the operational scheme employed by Admiral Doenitz. German victory during the spring offensive clearly demonstrated numerous operational successes, a reasonably well conceived operational plan, and proof positive of the potential for a larger scale victory, yet history recorded Germany's ultimate defeat in the Battle of the Atlantic. This analysis identified three significant flaws which led to the German demise; first, strategic guidance and operational means were inadequately reconciled.*
NPS/DKL Location: MICROFORM ADA264185

Grossmith, Frederick. **The sinking of the Laconia: a tragedy in the battle of the Atlantic.** Stamford: P.Watkins, c1994. 236 p.

Grove, Eric J., ed. **The defeat of the enemy attack on shipping, 1939-1945.** Aldershot, Hant; Brookfield, VT: Ashgate for the Navy Records Society, c1997. (Publications of the Navy Records Society; vol. 137). 380 p.

Hadley, Michael L. **U-boats against Canada: German submarines in Canadian waters.** Kingston: McGill-Queen's University Press, c1985. 360 p.

_____. **U-Boote gegen Kanada: Unternehmungen deutscher U-Boote in kanadischen Gewässern** [übertr. von Hans und Hanne Meckel]. Herford; Bonn: Mittler, 1990. 352 p.
German translation of U-Boats against Canada

_____. **U-Boote gegen Kanada: Unternehmungen deutscher U-Boote in kanadischen Gewässern** [Aus dem Engl. von Hans und Hanne Meckel]. Ungekürzte Ausg. Berlin: Ullstein, 1997. 568 p.
German translation of U-boats against Canada

Hahn, Gunther Ernst. **Deutsche Netzleger.** Wolfersheim-Berstadt: Podzun-Pallas-Verlag, 1996. (Marine-Arsenal; Bd. 37). 48 p.

Hardegen, Reinhard. **"Auf gefechtsstationen!": U-boote im einsatz gegen England und Amerika.** Leipzig: Boreas, 1943. 227 p.

Hartmann, Werner. **Feind im fadenkreuz, U-boot auf jagd im Atlantik**. Berlin, Die Heimbucherei, [1942]. 232 p.

Hasslinger, Karl M. **The U-Boat War in the Caribbean: Opportunities Lost**. Newport, RI: Naval War College; Available from National Technical Information Service, Springfield, VA, 1995. (ADA297938). 22 p.
Abstract: *This paper reviews the specific segments of the Battle of the Atlantic that were conducted in and around the Caribbean Sea. The background information explores Germany's political goals and policies in the years prior to the second world war, and the military situation that resulted. The Battle of the Atlantic is reviewed to determine the reasons for sending U-boats to the Caribbean theater, which was at the effective limit of their operational endurance. Further, the operational art aspects of the use of U-boats in the Caribbean theater and the results they achieved are examined in detail. The subsequent withdrawal of U-boats from the Caribbean after only eleven months in the theater is specifically evaluated in light of the personal leadership and operational art abilities of the Command in Chief of the U-boat Arm, Admiral Karl Doenitz. The paper's conclusion is an evaluation of the title question. Despite the acknowledged tactical success of sinking 400 merchant ships, with the loss of only seventeen U-boats, the author concluded that the Germans did not exploit all available opportunities that may have allowed them to achieve an even greater operational success in the prosecution of the Battle of the Atlantic.*
NPS/DKL Location: MICROFORM ADA297938

Herlin, Hans. **Les damnés de l'Atlantique**. Paris, Éditions France-Empire, [1960]. 300 p.
French translation of Verdammter Atlantik.

_____. **Verdammter Atlantik: Schicksale deutscher U-Boot-Fahrer: Tatsachenbericht**. Sonderausg. Dusseldorf: Econ, c1981. 317 p.
Originally published: Hamburg: Nannen Verlag, [1960, c1959].

Herzog, Bodo. **Kapitanleutnant Otto Steinbrinck; die Geschichte des erfolgreichsten U-Boot-Kommandanten in den Gewassern um England**. Krefeld, H. Ruhl, [1963]. 223 p.

_____. **U-Boote im Einsatz. U-Boats in action. 1939-1945. Eine Bilddokumentation** [translated by von Sigrun und Ulrich Elfrath. Dorheim, Podzun-Verlag, [1970]. 256 p.

_____ und Gunter Schomaekers. **Ritter der Tiefe--Graue Wölfe; die erfolgreichsten U-Boot-Kommandanten der Welt des Ersten und Zweiten Weltkrieges**. München, Wels, Verlag Welsermuhl, [1965]. 563 p.

Hessler, Gunter, comp. and Andrew J. Withers. **The U-boat war in the Atlantic, 1939-1945: German naval history**. facsimile edition. London: H.M.S.O., 1989. (German naval history). 3 v. in 1.
NPS/DKL Location: GENERAL D781 .U33 1989

Hickam, Homer H. **Torpedo junction: U-boat war off America's East Coast, 1942**. Annapolis, MD: Naval Institute Press, c1989. 367 p.

Hirschfeld, Wolfgang. **Das letzte Boot: Atlantik farewell**. München: Universitas, c1989. 332 p.

_____. **Feindfahrten: das Logbuch eines U-Boot-Funkers**. Wien: Neff, c1982. 384 p.

_____. **Feindfahrten: das Logbuch eines U-Boot-Funkers**. Berlin; Darmstadt; Wien: Dt. Buch-Gemeinschaft, [1984]. 384 p.

_____. **Feindfahrten: das Logbuch eines U-Boot-Funkers**. Gütersloh: Prisma-Verlag, 1986. 384 p.

_____. **Feindfahrten: das Logbuch eines U-Boot-Funkers**. Klagenfurt: Kaiser, c1991. 383 p.

_____. **Hirschfeld: the secret diary of a U-boat**. London: Cassell, 2000. (Cassell military paperbacks). 255 p.

_____. **Hirschfeld: the story of a U-boat NCO, 1940-1946**. Annapolis, MD: Naval Institute Press, c1996. 253 p.
Originally published: London: Leo Cooper, 1996.

Hogel, Georg. **Embleme, Wappen, Malings deutscher U-Boote 1939-1945**. Herford: Koehlers, c1987. 316 p.

_____. **Embleme, Wappen, Malings deutscher U-Boote 1939-1945**. 3. erw. Aufl. Hamburg: Koehler, c1996. 227 p.

_____. **U-boat emblems of World War II**. Atglen, PA: Schiffer, c1999. (Schiffer military history). 227 p.
English translation of Embleme, Wappen, Malings deutscher U-Boote 1939-1945.
NPS/DKL Location: GENERAL V859.G3 H6413 1999

How, Douglas. **Night of the Caribou**. Hantsport, N.S.: Lancelot Press, 1988. 153 p.

Hoyt, Edwin Palmer. **The death of the U-boats**. New York: McGraw-Hill, c1988. 248 p.
NPS/DKL Location: GENERAL D781 .H678 1988

_____. **The sea wolves: Germany's dreaded U-boats of WW II**. New York, Lancer Books, [1972]. 160 p.

_____. **The U-boat wars**. New York: Stein & Day, 1986, c1984. 242 p.
Originally published: New York: Arbor House, c1984.

_____. **U-boats: a pictorial history**. New York: McGraw-Hill, c1987. 289 p.
NPS/DKL Location: GENERAL V859.G3 H68 1987

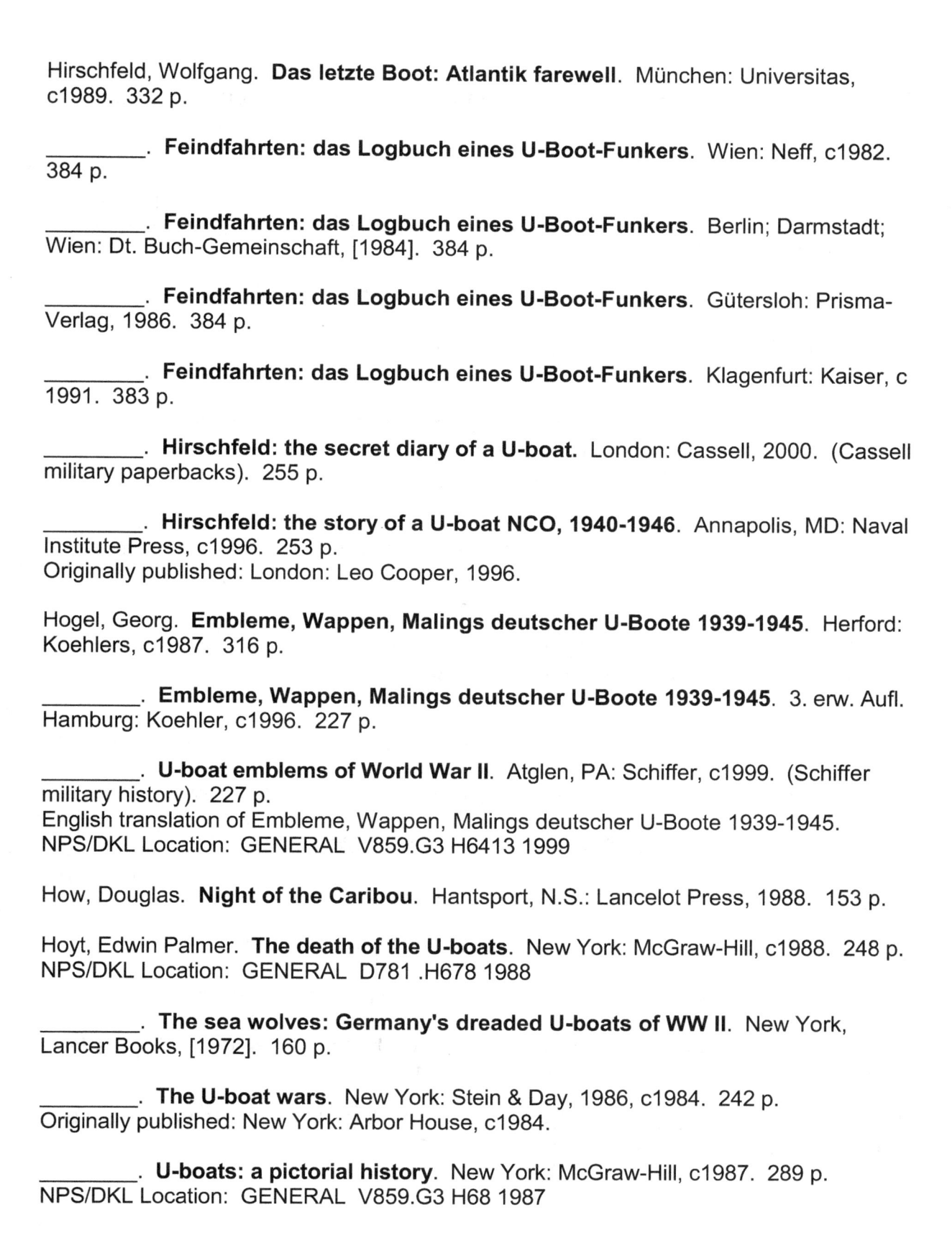

_____. **U-boats offshore: when Hitler struck America**. New York: Stein & Day, c1978. 278 p.
NPS/DKL Location: GENERAL D781 .H84

_____. **U-boats offshore: when Hitler struck America**. 1st Scarborough House pbk. ed. Chelsea, MI: Scarborough House; Chicago, IL: Distributed by Independent Publishers Group, 1990. 278 p.

Ireland, Bernard. **Battle of the Atlantic**. Annapolis, MD: Naval Institute Press, c2003. 232 p.
Originally published: Barnsley [U.K.]: Leo Cooper, 2003
NPS/DKL Location: GENERAL D770 .I735 2003.

Jackson, Carlton. **Allied secret: the sinking of the HMT Rohna**. Norman, OK: University of Oklahoma Press, 2001. 207 p.
Originally published: Annapolis, MD: Naval Institute Press, c1997, as Forgotten tragedy: the sinking of HMT Rohna.

Jackson, Robert. **Kriegsmarine: the illustrated history of the German Navy in WW II**. London: Aurum, 2001. 176 p.
Also published: Osceola, WI: MBI Pub. Co., 2001.

Johr, Barbara, Hartmut Roder and Thomas Mitscherlich. **Der Bunker: ein Beispiel nationalsozialistischen Wahns: Bremen-Farge 1943-45**. Bremen: Edition Temmen, 1989. 68 p.

Jones, Geoffrey Patrick. **Autumn of the U-boats**. London: W. Kimber, 1984. 224 p.

_____. **Defeat of the wolf packs**. London: William Kimber, 1986. 223 p.

_____. **Defeat of the wolf packs**. Novato, CA: Presidio, 1987, c1986. 223 p.

_____. **The month of the lost U-boats**. London: Kimber, 1977. 207 p.

_____. **Submarines versus U-boats**. London: W. Kimber, 1986. 224 p.

_____. **U-boat aces and their fates**. London: Kimber, 1988. 256 p.

Jouan, René. **La Marine allemande dans la seconde guerre mondiale, d'aprés les Conférences navales de Führer**. Paris, Payot, 1949. 304 p.

Just, Paul. **Vom Seeflieger zum Uboot-Fahrer: Feindflüge u. Feindfahrten 1939-1945**. Aufl. Stuttgart: Motorbuch-Verlag, 1979. 220 p.

Kaden, Wolfgang. **Auf Ubootjagd gegen England**. 2. aufl. Leipzig, Hase & Koehler, [1941, reprinted 1942]. 203 p.

Kahn, David. **Seizing the enigma: the race to break the German U-boat codes, 1939-1943**. Boston: Houghton Mifflin Co., 1991. 336 p.
NPS/DKL Location: GENERAL D810.C88 K34 1991

Kaplan, Philip and Jack Currie. **Wolfpack: U-boats at war, 1939-1945**. Annapolis, MD: Naval Institute Press, c1997. 238 p.
Originally published: London: Aurum Press, 1997.
NPS/DKL Location: GENERAL D781 .K37 1997

Karzhavin, Boris Aleksandrovich and Gunter Fuhrmann. **Nemetskaia podvodnaia lodka U-250: novye dokumenty I fakty**. Sankt-Peterburg: Nevskoe vremia, 1996. 71 p.
Original Russian text was published in 1994 in a German translation under title: Das Deutsche Unterseeboot U-250. The present edition was translated from the 1984 German translation into Russian.

Keatts, Henry and George Farr. **U-boats**. Houston, TX: Pisces Books, c1994. 224 p.
Originally published: Kings Point, NY: American Merchant Marine Museum Press, United States Merchant Marine Academy; Eastport, NY: Distributed by Fathom Press, c1986.

Kelshall, Gaylord. **The U-boat war in the Caribbean**. Annapolis, MD: Naval Institute Press, c1994. 514 p.
Originally published: Port of Spain, Trinidad, West Indies: Paria Pub. Co., 1988.
NPS/DKL Location: GENERAL D781 .K45 1994

_____. **U-Bootkrieg in der Karibik 1942 – 1945** [translated by Hans-Jürgen Steffen]. Hamburg; Berlin; Bonn: Mittler, 1999. 336 p.
German translation of The U-boat war in the Caribbean.

Kemp, Paul. **U-boats destroyed: German submarine losses in the World Wars**. Kemp. Annapolis, MD: Naval Institute Press, 1997. 288 p.
NPS/DKL Location: GENERAL D781 .K46 1997

King, Joyce E. **Admiral Karl Doenitz: A Legacy of Leadership**. Newport, RI: Naval War College; Available from National Technical Information Service, Springfield, VA, 1995. (ADA297842). 23 p.
Abstract: *A study of the operational leadership exhibited by ADM Karl Doenitz, Commander in Chief, Submarines, German Navy, during World War II. An examination of his planning, preparation and conduct of the U-boat war in the Atlantic Theater of Operations against the British, and later, the Americans. His objective of sinking the merchant fleet of the British nearly brought Britain to defeat. He displayed great talent in his dedicated fight for resources, innovative tactics of using wolfpacks, his intensive training program and unique command and control system, of these innovations enabled him to maximize use of his limited resources in the optimum way possible to achieve the strategic and operational objectives in the theater. He was also a master of operational maneuver and shifted his focus and his assets within his theater to take advantage of allied vulnerabilities. His personal character traits served him well as he*

inspired trust and unparalleled loyalty from his subordinates. His operational thinking and practice of operational art throughout this campaign remain a relevant model of operational leadership.
NPS/DKL Location: MICROFORM ADA297842

Kludas, Arnold. **Technikmuseum Wilhelm Bauer**. Bremerhaven: A. Kludas, 1984. 44 p.

Koop, Gerhard. **Kampf und Untergang der deutschen U-Boot-Waffe: eine Bilanz in Wort und Bild aus der Sicht des Gegners**. Bonn: Bernard & Graefe, c1998. 223 p.

Korganoff, Alexandre. **Le Mystère de Scapa Flow**. Paris, Arthaud: 1969. 272 p.

_____. **The phantom of Scapa Flow** [translated from the French by W. Strachan and D. M. Strachan]. London: Allan, 1974. 235 p.
English translation of Le Mystère de Scapa Flow.

_____. **Prien gegen Scapa Flow: Tatsachen, Geheimnisse, Legenden** [Übers. ins Dt. besorgte Hans u. Hanne Meckel]. 1. Aufl.. Stuttgart: Motorbuch-Verlag, 1977. 228 p.
German translation of The phantom of Scapa Flow.

_____. **Prien gegen Scapa Flow: Tatsachen, Geheimnisse, Legenden** [Übers. ins Dt. besorgte Hans u. Hanne Meckel]. 3. Aufl.. Stuttgart: Motorbuch-Verl., 1989. 228 p.

_____. **Prien proti Scapa Flow** [Czech transl. by V. Pauer]. - Vyd. 1. - Plzen: Mustang, 1994. 135 p.
Czech translation of The phantom of Scapa Flow.

Kosich, Francis X. **Strategic Implications of the Battle of the Atlantic**. Carlisle Barracks, PA: Army War College; Available from National Technical Information Service, Springfield, VA, 2000. (ADA378290). 26 p.
Abstract: *The battle of the Atlantic is perhaps the most pivotal battle of the Second World War. In it, Germany's use of asymmetric warfare, mines and submarines, once again threatened Britain's economic survival. Although Hitler correctly identified Britain's economy as its center of gravity and had success in attacking it through 1943, he failed to demonstrate the strategic vision necessary to achieve a decisive victory in the Atlantic. Hitler failed because he did not grasp the impact that submarine warfare had in WWII, he wanted a quick, decisive victory like Poland, and he underestimated both the will of the British leadership and the industrial capacity of the United States. Allied success in this campaign enabled the successful prosecution of the war on the European continent through the sallyport of England. From an Allied perspective the Atlantic became the proving ground for the advancement of the carrier-based navy and emerging technology, gave Roosevelt a means with which to invigorate an isolationist society, and gave Britain the time she needed to develop and improve measures to fight the asymmetric threat confronting them. This study looks at the battle of the Atlantic in closer detail while examining Hitler's overarching strategic objectives and those in this decisive theater in an attempt to explain why Hitler allowed it to simply run its course.*
NPS/DKL Location: MICROFORM ADA378290
Electronic access: http://handle.dtic.mil/100.2/ADA378290

Kraft, Helmut J. **Submarinos alemanes en Argentina**. Córdoba: Editorial PUMA, 1998. 171 p.

Krause, Günter. **U-Boot-Alarm: zur Geschichte der U-Boot-Abwehr 1914-1945**. Berlin: Brandenburgisches Verlagshaus, c1998. 255 p.

Kühn, Volkmar [pseud.] -- see Kurowski, Franz.

Kurowski, Franz. [Also uses the following pseudonyms: K(arl) Alman, Heinrich Bernig, Rüdiger Greif, Franz K. Kaufmann, Volkmar Kühn, Jason Meeker, Gloria Mellina, Joh(anna) Schulz, Hermann Schulze-Dierschau, Heinrich Schulze-Dirschau, Franz Kurowski-Tornau].

_____. **An alle Wölfe, Angriff!: deutsche U-Boot-Kommandanten im Einsatz, 1939-1945**. Friedberg/H.: Podzun-Pallas, c1986. 525 p.

_____. **An alle Wölfe: Angriff!: U-Boote, Crews und Kommandanten im Zweiten Weltkrieg 1939 - 1945**. Herrsching: Pawlak, 1990. 525 p.

_____. **An alle Wölfe: Angriff!: U-Boote, Crews und Kommandanten im Zweiten Weltkrieg 1939 - 1945**. Utting: Ed. Dörfler im Nebel-Verl., [2000]. 525 p.

_____. **Angriff, ran, versenken; die U-Bootschlacht im Atlantik**. Rastatt, E. Pabel, [1965]. 326 p.

_____. **Angriff, ran, versenken!: Die U-Boot-Schlacht im Atlantik**. Rastatt: Pabel, c1975. 350 p.

_____. **Angriff, ran, versenken!: die U-Boot-Schlacht im Atlantik**. Rastatt: Moewig, 1998. 350 p.

_____. **Graue Wölfe in blauer See: Der Einsatz d. dt. U-Boote im Mittelmeer**. Rastatt: Pabel, 1967. 310 p.

_____. **Graue Wölfe in blauer See: der Einsatz der deutschen U-Boote im Mittelmeer; Tatsachenbericht**. 8. Aufl. München: Heyne, 1995. 315 p. Originally published: München: Heyne, 1980.

_____. **Günther Prien: der "Wolf" und sein Admiral**. Leoni am Starnberger See: Druffel, 1981. 288 p.

_____. **Günther Prien: ein deutsches U-Boot in der britischen Flottenbasis Scapa Flow**. Rastatt: E. Pabel, c1986. (Deutsche Wehrmacht im II. Weltkrieg). 108 p.

_____. **Jäger der sieben Meere. Die berühmtesten U-Boot-Kommandanten des II. Weltkriegs**. Stuttgart: Motorbuch, 1994. 508 p.

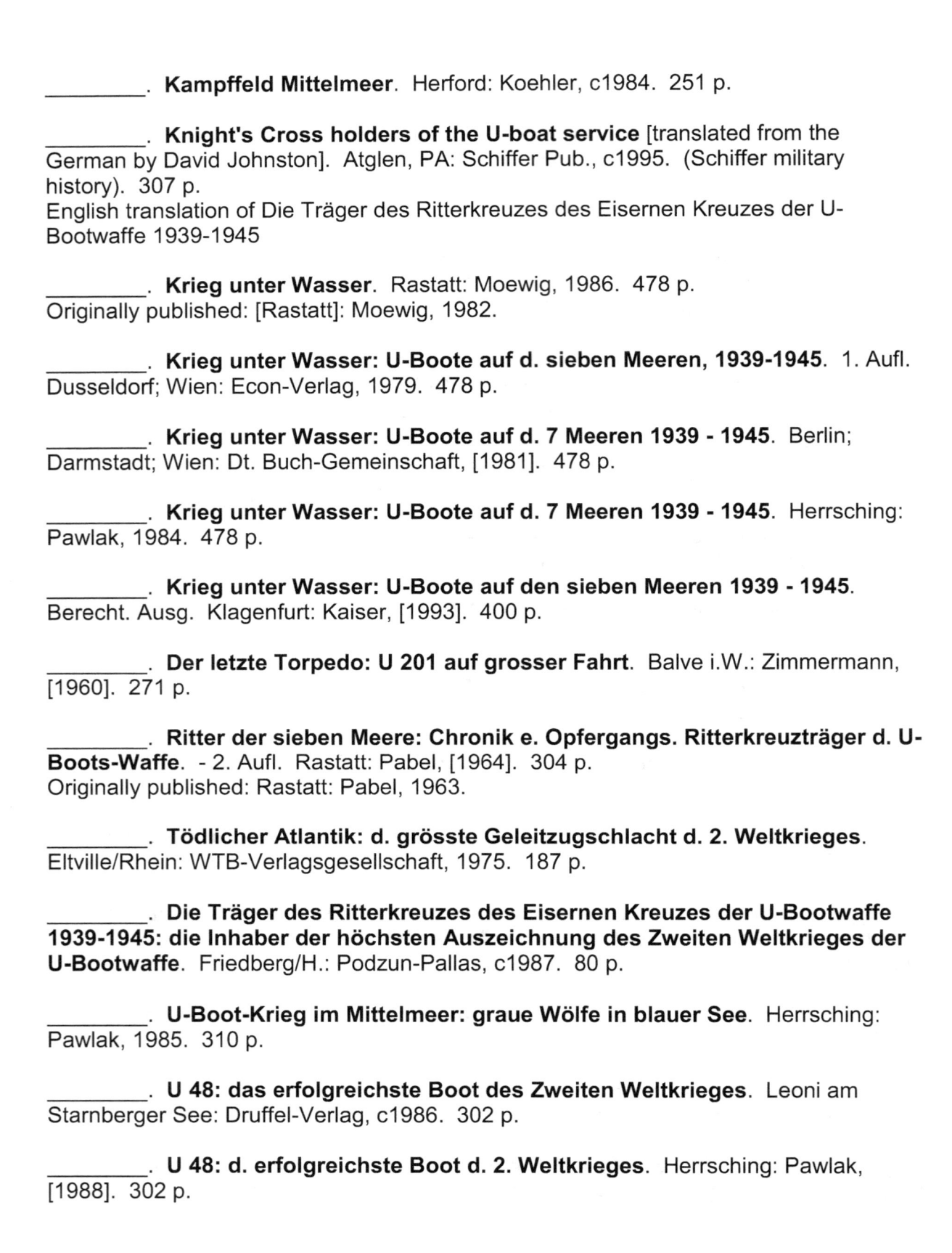

_____. **Kampffeld Mittelmeer**. Herford: Koehler, c1984. 251 p.

_____. **Knight's Cross holders of the U-boat service** [translated from the German by David Johnston]. Atglen, PA: Schiffer Pub., c1995. (Schiffer military history). 307 p.
English translation of Die Träger des Ritterkreuzes des Eisernen Kreuzes der U-Bootwaffe 1939-1945

_____. **Krieg unter Wasser**. Rastatt: Moewig, 1986. 478 p.
Originally published: [Rastatt]: Moewig, 1982.

_____. **Krieg unter Wasser: U-Boote auf d. sieben Meeren, 1939-1945**. 1. Aufl. Dusseldorf; Wien: Econ-Verlag, 1979. 478 p.

_____. **Krieg unter Wasser: U-Boote auf d. 7 Meeren 1939 - 1945**. Berlin; Darmstadt; Wien: Dt. Buch-Gemeinschaft, [1981]. 478 p.

_____. **Krieg unter Wasser: U-Boote auf d. 7 Meeren 1939 - 1945**. Herrsching: Pawlak, 1984. 478 p.

_____. **Krieg unter Wasser: U-Boote auf den sieben Meeren 1939 - 1945**. Berecht. Ausg. Klagenfurt: Kaiser, [1993]. 400 p.

_____. **Der letzte Torpedo: U 201 auf grosser Fahrt**. Balve i.W.: Zimmermann, [1960]. 271 p.

_____. **Ritter der sieben Meere: Chronik e. Opfergangs. Ritterkreuzträger d. U-Boots-Waffe**. - 2. Aufl. Rastatt: Pabel, [1964]. 304 p.
Originally published: Rastatt: Pabel, 1963.

_____. **Tödlicher Atlantik: d. grösste Geleitzugschlacht d. 2. Weltkrieges**. Eltville/Rhein: WTB-Verlagsgesellschaft, 1975. 187 p.

_____. **Die Träger des Ritterkreuzes des Eisernen Kreuzes der U-Bootwaffe 1939-1945: die Inhaber der höchsten Auszeichnung des Zweiten Weltkrieges der U-Bootwaffe**. Friedberg/H.: Podzun-Pallas, c1987. 80 p.

_____. **U-Boot-Krieg im Mittelmeer: graue Wölfe in blauer See**. Herrsching: Pawlak, 1985. 310 p.

_____. **U 48: das erfolgreichste Boot des Zweiten Weltkrieges**. Leoni am Starnberger See: Druffel-Verlag, c1986. 302 p.

_____. **U 48: d. erfolgreichste Boot d. 2. Weltkrieges**. Herrsching: Pawlak, [1988]. 302 p.

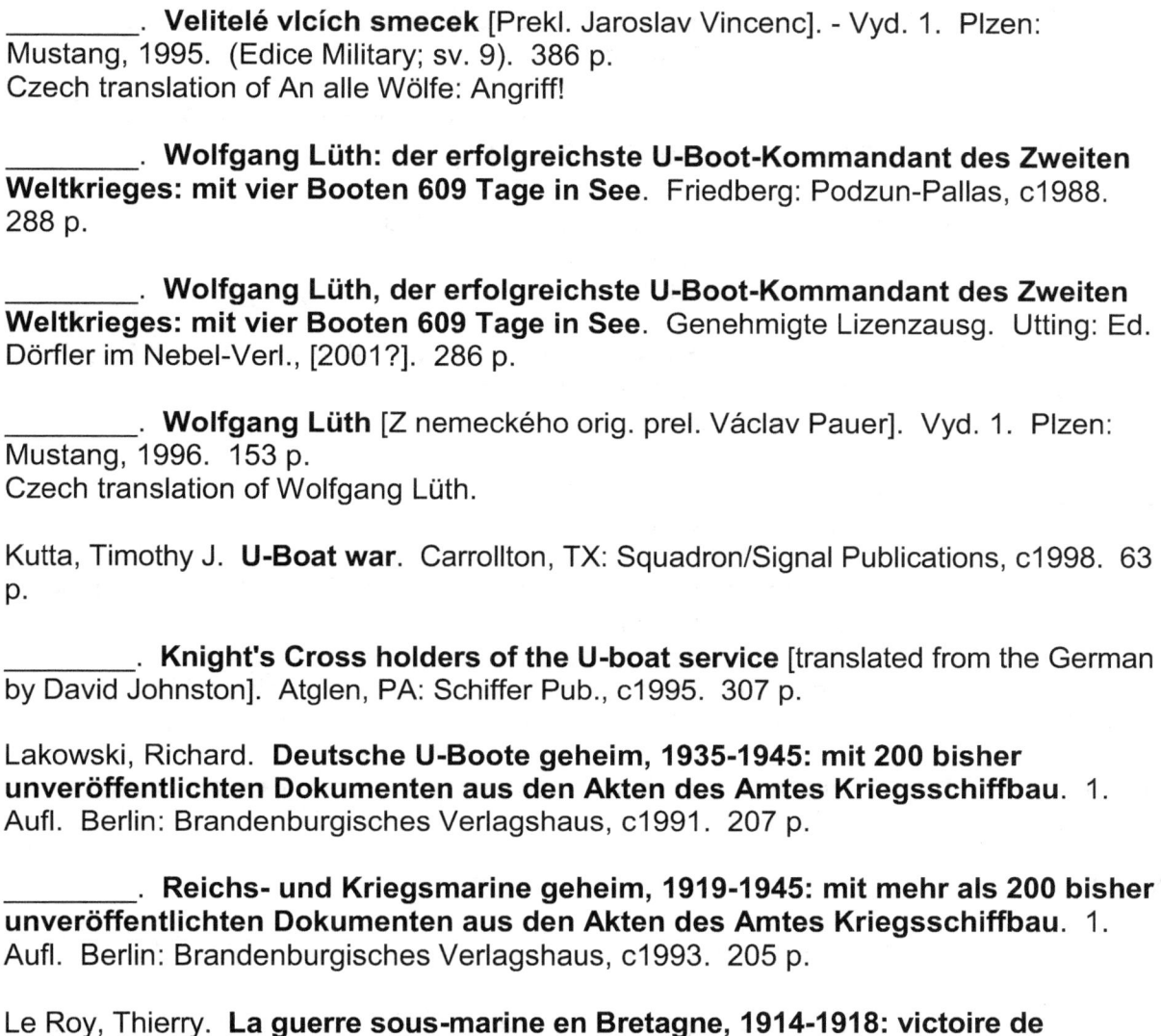

_____. **Velitelé vlcích smecek** [Prekl. Jaroslav Vincenc]. - Vyd. 1. Plzen: Mustang, 1995. (Edice Military; sv. 9). 386 p.
Czech translation of An alle Wölfe: Angriff!

_____. **Wolfgang Lüth: der erfolgreichste U-Boot-Kommandant des Zweiten Weltkrieges: mit vier Booten 609 Tage in See.** Friedberg: Podzun-Pallas, c1988. 288 p.

_____. **Wolfgang Lüth, der erfolgreichste U-Boot-Kommandant des Zweiten Weltkrieges: mit vier Booten 609 Tage in See.** Genehmigte Lizenzausg. Utting: Ed. Dörfler im Nebel-Verl., [2001?]. 286 p.

_____. **Wolfgang Lüth** [Z nemeckého orig. prel. Václav Pauer]. Vyd. 1. Plzen: Mustang, 1996. 153 p.
Czech translation of Wolfgang Lüth.

Kutta, Timothy J. **U-Boat war.** Carrollton, TX: Squadron/Signal Publications, c1998. 63 p.

_____. **Knight's Cross holders of the U-boat service** [translated from the German by David Johnston]. Atglen, PA: Schiffer Pub., c1995. 307 p.

Lakowski, Richard. **Deutsche U-Boote geheim, 1935-1945: mit 200 bisher unveröffentlichten Dokumenten aus den Akten des Amtes Kriegsschiffbau.** 1. Aufl. Berlin: Brandenburgisches Verlagshaus, c1991. 207 p.

_____. **Reichs- und Kriegsmarine geheim, 1919-1945: mit mehr als 200 bisher unveröffentlichten Dokumenten aus den Akten des Amtes Kriegsschiffbau.** 1. Aufl. Berlin: Brandenburgisches Verlagshaus, c1993. 205 p.

Le Roy, Thierry. **La guerre sous-marine en Bretagne, 1914-1918: victoire de l'aeronavale.** [Quimper]: T. Le Roy, [1990]. 254 p.

Lohmann, Walter. **Kameraden auf see, zwischen minen und torpedos.** Berlin, K. Curtius, [1943] 220 p.

Lundeberg, Philip Karl. **American anti-submarine operations in the Atlantic, May 1943-May 1945.** Cambridge, MA: Harvard University, 1954. 2 v.
Thesis (Ph.D.), Harvard University, 1954.

Luth, Wolfgang and Claus Korth. **Boot greift wieder an! Ritterkreuztrager erzahlen.** Berlin, E. Klinghammer, [c1944]. 320 p.

Marbach, Karl-Heinz. **Von Kolberg uber La Rochelle nach Berlin: ein langer Weg von Deutschland nach Europa.** Frankfurt am Main: Haag + Herchen, c1995. 275 p.

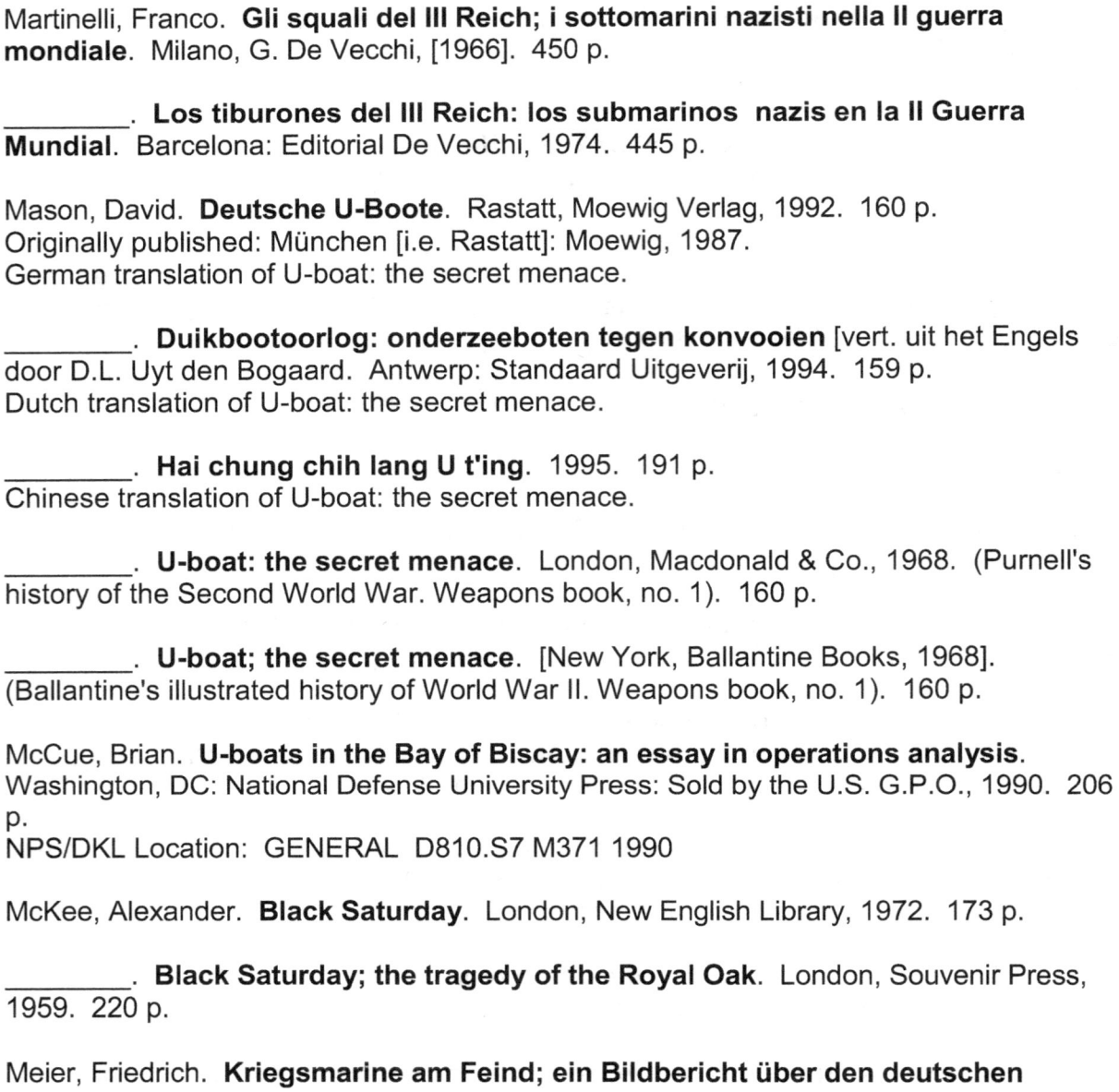

Martinelli, Franco. **Gli squali del III Reich; i sottomarini nazisti nella II guerra mondiale**. Milano, G. De Vecchi, [1966]. 450 p.

_____. **Los tiburones del III Reich: los submarinos nazis en la II Guerra Mundial**. Barcelona: Editorial De Vecchi, 1974. 445 p.

Mason, David. **Deutsche U-Boote**. Rastatt, Moewig Verlag, 1992. 160 p.
Originally published: München [i.e. Rastatt]: Moewig, 1987.
German translation of U-boat: the secret menace.

_____. **Duikbootoorlog: onderzeeboten tegen konvooien** [vert. uit het Engels door D.L. Uyt den Bogaard. Antwerp: Standaard Uitgeverij, 1994. 159 p.
Dutch translation of U-boat: the secret menace.

_____. **Hai chung chih lang U t'ing**. 1995. 191 p.
Chinese translation of U-boat: the secret menace.

_____. **U-boat: the secret menace**. London, Macdonald & Co., 1968. (Purnell's history of the Second World War. Weapons book, no. 1). 160 p.

_____. **U-boat; the secret menace**. [New York, Ballantine Books, 1968]. (Ballantine's illustrated history of World War II. Weapons book, no. 1). 160 p.

McCue, Brian. **U-boats in the Bay of Biscay: an essay in operations analysis**. Washington, DC: National Defense University Press: Sold by the U.S. G.P.O., 1990. 206 p.
NPS/DKL Location: GENERAL D810.S7 M371 1990

McKee, Alexander. **Black Saturday**. London, New English Library, 1972. 173 p.

_____. **Black Saturday; the tragedy of the Royal Oak**. London, Souvenir Press, 1959. 220 p.

Meier, Friedrich. **Kriegsmarine am Feind; ein Bildbericht über den deutschen Freiheitskampf zur See**. Berlin, E. Klinghammer, 1943. 238 p.

Meister, Jürg. **Der Seekrieg in den osteuropäischen Gewässern, 1941-45**. München, J. F. Lehmann, [1958]. 392 p.

Merten, Karl-Friedrich. **Nach Kompass: Lebenserinnerungen eines Seeoffiziers**. Berlin: E.S. Mittler, c1994. 511 p.

Metzler, Jost. **Alaaarm! U-Boot!: Die Feindfahrten von U 69 unter Kapitänleutnant Jost Metzler**. Rastatt: Pabel, 1966. 159 p.
Abridged edition of Die lachende Kuh.

_____. **Die lachende Kuh: Die e. spannenden Abenteuer gleichenden Erlebnisberichte d. Besatzung von U 69 während d. 2. Weltkrieges.** Ravensburg: Veitsburg-Verlag, 1954. 215 p.

_____. **The laughing cow: a U-boat captain's story** [translated by Mervyn Savill]. London, [Eng.]: W. Kimber, 1955. 217 p.
English translation of Die lachende Kuh.
NPS/DKL Location: BUCKLEY D782.U15 M5

_____ and Otto Mielke. **Sehrohr sudwarts!** Berlin, W. Limpert-Verlag, [1943]. 296 p.

Miller, David. **U-boats: the illustrated history of the raiders of the deep.** 1st ed. Washington, DC: Brassey's, c2000. 208 p.
NPS/DKL Location: GENERAL V859.G3 M555 2000

Milner, Marc. **The U-boat hunters: the Royal Canadian Navy and the offensive against Germany's submarines, 1943-1945.** Annapolis, MD: Naval Institute Press, c1994. 326 p.
NPS/DKL Location: GENERAL D784.C2 M55 1994

Ministry of Defence (Navy). **The U-boat war in the Atlantic, 1939-1945: German naval history** / facsimile edition with introduction by Andrew J. Withers. London: H.M.S.O., 1989. (German naval history). 528 p.

Mouton, Patrick. **L'or de Staline: 5 tonnes par 260 metres de fond.** Rueil-Malmaison: Editions du Pen Duick, c1984. 196 p.

Müller, Wolfgang and Reinhard Kramer. **Gesunken und verschollen: Menschen- und Schiffsschicksale, Ostsee 1945.** Herford: Koehlers Verlagsgesellschaft, c1994. 268 p.

Mulligan, Timothy. **Lone wolf: the life and death of U-boat Ace Werner Henke.** Westport, CT: Praeger, 1993. 247 p.
NPS/DKL Location: GENERAL CT8.E6 M84 1993

_____. **Lone wolf: the life and death of U-boat Ace Werner Henke.** Norman: University of Oklahoma Press, c1995. 247 p.
Original published: Westport, CT: Praeger, 1993.

_____. **Die Männer der deutschen U-Bootwaffe 1939** [Dt. übertr. Von Wolfram Schürer]. 1. Aufl. Stuttgart: Motorbuch-Verl., 2001. 368 p.
German translation of Neither sharks nor wolves.

_____. **Neither sharks nor wolves: the men of Nazi Germany's U-boat arm, 1939-1945.** Annapolis, MD: Naval Institute Press, c1999. 340 p.

Originally published: London: Chatham, 1999.
NPS/DKL Location: GENERAL D781 .M85 1999

Naisawald, L. VanLoan. **In some foreign field: four British graves and submarine warfare on the North Carolina Outer Banks**. 3rd rev. ed. Raleigh: Division of Archives and History, North Carolina Dept. of Cultural Resources, c1997. 99 p.
NPS/DKL Location: GENERAL D782.B4 N3 1997

Neary, Steve. **The enemy on our doorstep: the German attacks at Bell Island, Newfoundland, 1942**. St. John's, NF, Canada: Jesperson Press, 1994. 140 p.

Niestle, Axel. **German U-boat losses during World War II: details of destruction**. Annapolis, MD: Naval Institute Press, c1998. 305 p.
NPS/DKL Location: GENERAL D781 .N54 1998

Nofi, Albert A., ed. **The War against Hitler: military strategy in the West**. New York: Hippocrene Books, c1982. 273 p.
Portions of this work appeared previously published in Strategy & tactics magazine.

Õun, Mati. **Võitlused Läänemerel sügis 1941 ja 1942. aasta: järg legendidele, oletustele ja mõningasele tõele võitlustest Läänemerel, selle rannikul ja saartel**. Tallinn: Olion, 1997. 231 p.

_____. **Võitlused Läänemerel suvi 1941: legende, oletusi ja veidi tõttki võitlustest Läänemerel, ta rannikul ja saartel**. Tallinn: M. Õun, 1996. 143 p.

Paterson, Lawrence. **The First U-Boat Flotilla**. Annapolis, MD: Naval Institute Press, 2002. 320 p.
NPS/DKL Location: GENERAL D781 .P38 2002

Peillard, Leonce. **L'affaire du Laconia, 12 septembre 1942**. Ed. rev. et augm. Paris: R. Laffont, c1988. 268 p.
Originally published: Paris, R. Laffont, [1961].

_____. **The Laconia affair** [translated from the French by Oliver Coburn]. [1st American ed.]. New York: Putnam, [1963]. 270 p.
English translation of L'affaire du Laconia, 12 septembre 1942.
British edition published: J. Cape, 1963 as U-Boats to the rescue.
NPS/DKL Location: GENERAL D772 .L23 P3

Prien, Gunther. **I sank the Royal Oak** [translated by Comte de la Vatine]. London, Gray's Inn Press, [1954]. 196 p.
English translation of Mein Weg nach Scapa Flow.
NPS/DKL Location: BUCKLEY D811 .P9

_____. **U-boat commander** [translated by Georges Vatine]. London, New York, Wingate-Baker, 1969. 159 p.
Originally published: Berlin: Deutscher Verlag, 1940, as Mein Weg nach Scapa Flow.

Quinn, John and Alan Reilly. **Covering the approaches: the war against the U-Boats: Limavady and Ballykelly´s role in the Battle of the Atlantic.** Coleraine, N. Ireland: Impact Printing, c1996. 132 p.

Reich, Kurt. **Dem Tommy entwischt. Der geist unserer U-boot-waffe.** Berlin, Deutscher verlag, [c1942]. 205 p.

Reintjes, Karl Heinrich. **Quo vadis Germania?: der Weg vom Kaiseradler zum Bundesadler.** Melle: E. Knoth, c1991. 199 p.

_____. **Quo vadis Germania?: from Weimar to Bonn.** [translated [and supplemented] by Jack Puzey]. 1st English ed. London: Janus, c1997. 237 p.

_____. **U 524: das Kriegstagebuch eines U-Bootes.** Melle: Knoth, 1994. 227 p.

Roberts, Raymond J. **Survivors of the Leopoldville disaster.** Bridgman, MI: R.J. Roberts, [c1995]. 112 p.

Robertson, Terence J. **The Golden Horseshoe.** London, Evans Bros., [1955]. 210 p.

_____. **The Golden Horseshoe.** Cadet edition. London, England: Evans Brothers, 1966. 159 p.

_____. **The Golden Horseshoe--Story of Otto Kretschmer.** Stroud, Gloucestershire: Tempus, 2000. 192 p.

_____. **Night raider of the Atlantic; the saga of the German submarine "The Golden Horseshoe" and her daring commander, Otto Kretschmer.** [1st ed.]. New York, Dutton, 1956. 256 p.
NPS/DKL Location: BUCKLEY D782.U16 R6

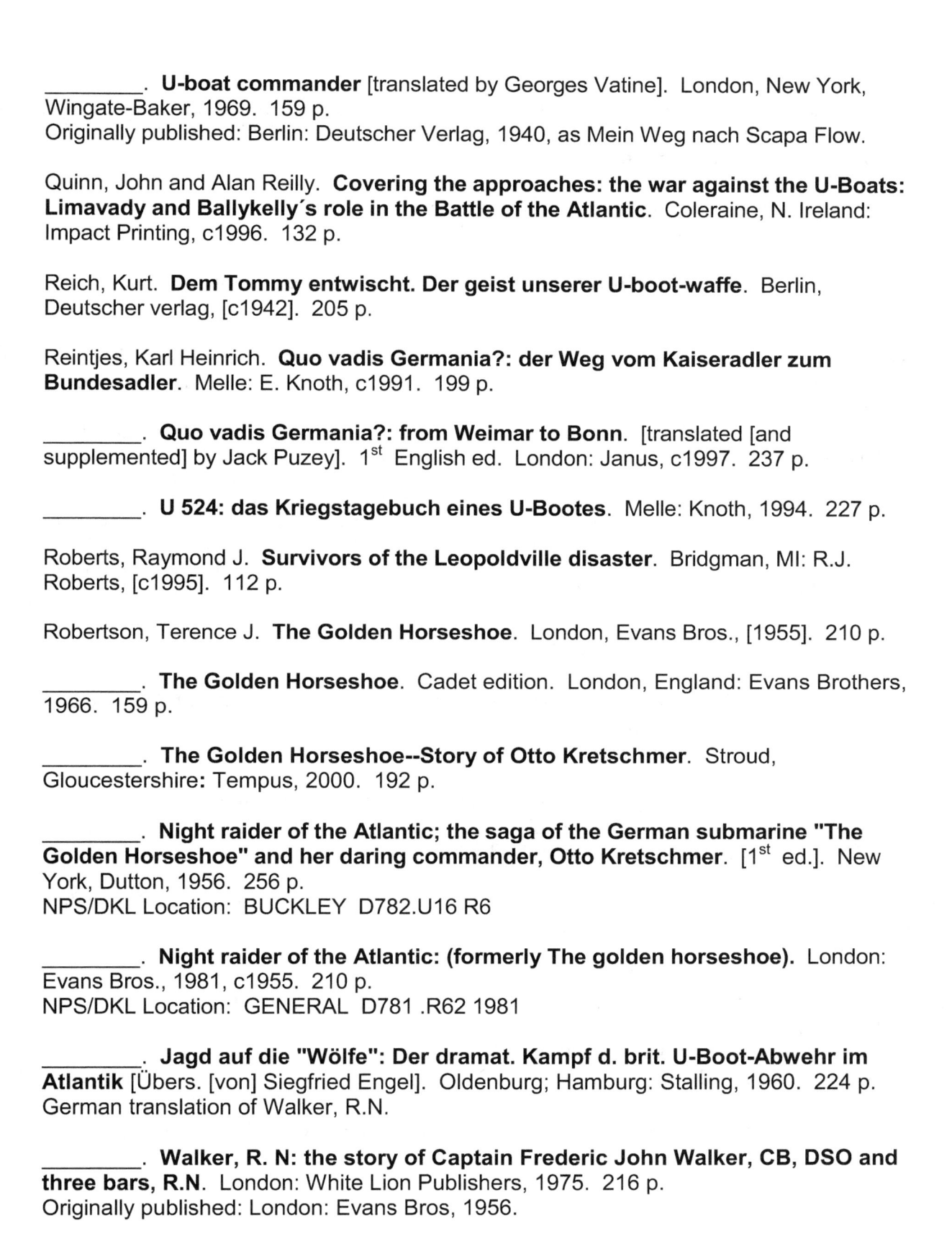

_____. **Night raider of the Atlantic: (formerly The golden horseshoe).** London: Evans Bros., 1981, c1955. 210 p.
NPS/DKL Location: GENERAL D781 .R62 1981

_____. **Jagd auf die "Wölfe": Der dramat. Kampf d. brit. U-Boot-Abwehr im Atlantik** [Übers. [von] Siegfried Engel]. Oldenburg; Hamburg: Stalling, 1960. 224 p.
German translation of Walker, R.N.

_____. **Walker, R. N: the story of Captain Frederic John Walker, CB, DSO and three bars, R.N.** London: White Lion Publishers, 1975. 216 p.
Originally published: London: Evans Bros, 1956.

Roskill, Stephen Wentworth. **Das Geheimnis um U 110** [Übers. aus d. Engl.: Dietrich Niebuhr]. Frankfurt a.M.: Verl. f. Wehrwesen Bernard & Graefe, 1960. 147 p.
German translation of The Secret Capture.

_____. **The Secret Capture**. London: Collins, 1959. 156 p.
Capture of the U-110, recommissioned as HMS Graph.

Ruge, Friedrich. **Im Kustenvorfeld: Minensuchen, Geleit, Ubootsjagd, Vorpostendienst**. Munchen: Lehmann, 1974. 143 p.

Russell, Jerry C. **Ultra and the Campaign Against the U-Boats in World War II**. Carlisle Barracks, PA: Army War College; Available from National Technical Information Service, Springfield, VA, 1980. (ADA089275). 45 p.
Abstract: *The problem addressed is the extent to which the United States Navy used Ultra, or Special Intelligence, in the campaign against the German U-boats. Information was gathered through published and unpublished sources. Through a chronological approach, United States Navy involvement is traced from entry into the war until its conclusion. Many factors are involved in the final outcome of the war and Ultra is only one. The Battle of the Atlantic was long and gruesome rather than short and spectacular. The United States Navy used Ultra along with technology, tactics, brilliant leadership and courageous men at sea to win the Battle of the Atlantic in World War II. The lessons for the future are clear. If the United States intends to oppose the Soviet submarine force at sea anywhere in the world, then we must maintain the lead in intelligence, tactics and technology. Further, and most importantly, we must strive to regain superiority of forces in those ocean areas where our interests are at stake.*
NPS/DKL Location: MICROFORM ADA089275

Savas, Theodore P., ed. **Silent hunters: German U-boat commanders of World War II**. Campbell, CA: Savas Pub. Co., c1997. 215 p.
NPS/DKL Location: ON-ORDER

Scalia, Joseph M. **Germany's last mission to Japan: the failed voyage of U-234**. Annapolis, MD: Naval Institute Press, c2000. 250 p.
NPS/DKL Location: GENERAL D781 .S43 2000

Schaeffer, Heinz. **66 Tage unter Wasser: die geheimnisumwobene U-Boot-Fernfahrt nach Argentinien**. Rastatt: E. Pabel, c1986. (Die Deutsche Wehrmacht im II. Weltkrieg). 125 p.

_____ and Nicholas Monsarrat. **U-boat 977**. [1st American ed.] New York, Norton, [1953, c1952]. 260 p.
Originally published: London: W. Kimber, [1952].
NPS/DKL Location: BUCKLEY D782.U2 S3

Schepke, Joachim. **U-Boot-Fahrer von Heute**. Berlin, Deutscher Verlag [c1940]. 131 p.

Schmeelke, Karl-Heinz and and Michael Schmeelke. **Deutsche U-Bootbunker gestern und heute**. Wölfersheim-Berstadt: Podzun-Pallas, c1996. (Waffen-Arsenal: Special; Bd. 15). 48 p.

_____ and _____. **German U-boat bunkers yesterday and today** [translated from the German by Ed Force]. Atglen, PA: Schiffer Pub. Co., c1999. (Schiffer military history). 48 p.
English translation of Deutsche U-Bootbunker gestern und heute.

Schulz, Joh [pseudo.] -- see Kurowski, Franz.

Schulz, Werner. **Im Kleinst-U-Boot: aus dem Nachlass eines "Seehund"-Fahrers**. 1. Aufl. Berlin: Brandenburgisches Verlagshaus, 1995. 215 p.

_____. **Über dem nassen Abgrund: als Kommandant und Flottillenchef im U-Boot-Krieg**. 3. Aufl.. Berlin; Bonn; Herford: Mittler, 2000. 232 p.
Originally published: Berlin; Bonn; Herford: E.S.Mittler, 1994.

Schütze, Hans G. **Operation unter Wasser**. Herford: Koehler, c1985. 249 p.

Sellwood, Arthur V. **Dynamite for hire; the story of Hein Fehler**. London, W. Laurie, [1956]. 264 p.

Sharpe, Peter. **U-Boat fact file, 1935-1945**. Earl Shilton: Midland Publishing Ltd., 1998. 224 p.

Showell, Jak P. Mallmann. **Deutsche U-Boote an feindlichen Küsten: 1939 - 1945; Kommandounternehmen, Spionage und Sabotage, Versorgungsfahrten** [Ins Dt. übertr. von Wolfram Schürer. Dt. Bearb.: Helma und Wolfram Schürer]. 1. Aufl.. Stuttgart: Motorbuch-Verl., 2002. 191 p.
German translation of U-boats at war.

_____. **Enigma U-boats**. Annapolis, MD: Naval Institute Press, c2000. 192 p.
Originally published: Shepperton: Ian Allan, 2000 as Enigma U-boats: breaking the code.
NPS/DKL Location: GENERAL D810.C88 S56 2000

_____. **Hitler's U-Boat Bases**. Annapolis, MD: Naval Institute Press, 2002. 224 p.

_____. **U-Boat Command and the Battle of the Atlantic**. London: Conway Maritime Press, 1989. 224 p.

_____. **U-boat commanders and crews, 1935-45**. Ramsbury, Marlborough, Wiltshire: Crowood Press, 1998. 144 p.

_____. **U-Boat warfare: the evolution of the Wolf Pack**. Annapolis, MD: Naval Institute Press, 2002. 160 p.

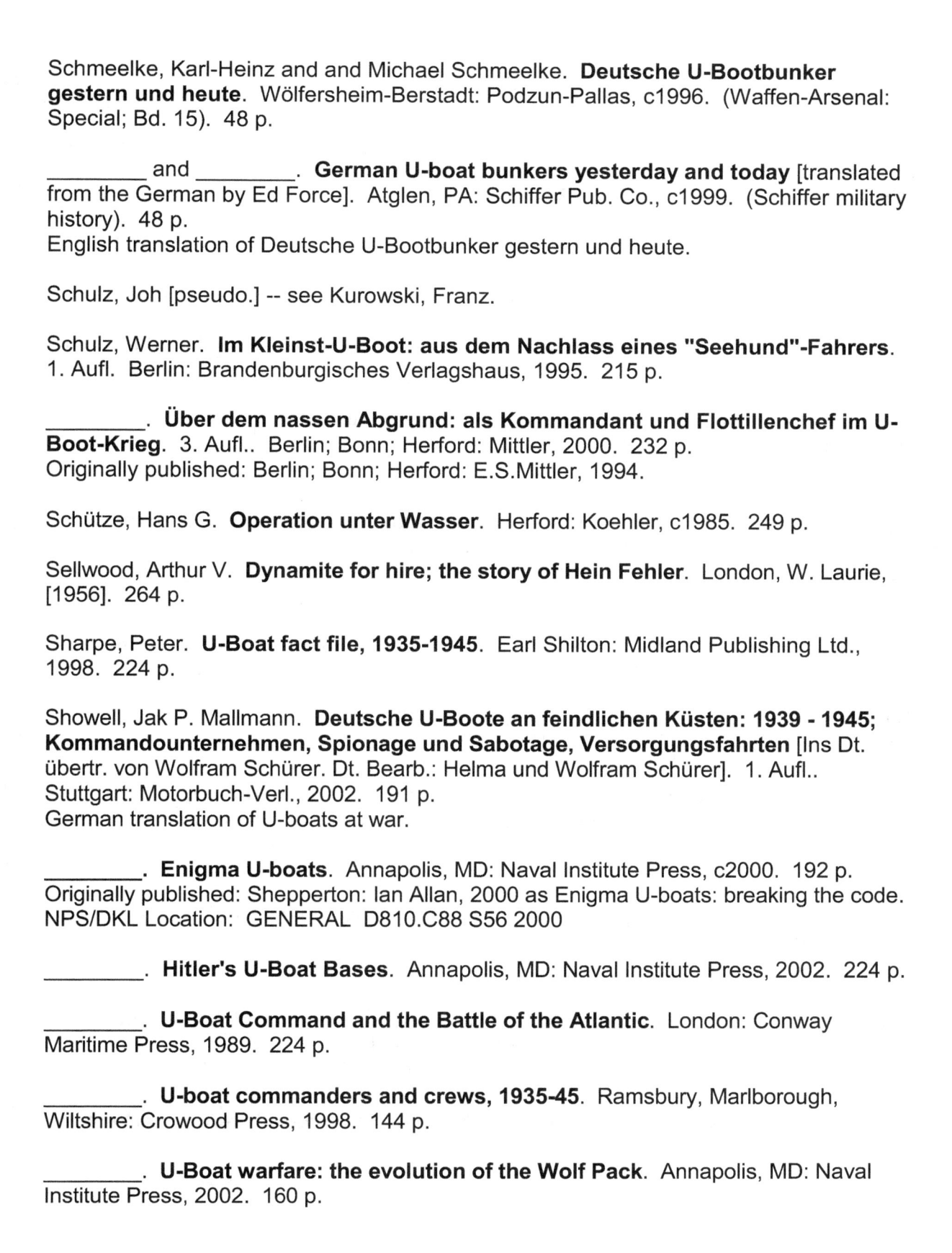

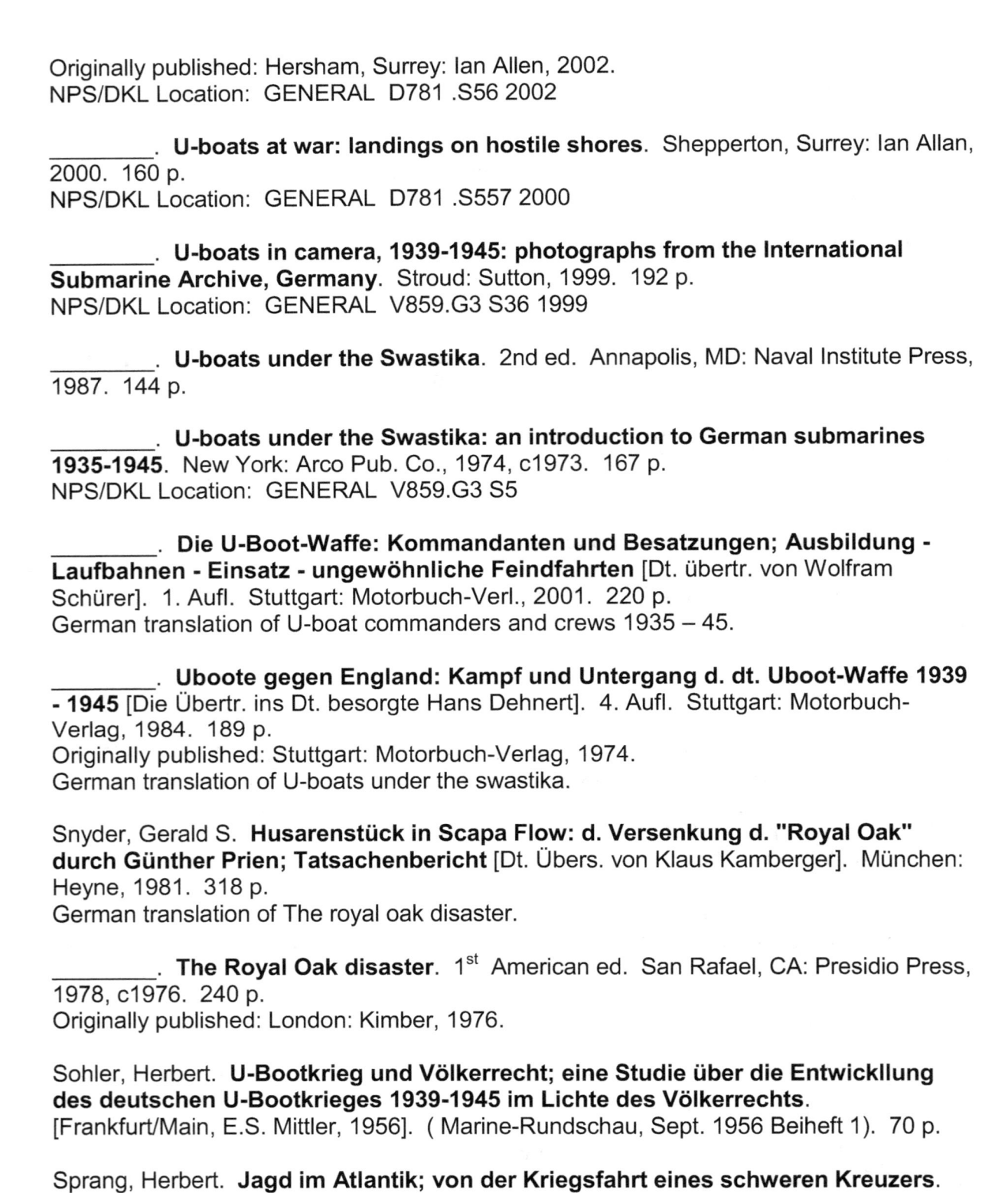

Originally published: Hersham, Surrey: Ian Allen, 2002.
NPS/DKL Location: GENERAL D781 .S56 2002

_____. **U-boats at war: landings on hostile shores**. Shepperton, Surrey: Ian Allan, 2000. 160 p.
NPS/DKL Location: GENERAL D781 .S557 2000

_____. **U-boats in camera, 1939-1945: photographs from the International Submarine Archive, Germany**. Stroud: Sutton, 1999. 192 p.
NPS/DKL Location: GENERAL V859.G3 S36 1999

_____. **U-boats under the Swastika**. 2nd ed. Annapolis, MD: Naval Institute Press, 1987. 144 p.

_____. **U-boats under the Swastika: an introduction to German submarines 1935-1945**. New York: Arco Pub. Co., 1974, c1973. 167 p.
NPS/DKL Location: GENERAL V859.G3 S5

_____. **Die U-Boot-Waffe: Kommandanten und Besatzungen; Ausbildung - Laufbahnen - Einsatz - ungewöhnliche Feindfahrten** [Dt. übertr. von Wolfram Schürer]. 1. Aufl. Stuttgart: Motorbuch-Verl., 2001. 220 p.
German translation of U-boat commanders and crews 1935 – 45.

_____. **Uboote gegen England: Kampf und Untergang d. dt. Uboot-Waffe 1939 - 1945** [Die Übertr. ins Dt. besorgte Hans Dehnert]. 4. Aufl. Stuttgart: Motorbuch-Verlag, 1984. 189 p.
Originally published: Stuttgart: Motorbuch-Verlag, 1974.
German translation of U-boats under the swastika.

Snyder, Gerald S. **Husarenstück in Scapa Flow: d. Versenkung d. "Royal Oak" durch Günther Prien; Tatsachenbericht** [Dt. Übers. von Klaus Kamberger]. München: Heyne, 1981. 318 p.
German translation of The royal oak disaster.

_____. **The Royal Oak disaster**. 1st American ed. San Rafael, CA: Presidio Press, 1978, c1976. 240 p.
Originally published: London: Kimber, 1976.

Sohler, Herbert. **U-Bootkrieg und Völkerrecht; eine Studie über die Entwickllung des deutschen U-Bootkrieges 1939-1945 im Lichte des Völkerrechts**.
[Frankfurt/Main, E.S. Mittler, 1956]. (Marine-Rundschau, Sept. 1956 Beiheft 1). 70 p.

Sprang, Herbert. **Jagd im Atlantik; von der Kriegsfahrt eines schweren Kreuzers**. Berlin, Steiniger, [1942]. 158 p.

Standard Oil Company and Panama Transport Company. **Ships of the Esso fleet in World War II**. New Jersey: Standard Oil Company, 1946. 530 p.
Records of 135 ocean tankers of the Standard Oil Company (New Jersey) and the Panama Transport Company.
NPS/DKL Location: GENERAL D773 .S7

Stempniewicz, Czeslaw. **Bomby na Nowy Jork**. Wyd. 1. Warszawa: Krajowa Agencja Wydawnicza, 1979. 104 p.

Stern, Robert Cecil. **Battle beneath the waves: the U-boat war**. London: Arms & Armour, c1999. 224 p.
NPS/DKL Location: GENERAL D781 .S76 1999

_____. **Type VII U-boats**. London: Brockhampton Press, 1998. 160 p.
Originally published: Annapolis, MD: Naval Institute Press, c1991.

_____. **U-boats of World War Two**. London; New York: Arms & Armour Press; [New York, NY: Distributed in the USA by Sterling Pub. Co., 1988-]. (Warships illustrated; no. 13). v.

Stewart, James W. **Sinking of the Zam Zam: diary of James W. Stewart, with the British American Ambulance Corps, January-September 1941**. [Lahaska, Pa.]: D.A. & M.P. Stewart, c1993. 230 p.

Syrett, David, ed. **The Battle of the Atlantic and signals intelligence: U-boat situations and trends, 1941-1945**. Aldershot, Hants, England; Brookkfield, VT, USA: Published by Ashgate for the Navy Records Society, c1998. (Publications of the Navy Records Society; vol. 139). 628 p.
NPS/DKL Location: GENERAL D810.C88 B37 1998

_____. **The defeat of the German U-boats: the Battle of the Atlantic**. Columbia, S.C.: University of South Carolina Press, c1994. (Studies in maritime history). 344 p.
NPS/DKL Location: GENERAL D780 .S96 1994

Tarrant, V. E. **The last year of the Kriegsmarine: May 1944 - May 1945**. Annapolis, MD: Naval Institute Press, c1994. 256 p.
"The decline and destruction of the few remaining capital ships ... is fully detailed, as is the remarkable history of the U-boat service, which continued to fight at sea right up to the cease-fire in May." -- Dustjacket
NPS/DKL Location: GENERAL D770 .T35 1994

_____. **The last year of the Kriegsmarine: May 1944 - May 1945**. Pbk. ed. London: Arms & Armour; New York, NY: Sterling [distributor], 1996. 256 p.

Taylor, Theodore. **Fire on the beaches**. W.W. Norton, [1958]. 248 p.
NPS/DKL Location: BUCKLEY D780 .T2

Terraine, John. **The U-boat wars, 1916-1945**. New York: Putman, 1989. 841 p.
Simultaneously published: London: L. Cooper, 1989, as Business in great waters: the U-boat wars, 1916-1945.
NPS/DKL Location: GENERAL D591 .T47 1989

Topp, Erich. **Fackeln über dem Atlantik: Lebensbericht eines U-Boot-Kommandanten**. Herford: Mittler, c1990. 287 p.

_____. **The odyssey of a U-boat commander: recollections of Erich Topp** [translated by Eric C. Rust]. Westport, CT: Praeger, 1992. 242 p.
English translation of Fackeln uber dem Atlantik.

Tojo Ramallo, José Antonio. **Lobos acosados: historia de los submarinos alemanes hundidos frente a las costas de Galicia durante la Segunda Guerra Mundial**. Santiago de Compostela: Laverde Ediciones, [2000]. 297 p.
Cuadro cronológico y relación de marinos alemanes fallecidos frente a las costas de Galicia.

United States. National Archives and Records Administration. **Records relating to U-boat warfare, 1939-1945**. Washington, DC: NARA, 1985. (Guides to the microfilmed records of the German Navy, 1850-1945; no. 2). 263 p.
NPS/DKL Location: FEDDOCS AE 1.112/2:2

United States. Naval History Division. **Records of the German Navy, 1850-1945**. Washington, DC: National Archives and Records Service, [1978?]. (National Archives microfilm publications. T; microcopy no. 1022). 4,268 microfilm reels.

Vaeth, Joseph Gordon. **Blimps & U-boats: U.S. Navy airships in the battle of the Atlantic**. Annapolis, MD: Naval Institute Press, c1992. 205 p.

Valvatne, Sigurd. **Med norske ubater i kamp**. [1. oppl.]. [Bergen], J.W. Eide, [1954]. 278 p.

Vause, Jordan. **U-boat ace: the story of Wolfgang Lüth**. Annapolis, MD: Naval Institute Press, 2001. (Bluejacket Paperback Book Series). 264 p.
Originally published: Annapolis: Naval Institute Press, c1990.

_____. **Wolf: U-boat commanders in World War II**. Annapolis, MD: Naval Institute Press, c1997. 249 p.
NPS/DKL Location: GENERAL D781 .V38 1997

Villeneuve-Trans, R. de. **La liberte des mers. Le blocus de l'Allemagne, la guerre sous-marine**. Paris, A. Pedone, 1917. 101 p.

Waddington, C. H. **O.R. in World War 2: operational research against the U-boat.**
London, Elek, 1973. (Histories of science series). 253 p.
NPS/DKL Location: GENERAL V214 .W23

Warnock, A. Timothy. **The U.S. Army Air Forces in World War II: air power versus U-boats: confronting Hitler's submarine menace in the European theater.**
[Washington, DC?]: Air Force History and Museums Program; [Supt. of Docs., U.S. G.P.O., distributor], 1999. 24 p.
NPS/DKL Location: FEDDOCS D 301.82:SU 1

_____. **The U.S. Army Air Forces in World War II: the battle against the U-boat in the American Theater.** [Washington, D.C.?]: Center for Air Force History, [1993?]. 29 p.

Warwick, Colin. **Really not required: memoirs 1939 to 1946.** Edinburgh: The Pentland Press Limited, c1997. 289 p.

Watts, Anthony John. **The U-boat hunters.** London: Macdonald & Jane's, 1976. 192 p.
NPS/DKL Location: GENERAL D780 .W3

Weaver, H. J. **Nightmare at Scapa Flow: the truth about the sinking of HMS Royal Oak.** Peppard Common, Oxfordshire: Cressrelles, 1980. 191 p.

Werner, Herbert A. **Die eisernen Särge** [Vom Verf. autoris. dt. Übers.]. 2. Aufl. Hamburg: Hoffmann und Campe, 1973. 399 p.
Originally published: Hamburg: Hoffmann u. Campe, 1970.
German translation of Iron Coffins.

_____. **Die eisernen Särge** [Vom Verf. autoris. dt. Übers.]. Stuttgart: Dt. Bücherbund, [1974]. 398 p.
German translation of Iron Coffins.

_____. **Die eisernen Särge.** 17. Aufl. München: Heyne, 1995. 399 p.
12. Aufl. Published: München: Heyne, 1987.
German translation of Iron Coffins.

_____. **Iron coffins; a personal account of the German U-boat battles of World War II.** [1st ed.]. New York, Holt, Rinehart & Winston, [1969]. 329 p.
NPS/DKL Location: GENERAL D780 .W45 1969

_____. **Iron coffins: a personal account of the German U-boat battles of World War II.** London: Barker, 1970. 329 p.
Originally published: New York: Holt, Rinehart & Winston, 1969.

_____. **Iron coffins: a personal account of the German U-boat battles of World War II.** Paperback library ed. New York: Paperback Library, 1971. 431 p.

Originally published: New York: Holt, Rinehart & Winston, 1969.

_____. **Iron coffins: a personal account of the German U-boat battles of World War II**. London: Pan Books, 1972. 342 p.
Originally published: New York: Holt, Rinehart & Winston, 1969.

_____. **Iron coffins: a personal account of the German U-boat battles of World War II**. Specially illustrated ed. New York: Bantam, 1978, c1969. 417 p.
Originally published: New York: Holt, Rinehart & Winston, 1969.

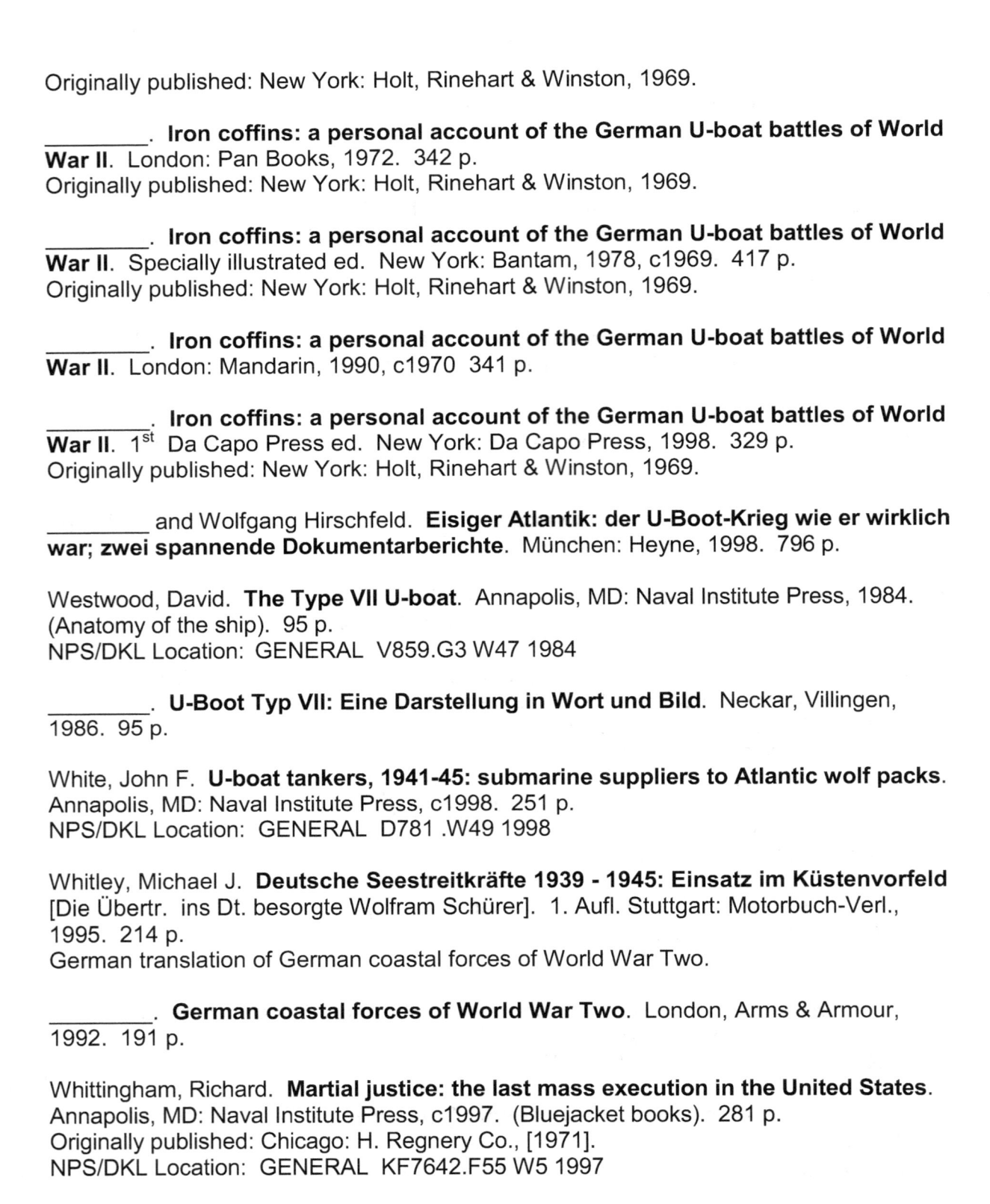

_____. **Iron coffins: a personal account of the German U-boat battles of World War II**. London: Mandarin, 1990, c1970 341 p.

_____. **Iron coffins: a personal account of the German U-boat battles of World War II**. 1st Da Capo Press ed. New York: Da Capo Press, 1998. 329 p.
Originally published: New York: Holt, Rinehart & Winston, 1969.

_____ and Wolfgang Hirschfeld. **Eisiger Atlantik: der U-Boot-Krieg wie er wirklich war; zwei spannende Dokumentarberichte**. München: Heyne, 1998. 796 p.

Westwood, David. **The Type VII U-boat**. Annapolis, MD: Naval Institute Press, 1984. (Anatomy of the ship). 95 p.
NPS/DKL Location: GENERAL V859.G3 W47 1984

_____. **U-Boot Typ VII: Eine Darstellung in Wort und Bild**. Neckar, Villingen, 1986. 95 p.

White, John F. **U-boat tankers, 1941-45: submarine suppliers to Atlantic wolf packs**. Annapolis, MD: Naval Institute Press, c1998. 251 p.
NPS/DKL Location: GENERAL D781 .W49 1998

Whitley, Michael J. **Deutsche Seestreitkräfte 1939 - 1945: Einsatz im Küstenvorfeld** [Die Übertr. ins Dt. besorgte Wolfram Schürer]. 1. Aufl. Stuttgart: Motorbuch-Verl., 1995. 214 p.
German translation of German coastal forces of World War Two.

_____. **German coastal forces of World War Two**. London, Arms & Armour, 1992. 191 p.

Whittingham, Richard. **Martial justice: the last mass execution in the United States**. Annapolis, MD: Naval Institute Press, c1997. (Bluejacket books). 281 p.
Originally published: Chicago: H. Regnery Co., [1971].
NPS/DKL Location: GENERAL KF7642.F55 W5 1997

Wiggins, Melanie. **Torpedoes in the Gulf: Galveston and the U-boats, 1942-1943**. 1st ed. College Station: Texas A&M University Press, c1995. (Texas A&M University military history series; 40). 265 p.
NPS/DKL Location: GENERAL D781 .W54 1995

_____. **U-boat adventures: firsthand accounts from World War II**. Annapolis, MD: Naval Institute Press, c1999. 250 p.
Contents: 1. Jurgen Wattenberg, U-162; 2. Georg Hogel, U-30, U-110; 3. Otto Kretschmer, U-35, U-23, U-99; 4. Josef Erben, U-128; 5. Hans Burck, U-67; 6. Ernst Gothling, U-26; 7. Hermann Frubrich, U-845; 8. Hermann Wien, U-180; 9. Otto Dietz, U-180, U-505; 10. Tom Poser, U-222; 11. Wolfgang von Bartenwerffer, U-536; 12. U-682 and U-735; 13. Oskar Kusch, U-154; 14. Siegfried Koitschka, U-616; 15. Hans Georg Hess, U-995; 16. Walter Tegtmeier, U-597, U-718, U-997; 17. Josef Matthes, U-963; 18. Helmut Wehlte, U-903, and Gunther Thiemrodt, U-1205; 19. Peter Petersen, U-518; 20. Wolfgang Heibges, U-999; 21. Peter Marl, U-103, U-196, U-195.
NPS/DKL Location: GENERAL D781 .W55 1999

Wilcox, Robert K. **Japan's secret war**. 1st ed. New York: Morrow, c19985. 236 p.

_____. **Japan's secret war: Japan's race against time to build its own atomic bomb**. New York: Marlowe & Co.: Distributed by Publishers Group West, c1995. 268 p.
Updated to include newly declassified information: U-234 captured with 10 containers marked "Japanese Army ," containing more than 1,000 pounds of unranium oxide.
NPS/DKL Location: GENERAL QC773.3.J3 W55 1995

Windt Lavandier, Cesar de. **La Segunda Guerra Mundial y los submarinos alemanes en el Caribe**. San Pedro de Macoris, Republica Dominicana: Universidad Central del Este, 1982. 362 p.

_____. **La Segunda Guerra Mundial y los submarinos alemanes en el Mar Caribe**. 2. ed. ampliada y corr. Santo Domingo, República Dominicana: Editora Amigo del Hogar, 1997. 414 p.
Rev. edition of: La Segunda Guerra Mundial y los submarinos alemanes en el Caribe, 1982.

Woltereck, Heinz. **Seekrieg im Osten; der kampf der deutschen Kriegsmarine gegen die Sowjets**. Leipzig, Quelle & Meyer, 1943. 252 p.

Wynn, Kenneth G. **U-boat operations of the Second World War**. Annapolis, MD: Naval Institute Press, c1997-[c1998]. 2 v.
Contents: v. 1. Career histories, U1-U510; v. 2. Career histories, U511-UIT25.
NPS/DKL Location: GENERAL D781 .W96 1997

SUBMARINES -- GREEK -- GENERAL

Tzamtzes, A. I. **Ta Hellenika hyperokeania 1907-1977: to chroniko mias epoches tes neoteres emporikes mas nautilias pou ekleise oristika**. [Athena]: Ekdoseis Miletos, [1997?]. 159 p.

SUBMARINES – INDIAN -- GENERAL

Koteswara Rao, M. V. R. **Environment in submarines**. New Delhi: Defence Research & Development Organisation, Ministry of Defence, c2001. (DRDO monographs/special publications series). 124 p.

SUBMARINES – ISRAELI -- GENERAL

Eldar, Maik. **Dakar: ve-sipurah shel shayetet ha-tsolelot**. Tel-Aviv: Modan: Aryeh Nir, c1997. 300 p.
Title on t.p. verso: Dakar and the story of the Israeli submarines.

Erell, Shlomo. **Diplomatyah be-ma'amake ha-yam**. Or Yehudah: Hed artsi, c2000. 191 p.
Title on t.p. verso: Submarine diplomacy.

Husayn, Jamal al-Din. **Sirr al-ghawasah Dakar: safahat mansiyah min tarikh al-sira maa Israil**. [Cairo: s.n., 1998]. 131 p.

SUBMARINES -- ITALIAN -- GENERAL

Giorgerini, Giorgio. **Uomini sul fondo: storia del sommergibilismo italiano dalle origini a oggi**. 1. ed. Milano: A. Mondadori, 1994. 711 p.

Pollina, Paolo M. and Aldo Cocchia. **I sommergibili italiani, 1895-1962**. Roma, 1963. 300 p.

SUBMARINES -- ITALIAN -- WWI

Bravetta, Ettore. **L'Insidia sottomarina e come fu debellata: con notizie sul recupero delle navi affondate**. Milano: Ulrico Hoepli, 1919. 461 p.

Fraccaroli, Aldo. **Italian warships of World War I**. Shepperton, Allan, 1970. 304 p.

Milanesi, Guido. **Sommergibili; il "Monge"--l' "H. 3"--l' "U. C. 12"--i Nostri ... Pubblicazione dell'Ufficio speciale del Ministero della marina.** Milano, Alfieri & Lacroix, [c1917]. 64 p.

_____. **Submarines: the "Monge"--the "H.3"--the "U.C.12"--Ours ... Publications of the Special department of the Ministry of marine.** Milan, Alfieri & Lacroix, [1916?]. 56 p.

SUBMARINES -- ITALIAN – WWII

Cocchia, Aldo. **The hunters and the hunted; adventures of Italian naval forces** [translated by M. Gwyer]. Annapolis, U.S. Naval Institute, [1958]. 179 p.
English translation of Sommergibili all'attacco.

_____. **The hunters and the hunted: adventures of Italian naval forces** [translated by M. Gwyer]. New York: Arno Press, 1980, c1958. (Navies and men). 178 p.
Reprint of the edition published by United States Naval Institute, Annapolis, MD, 1958.
English translation of Sommergibili all'attacco.

_____. **Submarines attacking; adventures of Italian naval forces** [translated by Margaret Gwyer]. London, W. Kimber, [1956]. 204 p.
English translation of Sommergibili all'attacco.
NPS/DKL Location: GENERAL D784.I8 C6

Italy. Marina. Stato maggiore. Ufficio storico. **Sommergibili in guerra: episodi in Mediterraneo.** [Roma: Edizione "Rivista marittima", 1956]. 215 p.

Lazzero, Ricciotti G. **La Decima Mas: [la compagnia di venture del "principe nero]."** 1a ed. Milano: Rizzoli, 1984. 367 p.

Lelj, Massimo. **Torpediniere.** Palermo: Flaccovio, [1942]. 165 p.

Nelli, R. B. **Eroismo italiano sotto i mari. 1940-1943.** Milano, G. De Vecchi, 1968. 574 p.

SUBMARINES -- JAPANESE -- GENERAL

Iijima, Eiichi. **Dairoku Sensuitei fujsezu: Soseki, Sakuma Teicho, Hirose Chusa.** Tokyo: Sozosha, Heisei 6 [1994]. 231 p.

Ogawa Kazuhisa. **Gensen kairo: daisanji sekai taisen wa Nihonkai kara hajimaru.** Tokyo: Kodansha, Showa 59 [1984]. 282 p.

SUBMARINES -- JAPANESE -- WWI

Fukuda, Ichiro. **Sensuikan no hanashi**. 3 [1928]. 392 p.

_____. **Sensuikansen**. Tokyo: Kaigunkenkyusha, 4 [1929]. 360 p.

Ikezaki, Chuko. **Nihon sensuikan**. 4 [1929]. 158 p.

Mori, Kenkichi. **The submarine in war: a study of relevant rules and problems: num inter arma leges silent?** Tokyo: Maruzen, 1931. 185 p.
Issued under the auspices of the Keimeikai.

SUBMARINES -- JAPANESE -- WWII

Boeicho Boei Kenshujo Senshibu. **Sensuikanshi**. Tokyo: Asagumo Shinbunsha, Showa 54 [1979]. (Senshi sosho 98). 491 p.

Boyd, Carl and Akihiko Yoshida. **The Japanese submarine force and World War II**. Annapolis, MD: Naval Institute Press, c1995. 272 p.
NPS/DKL Location: GENERAL D783.6 .B89 1995

_____ and _____. **The Japanese submarine force and World War II**. Annapolis, MD: Naval Institute Press, 2002. (Bluejacket books). 296 p.

Burlingame, Burl. **Advance force-- Pearl Harbor: the Imperial Navy's underwater assault on America**. Kailua, Hawaii: Pacific Monograph, c1992. 480 p.
NPS/DKL Location: GENERAL D767.92 .B87 1992

_____. **Advance force Pearl Harbor**. 1st Naval Institute Press ed. Annapolis, MD: Naval Institute Press, 2002. 481 p.

Carpenter, Dorr and Norman Polmar. **Submarines of the Imperial Japanese Navy**. Annapolis, MD: Naval Institute Press, c1986. 177 p.
Originally published: London: Conway Maritime Press, 1986.

Fukuda, Ichiro. **Sensuikan**. 17-18 [1942-1943]. 2 v.

Fukumitsu, Minoru. **Awa Maru jiken**. 48 (1973). 211 p.

Fukunaga, Kyosuke. **Sensuiakan**. Tokyo: Arusu, Showa 16 [1941] (Showa 18 [1943] printing). (Arusu bunka sosho; 6). 98 p.

Hara, Genji. **I 17-sen funsenki**. Shohan. Tokyo: Asahi Sonorama, 1988. (Koku senshi shirizu; 100). 466 p.

Hashimoto, Iko. **Igo gojuhachi kitoseri**. 50 [1975]. 254 p.

Hashimoto, Mochitsura. **Sunk; the story of the Japanese submarine fleet, 1941-1945**. [1st American ed.]. New York, Holt, [1954]. 276 p.
English translation of I-go 58 kitoseri.
NPS/DKL Location: GENERAL D784.J3 H3

_____. **Sunk; the story of the Japanese submarine fleet, 1942-1945** [translated by E. H. M. Colegrave]. London: Cassell, 1954. 218 p.
English translation of I-go 58 kitoseri.

_____. **Sunk; the story of the Japanese submarine fleet, 1942-1945** [translated by E. H. M. Colegrave]. London: Hamilton, 1955. 192 p.
English translation of I-go 58 kitoseri.

Higashinaka, Mitsuo. **Itsuwari no patonashippu: Nichi-Bei anpo-- jiji kokkoku**. Shohan. Osaka-shi: Seifudo Shoten, 1989. 246 p.

Hori, Motoyoshi. **Sensuikan**. 1959. 312 p.

_____. **Sensuikan**. 1980. 409 p.

Imaizumi, Kijiro. **Tayuminaki shingeki**. 45 [1970]. 501 p.

Inaba, Michimune. **Kaitei juichimankairi**. 1980. 314 p.

_____. **Shinro higashi e**. [31 i.e. 1956]. 212 p.

Isamukai, Kankokai. **Isamusen omoide no ki**. Osaka-shi: Isamukai Kankokai, Showa 59 [1984]. 436 p.

Ishikawa, Kotaro. **Sensuikan I-16-go tsushinhei no nisshi**. Tokyo: Soshisha, 1992. 223 p.

Itakura, Hidenobu. **Itakura kancho sensuikan senki**. 36 [1961]. 262 p.

Itakura, Mitsuma. **Aa Igo sensuikan**. 1979. 262 p.

_____. **Dongame kancho seishunki: Igo fuchin sensui kancho no kiroku**. Tokyo: Kojinsha, Showa 59 [1984]. 238 p.

_____. **Zoku Aa Igo sensuikan**. 55 [1980]. 222 p.

Iura, Shojiro. **Sensuikantai**. 28 [1953]. 269 p.

_____. **Sensuikantai**. Shinsoban shohan. Tokyo: Asahi Sonorama, Showa 58 [1983]. (Koku senshi shirizu; 25). 398 p.

Jenkins, David. **Battle surface!: Japan's submarine war against Australia 1942-44**. Milsons Point, N.S.W: Random House Australia, 1992. 304 p.

Kaigunsho Kuroshiokai. **Kesshi senko juyushi**. 18 [1943]. 306 p.

Kimata, Jiro. **Nihon sensuikan senshi**. Shohan. Tokyo: Tosho Shuppansha, 1993. 925 p.

_____. **Rengo kantai sensuikan**. 52 [1977]. 253 p.

Koro Keikaishi Hensankai. **Nihon no sokai: koro keikai gojunen no ayumi**. Tokyo: Kokusho Kankokai, Heisei 4 [1992]. 205 p.

Kurihara, Ryuichi. **Kohyoteki**. 49 [1974]. 235 p.

Kurosaki, Teijiro. **Muteki sensuikan**. [17 i.e. 1942]. 218 p.

Kurzman, Dan. **Fatal voyage: the sinking of the USS Indianapolis**. New York: Atheneum, 1990. 331 p.
NPS/DKL Location: GENERAL D774.I5 K87 1990

_____. **Fatal voyage: the sinking of the USS Indianapolis**. 1st Broadway Books paperback ed. Broadway Books, 2001. 352 p.
Originally published: New York: Atheneum, 1990.

Lech, Raymond B. **The tragic fate of the U.S.S. Indianapolis: the U.S. Navy's worst disaster at sea**. 1st Cooper Square Press ed. New York: Cooper Square Press, 2001. 309 p.
Originally published: New York: Stein & Day, 1982, as All the drowned sailors.

Lewis, Tom. **Sensuikan I-124: a history of the Imperial Japanese Navy fleet submarine sunk in Northern Territory waters**. Darwin: Tall Stories Publications, c1997. 131 p.

Lind, L. J. **Toku-tai: Japanese submarine operations in Australian waters**. [Kenthurst, NSW]: Kangaroo Press, 1992. 160 p.

Mikesh, Robert C. **Aichi M6AI Seiran: Japan's submarine-launched Panama Canal bomber**. [Boylston, Mass.: Monogram Aviation Publications], c1975. 32 p.

Milligan, Christopher and John Foley. **Australian hospital ship Centaur: the myth of immunity**. Hendra, Queensland: Nairana Publications, 1993. 340 p.

Monna, Takaichiro. **Kaitei no shonen hikohei: Kaigun saigo no tokko, fukuryotai no kiroku**. Tokyo: Kojinsha, 1992. 261 p.

Nihon Kaigun Sensuikanshi Kankokai. **Nihon Kaigun sensuikanshi**. Tokyo: Nihon Kaigun Sensuikanshi Kankokai, 1979. 922 p.

Nozawa, Tadashi. **Nippon no kubo sensuikan**. 49 [1974]. 156 p.

Oka, Yukio. **Kaerazaru wakaki eiyutachi no densetsu**. Tokyo: Kojinsha, 1990. (Shogen Showa no senso. Ribaibaru senki korekushon; [4]). 350 p.

Orita, Zenji and Joseph D. Harrington. **I-boat captain**. Canoga Park, CA: Major Books, c1976. 336 p.

Sakamoto, Kaneyoshi. **Nihon sensuikan senshi**. Tokyo: Tosho Shuppansha, 1979. 258 p.

Sano, Hirokazu. **Tokushu senkotei**. 50 [1975]. 277 p.

Sasaki, Hanku and Kijiro Imaizumi. **Chinkon no umi**. 43 (1968). 291 p.

Sato, Goro. **Atarashii genri no suichu yusosen**. 37 [i. e. 1962]. ?? p.

Sato, Tsugio. **Maboroshi no sensui kubo**. Shohan. Tokyo: Tosho Shuppansha, 1989. 273 p.

Shiba, Kinpei. **Special attack flotilla**: [bushido in the War of Greater East Asia]. Tokyo, Japan: Japan Times, 1942. 82 p.

Shiga, Hiroshi. **Gyoraitei no futari: wakaki doki no sakura no shogai**. Tokyo: Kojinsha, Showa 62 [1987]. 221 p.

Shrader, Grahame F. **The Phantom War in the Northwest and an account of Japanese submarine operations on the West Coast, 1941-1942**. [Edmonds, WA, 1970, c1969]. 60 p.

Smith, Alan E. **Three minutes of time: the torpedoing of the Australian hospital ship Centaur**. [Tugun, Qld.]: Lower Tweed River Historical Society, c1991. 78 p.

Stanton, Doug. **In harm's way: the sinking of the USS Indianapolis and the extraordinary story of its survivors**. 1st ed. New York: H. Holt, 2001. 333 p.

Sugawara, Kumao. **I Nijuissen [I-21] funsen no kiroku**. [44 i.e. 1969]. 218 p.

Toyoda, Jo. **Getsumei no wanko**. 49 [1974]. 233 p.

Tsuchiya, Ken'ichi. **Junchu Daiichiji Dainiji Tokubetsu Kogekitai**. Tokyo: Shun'yodo, Showa 18 [1943]. 283 p.

United States. Office of Naval Intelligence. **Japanese naval vessels of World War Two as seen by U.S. Naval Intelligence**. Annapolis, MD: Naval Institute Press, 1987. 1 v.
NPS/DKL Location: REFERENCE VA653.A22 J37 1987

_____. **Japanese naval vessels of World War Two as seen by U.S. Naval Intelligence**. Poole, Dorset: Arms & Armour Press; [New York, NY, U.S.A.: Sterling Pub. Co., distributor], 1987. ca. 500 p.
Originally published 1942-1944 by the U.S. Division of Naval Intelligence as ONI 41-421, Index to all Japanese naval vessels; ONI 41-42, Japanese naval vessels; ONI 220J, Japanese submarines; and ONI 225J, Japanese landing operations and equipment.

Ushijima, Hidehiko. **Kyugunshin wa katarazu**. 1976. 257 p.

Webber, Bert. **Panic! at Fort Stevens: Japanese navy shells Fort Stevens, Oregon in World War -II: documentary**. Medford, OR: Webb Research Group Publishers, c1995. 94 p.

_____. **Silent siege: Japanese attacks against North America in World War II**. [1st ed.]. Fairfield, WA: Ye Galleon Press, c1984. 396 p.

_____. **Silent siege-II: Japanese attacks on North America in World War II = Chinmoku no hangeki: Nihongun no Beihondo kogeki sakusen kiroku**. [2nd ed.]. Medford, OR: Webb Research Group, c1988. 397 p.

_____. **Silent siege-III: Japanese attacks on North America in World War II: ships sunk, air raids, bombs dropped, civilians killed: documentary = Chinmoku no hangeki**. [3rd ed.]. Medford, OR: Webb Research Group, c1992. 301 p.
NPS/DKL Location: GENERAL D768 .W44 1992

Yamaoka, Sohachi. **Tatako sensuikan**. Tokyo: Shunkodo, Showa 19 [1944]. 227 p.

Yoshida, Kikuyoshi. **Igo dai 27 sensuikan no katsuyaku: ango-cho no sensuikan senki**. Tokyo: Senshi Kankokai: Hatsubai Seiunsha, Heisei 7 [1995]. 212 p.

SUBMARINES – KOREAN -- GENERAL

Korea (South). Kukpangbu. **The submarine intrusion incident: North Korea's continuing provocation and the ROK's response**. Seoul: Ministry of National Defense, Republic of Korea, 1997. 22 p.

SUBMARINES -- PERUVIAN -- GENERAL

Ortiz Sotelo, Jorge. **Apuntes para la historia de los submarinos peruanos**. Lima: Biblioteca Nacional del Peru, Fondo Editorial; Asociacion de Historia Maritima y Naval Iberoamericana, 2001. 259 p.

SUBMARINES -- POLISH – WWII

Biskupski, Stanislaw. **ORP: "Orzel" zaginal**. [Wyd. 5]. Warszawa, Nasza Ksiegarnia, [1975]. 217 p.

Pertek, Jerzy. **Dzieje ORP "Orzel."** Wyd. 1. Gdynia: Wydawn. Morskie, 1961. 150 p.

_____. **Dzieje ORP "Orzel."** [Wyd. 2]. [Gdynia] Wydawn. Morskie [1966]. 154 p.

Zukowski, Olgierd. **Stalowe rekiny: Krótka historja i opis lodzi podwodnych**. Warszawa: WINW, 1931. 88 p.

SUBMARINES -- RUSSIAN/SOVIET – GENERAL

Achkasov, V. I and N. B. Pavlovich. **Soviet naval operations in the Great Patriotic War [translated from Russian by the U.S. Naval Intelligence Command Translation Project and members of the Naval Intelligence Command Translations Unit 0166]**. Annapolis, MD: Naval Institute Press, c1981. 393 p.
English translation of Sovetskoe voenno-morskoe iskusstvo v Velikoi Otechestvennoi voine.

Azarov, Vsevolod Borisovich. **Komandir "S-7": o Geroe Sovetskogo Soiuza S.P. Lisine**. Moskva: Izd-vo polit. lit-ry, 1986. (Geroi Sovetskoi Rodiny). 108 p.

Basov, Aleksei Vasil'evich. **Flot v Velikoi Otechestvennoi voine, 1941-1945: Opyt operativ.-strateg. primeneniia**. Moskva: Nauka, 1980. 304 p.

Bazilevskii, Sergei Aleksandrovich. **U kolybeli podvodnogo flota: zapiski konstruktora**. Sankt-Peterburg: Sankt-Peterburgskoe morskoe biuro mashinostroeniia "Malakhit", 1996. (Podvodnoe korablestroenie--proshloe, nastoiashchee, budushchee; Vyp. 8). 163 p.

Berg, John. **Sovjets ubåtar [fackgranskning av övers., bearb. och kompl. för svenska förhållanden: Frank Rosenius]**. Stockholm: Bonnier fakta, 1985. 80 p.
Swedish translation of Naerbilde av Sovjets ubåter.

_____ and John E. Moore. **The Soviet submarine fleet: a photographic survey**. London: Jane's, 1985. 80 p.

English translation of Naerbilde av Sovjets ubater, Oslo, 1984.
NPS/DKL Location: GENERAL V859.S65 B37 1985

Bolgarov, Nikolai Pavlovich. **Rasskazy o podvodnoi lodke**. Mosvka: Voen. Izd-vo, 1960. 230 p.

Breemer, Jan S. **Soviet submarines: design, development, and tactics**. Coulsdon, Surrey, U.K.; Alexandria, VA: Jane's Information Group, 1989. 187 p.
NPS/DKL Location: GENERAL V859.S65 B73 1989

Burleson, Clyde W. **The Jennifer Project**. 1st Texas A&M University Press ed. College Station: Texas A&M University Press, c1997. 183 p.
Originally published: New York: Prentice Hall, N.Y., 1977.
NPS/DKL Location: INTELL VB230 .B97 1997

Burns, Thomas S. **The secret war for the ocean depths: Soviet-American rivalry for mastery of the seas**. 1st ed. New York: Rawson Associates Publishers, c1978. 334 p.
NPS/DKL Location: GENERAL VA50 .B93

Burov, Viktor Nikolaevich. **Otechestvennoe voennoe korablestroenie v tret'em stoletii svoei istorii**. S.-Peterburg: Sudostroenie, 1995. 599 p.

Bykhovskii, Izrail' Adol'fovich. **Geroicheskaia "Pantera"**. [3., ispr. i. dop. izd.]. [Kaliningrad], Kaliningradskoe knizhnoe izd-vo, 1966. 317 p.

Chemesov, Oleg Grigor'evich. **V glubinakh Barentseva moria**. Moskva, Voen. izd-vo, 1965. (Geroicheskoe proshloe nashei rodiny). 109 p.

Cherkashin, Nikolai. **Kak pogibaiut submariny: khronika odnoi katastrofy**. Moskva: "Andreevskii flag", 1995. (Vakhtennyi zhurnal. Avariinaia trevoga).
71 p.

_____. **Povsednevnaia zhizn' rossiiskikh podvodnikov**. Moskva: Molodaia gvardiia, 2000. (Zhivaia istoriia Povsednevnaia zhizn' chelovechestva). 556 p.

Dmitriev, Vladimir Ivanovich. **Atakuiut podvodniki**. Izd. 2-e, pererab. i dop. Moskva, Voenizdat., 1973. 368 p.

_____. **Sovetskoe podvodnoe korablestroenie**. Moskva: Voennoe izdatel'stvo, 1990. 286 p.

Dobson, Christopher, John Miller, and Ronald Payne. **The cruelest night**. 1st American ed. Boston: Little, Brown, c1979. 223 p.
Wilhelm Gustloff sunk by the S-13 after leaving Gydnia (Gotenhafen) on January 30, 1945.

_____, _____ and _____. **The cruellest night: Germany's Dunkirk and the sinking of the Wilhelm Gustloff**. London: Hodder & Stoughton, c1979. 223 p.

_____, _____ and _____. **Die Versenkung der "Wilhelm Gustloff"** [Berecht. Übers. von Ilse Winger]. Wien; Hamburg: Zsolnay, 1979. 231 p.
German translation of The cruellest night.

_____, _____ and _____. **Die Versenkung der "Wilhelm Gustloff"** [Berecht. Übers. von Ilse Winger]. Berlin; Darmstadt; Wien: Dt. Buch-Gemeinschaft, [1981]. 231 p.

_____, _____ and _____. **Die Versenkung der "Wilhelm Gustloff": Tatsachenbericht** [Dt. Übers. von Ilse Winger]. Genehmigte, ungekürzte Taschenbuchausg. München: Heyne, 1981. 272 p.

_____, _____ and _____. **Die Versenkung der "Wilhelm Gustloff": Tatsachenbericht** [Dt. Übers. von Ilse Winger]. Genehmigte, ungekürzte Taschenbuchausg. München: Heyne, 1985. 272 p.

_____, _____ and _____. **Die Versenkung der "Wilhelm Gustloff"** [Aus dem Engl. von Ilse Winger]. Ungekürzte Ausg. Frankfurt/M; Berlin: Ullstein, 1995. 231 p.

Dunham, Roger C. **Spy sub: a top secret mission to the bottom of the Pacific**. Annapolis, MD: Naval Institute Press, c1996. 222 p.

Emel'ianov, Leonid Antonovich. **Sovetskie podvodnye lodki v Velikoi Otechestvennoi voine**. Moskva: Voen. izd-vo Ministerstva oborony SSSR, 1981. 189 p.

Favorov, Petr Alekseevich et al. **Anglo-russkii slovar po podvodnym lodkam i protivolodochnoi oborone**. Moskva: Voennoe izdatel'stvo, 1963. 260 p.

Gagin, V. **Sovetskie atomnye podvodnye lodki: [k 300-letiiu Rossiiskogo Flota]**. Voronezh: AO "Poligraf", 1995. (Seriia "Rossiia, prosnis'!"; vyp. 1). 31 p.

Gerdau, Kurt. **Goya, Rettung über See: die grösste Schiffskatastrophe der Welt**. Herford: Koehler, c1985. 175 p.
Sunk by the L-3 after leaving Hela near Danzig (Gdansk). April 16, 1945.

_____. **Ubena: Rettung über See: im Kielwasser des Krieges**. Herford: Koehlers Verlagsgesellschaft, c1985. 191 p.

Gheorghiu, Serban. **Tragedia navelor "Struma" si "Mefkure."** Constanta: Editura Fundatiei A. Saguna, 1998. 269 p.
Sinking of the refugee ship Struma in the Black Sea, February 23, 1942.

Grishchenko, Petr Denisovich. **Sol' sluzhby**. Leningrad: Lenizdat, 1979. 253 p.

Herrick, Robert Waring. **Final report on Soviet naval mission assignments**. Arlington, VA: Ketron, Inc., 1979-. (Report No.: KTR 119-79 - Soviet SSBN roles in strategic strike, KFR 234-79 - The SSBN-protection mission, KFR 235-79 - A protracted witholding role for Soviet SSBNs?). Frederick, MD: University Publications of America,Inc., 1986. (Soviet Union special studies, 1982-1985; 5-6). 2 35 mm microfilm reels. Submitted by Ketron, Inc. to the Assistant Director for Net Assessment (OP-090N), Navy Program Planning Office. N00014-77-C-0708

Huchthausen, Peter A., Igor Kurdin, and R. Alan White. **Hostile waters**. 1st ed. New York: St. Martin's Press, 1997. 303 p.
NPS/DKL Location: GENERAL VA575.K14 H83 1997

_____, _____ and _____. **In feindlichen Gewässern: das Ende von K 219** [Übers. von Jochen Zwikirsch]. 1. Aufl.. Hamburg; Berlin; Bonn: Mittler, 2000. 230 p.
German translation of Hostile waters.

_____, _____ and _____. **Mayim oyenim** [me-Anglit, Rut Vinshtin]. [Holon]: Kineret, 1999. 304 p.
Hebrew translation of Hostile waters.

Huan, Claude. **L'énigme des sous-marins soviétiques**. Paris: Éditions France Empire, [1959]. 297 p.

Ingebrigtsen, Jan. **En studie av Norskehavets strategiske betydning som funksjon av Sovjetunionens nordflåtes operasjoner**. Oslo: [s.n.], 1975. (Norsk utenrikspolitisk institut. NUPI-rapport; nr. 24). 64 p.

Iosseliani, IAroslav. **V bitvakh pod vodoi**. Moskva, Voen. izd-vo, 1959. (Voennye memuary). 270 p.

Izmailov, Semen. **Taina podvodnoi lodki: istoriia odnoi sistemy**. [Haifa?: s.n., 1995?]. 143 p.

Jordan, John. **Soviet submarines: 1945 to the present**. London: Arms & Armour; New York: Distributed by Sterling Pub., 1989. 192 p.
NPS/DKL Location: REFERENCE V859.S65 J67 1989

Kamaev, Valentin Sergeevich. **Na torpednom katere**. Leningrad: Lenizdat, 1977. 160 p.

Kaverin, Veniamin Aleksandrovich. **Der held fun sowetnhfarband**. 1944. 31 p.
The story of Izrail' Fisanovich, commander of the mini-sub M-172..

Khudiakov, Lev IUr'evich. **Podvodnye lodki XXI veka**. Sankt-Peterburg: "Malakhit", 1994. (Podvodnoe korablestroenie: proshloe, nastoiashchee, budushchee; vyp. 5). 60 p.

Kolyshkin, Ivan Aleksandrovich. **J'etais sous-marinier** [traduit du russe par Pierre Riffaud sous la redaction d'Oleg Tarassenkov]. Moscou, Editions du Progres, [1967]. 421 p.
French translation of V glubinakh poliarnykh morei.

_____. **Submarines in Arctic waters; memoirs** [translated from the Russian by David Skvirsky]. Moscow, Progress Publishers, [1966]. 252 p.
English translation of V glubinakh poliarnykh morei.

_____. **Submarines in Arctic waters; memoirs** [translated from the Russian by David Skvirsky]. New York: Universal Distributors Company, 1966. 253 p.
English translation of V glubinakh poliarnykh morei.

_____. **V glubinakh poliarnykh morei**. Moskva, Voen. izd-vo, 1964. (Voennye memuary). 327 p.

_____. **V glubinakh poliarnykh morei**. Izd. 2-e, ispr. i dop. Moskva, Voenizdat, 1970. (Voennye memuary). 328 p.

Konstantinov, Filipp Vasil'evich and . Igor' Petrovich Popov. **Pod periskopom**. Moskva, Voen. izd-vo, 1964. 84 p.

Kopenhagen, Wilfried. **Soviet and Russian nuclear submarines**. Atglen, PA: Schiffer Pub. 41 p.

Korzh, Viktor Emel'ianovich. **Zapas prochnosti**. Moskva, Voen. izd-vo, 1966. (Voennye memuary). 189 p.

Kozhevnikov, Valerii Aleksandrovich, P. Turmov, and G.IU. Illarionov. **Podvodnye lodki Rossii: istoriia i sovremennost'**. Vladivostok: Izd-vo DVGTU, 1995. 166 p.

_____, _____ and _____. **Podvodnye lodki Rossii. Kniga vtoraia: istoriia i sovremennost'**. 2nd ed. Vladivostok: Izd-vo "Ussuri", 1996. 287 p.

Kucher V.A. et al. **Russkie podvodnye lodki: istoriia sozdaniia i ispol'zovaniia 1834-1923 gg.: nauchno-istoricheskii spravchnik**. Sankt-Peterburg: Rubin, 1994. v.

Kuperstein, L. **Megilat Strumah; ketuvah va-`arukhah bi-yede L. Kupershtain**. Tel-Aviv: Hitahdut `ole Romanyah be-Erets-Yi´sra'el, 702 [1942]. 134 p.
Sinking of the refugee ship Struma in the Black Sea, February 23, 1942.

Kurushin, M. **Podvodnaia lodka "Kursk": rozhdenie, zhizn', versii gibeli, podrobnosti**. Moskva: "Olimp": Izd-vo AST, 2000. 157 p.

Leitenberg, Milton and Lawrence Freedman. **Soviet submarine operations in Swedish waters, 1980-1986**. New York: Praeger, 1987. (The Washington papers; 128). 199 p.
NPS/DKL Location: GENERAL VA593 .L45 1987

Livshits, I. A. and IU. A. PanteleeVA. **Podvodnoe korablestroenie v Rossii, 1900-1917; sbornik dokumentov**. [Leningrad], Sudostroenie, 1965. 402 p.

Martino, Ermanno and Alvise Gigante. **Submarines = Sottomarini e sommergibili**. Genoa: Intyrama, c1967. (Soviet fleet today; 1). 64 p.

McCormick, Gordon H. **Stranger than fiction: Soviet submarine operations in Swedish waters**. Santa Monica, CA: Rand Corporation, [1990]. 72 p.
"A Project of Air Force report prepared for the United States Air Force."

Minchenko, Sergei. **Zapiski komandira ekspeditsii osobogo naznacheniia: pravda o poiske, obsledovanii i pod"eme PL S-80**. Moskva: Andreevskii Flag, 1996. (Vakhtennyi zhurnal. Zhurnal boevykh deistvii). 107 p.

Mikhailovskii, A. P. **Rabochaia glubina: zapiski podvodnika**. Sankt-Peterburg: Nauka, 1996. 218 p.
Continues: Vertikal'noe vsplytie.

_____. **Vertikal'noe vsplytie: zapiski podvodnika**. Sankt-Peterburg: "Nauka", 1995. 533 p.
Continued by: Rabochaia glubina: zapiski podvodnika.

Mormul', Nikolai. **Katastrofy pod vodoi:Gibel' podvodnykh lodok v epokhu kholodnoi voiny**. Murmansk: Elteko, 1999. 572 p.

Mostseev, V. M., M.I. Khametov, L.A. Vlasov, I.N. Revenko, eds. **V studenykh glubinakh**. Moskva: Voenizdat, 1980. 344 p.

Ofir, Efraim. **Nirdafim le-lo motsa: sipurah shel "Strumah."** [Israel] A.K.M.O.R, [1999], 232 p.
Title on t.p. verso: With no way out: story of "Struma."

Ottar Sveinsson. **Utkall i Atlantshafi a jolanott**. Reykjavik: Islenska bokautgafan, 1999. 207 p.

Pavlov, A. S. **Voennye korabli Rossii 1997-1998 g.: spravochnik**. 5 vyp. [i.e. 5th ed.]. IAkutsk: A.S. Pavlov, 1997. 151 p.

Platonov, A. V. and V.M. Lur'e. **Komandiry sovetskikh podvodnykh lodok 1941-1945 gg.** Sankt-Peterburg: TSitadel', 1999.(Al'manakh "TSitadel'". Bol'shaia seriia). 122 p.

Polmar, Norman and Jurrien Noot. **Submarines of the Russian and Soviet navies, 1718-1990**. Annapolis, MD: Naval Institute Press, c1991. 370 p.
NPS/DKL Location: GENERAL V859 .S65 P64 1990

Rohwer, Jürgen. **Die Versenkung der jüdischen Flüchtlingstransporter Struma und Mefkure im Schwarzen Meer (Februar 1942, August 1944); Historische Untersuchung.** Frankfurt/M.: Bernard & Graefe, 1965. 153 p.
"Rohwer was the first to uncover the submarine records with the break-up of the Soviet Union. A Russian submarine commander, Lt Denezhko was given a "medal" for his "heroic" action of killing all these defenceless people."
http://www.dangoor.com/73page120.html

Romanov, D. A. **Tragediia podvodnoi lodki "Komsomolets": argumenty konstruktora.** Sankt-Peterburg: Assotsiatsiia izdatelei, 1993. 189 p.

Rubinshtain, Shim`on. **ha-Me'erah she-hitgashmah: `al kamah teragedyot ketanot be-tokh ha-teragedyah ha-gedolah she-shemah "Strumah".** Yerushalayim: Be-hotsa'at ha-Mehaber, [1997]. 13 p.
Sinking of the refugee ship Struma in the Black Sea, February 23, 1942.

Savateev, Petr Nikiforovich. **Udary iz morskikh glubin.** Moskva, Voen. izd-vo, 1961. (Geroicheskoe proshloe nashei rodiny). 133 p.

Schön, Heinz. **Die "Gustloff"-Katastrophe: Bericht eines Überlebenden über die grösste Schiffskatastrophe im Zweiten Weltkrieg.** 1. Aufl. Stuttgart: Motorbuch, 1984. 515 p.
Sunk by the S-13 after leaving Gydnia (Gotenhafen) on January 30, 1945.

_____. **Ostsee '45: Menschen, Schiffe, Schicksale; d. umfassende Dokumentarbericht über d. grösste Rettungswerk d. Seegeschichte; d. Rettung von mehr als 2 Millionen Menschen mit Schiffen d. Dt. Handels- u. Kriegsmarine 1944, 45 über d. Ostsee u.d. Schiffsuntergänge "Wilhelm Gustloff", "General Steuben", "Goya", "Cap Arcona" 1945 in d. Ostsee.** 1. Aufl. Stuttgart: Motorbuch-Verlag, 1983. 696 p.

_____. **SOS Wilhelm Gustloff: die größte Schiffskatastrophe der Geschichte.** 1. Aufl. Stuttgart: Motorbuch-Verl., 1998. 254 p.

Sellwood, A. V. **The damned don't drown: the sinking of the Wilhelm Gustloff.** Annapolis, MD: Naval Institute Press, 1996. (Bluejacket books). 159 p.
Originally published: London: A. Wingate, 1973.
Sunk by the S-13 after leaving Gydnia (Gotenhafen) on January 30, 1945.

Shakhov, Sergei Sergeevich. **Atakuet "Shchuka".** Moskva, Voen. izd-vo, 1967. (Rasskazyvaiut frontoviki). 94 p.

Shigin, Vladimir. **Nad bezdnoi.** Moskva: "Andreevskii flag", 1997. (Tainy minuvshego). 284 p.

Spassky, I. D., V.P. Semyonov, and Norman Polmar, eds. **Submarines of the tsarist navy: a pictorial history.** Annapolis, MD: Naval Institute Press, c1998. 94 p.
NPS/DKL Location: BUCKLEY V859.R8 S83 1998

Starikov, Valentin Georgievich. **Na boevom kurse / Literaturnaia obrabotka N. Mikhailovskogo.** [Leningrad]: Molodaia gvardiia, 1952. 173 p.

_____. **Na grani zhizni i smerti.** Izhevsk: Udmurtiia, 1972. 206 p.

Stefanick, Tom. **Strategic missile submarines and international security.** Washington, DC: American Association for the Advancement of Science, c1989. (AAAS publication; no. 89-20X). 33 p.

Stoecker, Sally W. **Life aboard a Soviet destroyer and a Soviet submarine.** Santa Monica, CA: Rand Corp., 1983. (Report No.: P-6910). 25 p.

Strizhak, Oleg. **Sekrety Baltiiskogo podplava: ne utait' v glubinakh flotskikh arkhivov.** Sankt-Peterburg: Pushkinskii fond, 1996. 252 p.

Suzin, Zvi. **Mi-tokh behirah.** [Israel]: Ts. Suzin, [1998]. 331 p.
Sinking of the refugee ship Salvador in the Sea of Marmara December 12, 1940.

Tributs, Vladimir Filippovich. **Podvodniki Baltiki atakuiut; voennye memuary.** [Leningrad], Lenizdat, 1963. 333 p.

Trusov, Grigorii Martynovich. **Podvodnye lodki v russkom i sovetskom flote.** Izd. 2., ispr. i dop. Pod red. V. IA. VeresoVA Leningrad, Gos. soiuznoe izd-vo sudostroit. promyshl., 1963. 439 p.

Tully, John M. **Russia's Submarine Force: determinants and prospects.** Monterey, CA: Naval Postgraduate School; Springfield, VA: Available from National Technical Information Service, 2001. (ADA392080). 85 p.
Thesis (M.A. in National Security Affairs), Naval Postgraduate School, 2001.
Abstract: *This thesis analyzes the factors likely to shape the future of the Russian submarine force. It examines key events affecting this force since the collapse of the Soviet Union in 1991, and explores the determinants of these events. The Russian Federation inherited a huge submarine fleet from the Soviet Union. Due to the changing conditions in the world and in Russia, its future status is in doubt. The thesis begins by analyzing the development and roles of the Soviet submarine force. It then considers the four most significant factors that have affected the submarine force since 1991: 1) Russia's poor economic performance, 2) Russia's changing national security requirements, 3) competition from the other military services for a limited defense budget, and 4) changes within the military and society . The thesis concludes that the Russian submarine force is unlikely to undergo a major revival. The most probable scenario*

involves a smaller and less capable force. The most influential factors may be Russia's economic performance and the military reform plans of Russian President Vladimir Putin and Defense Minister Sergey Ivanov.
NPS/DKL Location: THESIS T9315
Electronic access: http://theses.nps.navy.mil/Thesis_01jun_Tully.pdf
Electronic access: http://handle.dtic.mil/100.2/ADA392080

Tymoshchenko, Borys. **Pam'iatnyk: dokumental'na povist'**. L'viv: "Kameniar", 1995. 45 p.

Varner, Roy and Wayne Collier. **A matter of risk: the incredible inside story of the CIA's Hughes Glomar Explorer mission to raise a Russian submarine**. London: Hodder & Stoughton, 1979, c1978. 258 p.
Originally published: New York: Random House, c1978.
NPS/DKL Location: GENERAL VB230 .V28

Vinogradov, N. I. **Podvodnyi front**. Moskva: Voen. izd-vo, 1989. (Voennye memuary). 316 p.

Weir, Gary E. and Walter J. Boyne. **Rising tide**. New York: Basic Books, 2003. 304 p.
Contents: 1. Stalin's grand plan -- 2. Cruises and troubles -- 3. The Cuban Missile Crisis -- 4. The ever present danger -- 5. An uncertain nuclear beginning -- 6. A variety of intelligence gathering methods -- 7. Improving the breed -- 8. An insider's view of the mystery of the Kursk.

Zinger, Maks Emmanuilovich. **Geroi Sovetskogo soiuza I.A. Kolyshkin**. Moskva: Voennoe izd-vo Ministerstva vooruzhennykh sil SSSR, 1949. 87 p.

SUBMARINES -- RUSSIAN/SOVIET – NUCLEAR

Aleshin, Vasilii Sergeevich, Nikolai Mikhailovich Kuznetsov and Ashot Arakelovich Sarkisov. **Sudovye iadernye reaktory**. 1968. 491 p.

Antonov, Aleksandr Mikhailovich. **Pervoe pokolenie atomokhodov SKB-143**. Sankt-Petersburg: Sankt-Peterburgskoe morskoe biuro mashinostroeniia "Malakhit", 1996. (Podvodnoe korablestroenie--proshloe, nastoiashchee, budushchee; vyp. 6). 73 p.

_____, Walerie Marinin, and Nikolai Walujew. **Sowjetische-russische Atom-U-Boote: Gefahr aus der Tiefe**. Berlin: Brandenburgisches Verlagshaus, c1998. 144 p.

Bártl, Stanislav. **Stíny morských hlubin**. 1. vyd. Praha, CTK-Pragopress, t. Mír 1,, 1971. 211 p.

_____. **Stíny morských hlubin: dobrodružství atomových ponorek**. Praha: Mladá fronta, 1989. (Edice Archiv; sv. 59). 221 p.

Berezhnoi, S. S. **Atomnye podvodnye lodki VMF SSSR i Rossii**. Moskva: Naval Kollektsiia, 2001. (Naval Kollektsiia: Morskoi istoricheskii al'manakh; vyp.7). 80 p.

Bukalov, Valerii Mikhailovich and Aleksandr Abdugaparovich Narusbaev. **Proektirovanie atomnykh podvodnykh lodok**. 1964. 287 p.

_____. and _____ **Proektirovanie atomnykh podvodnykh lodok**. 1968. 334 p.

Burleson, Clyde W. **Kursk down!: the shocking true story of the sinking of a Russian nuclear submarine**. New York: Warner Books, c2002. 261 p.

Bykhosvkii, Izrail' Adol'fovich. **Atom-U-Boote** [Übers.: Eugen Arnold]. Berlin, Verlag des Ministeriums für Nationale Verteidigung, 1959. 119 p.
German translation of Atomnye podvodnye lodki.

_____. **Atomnye podvodnye lodki**. 1957.

_____. **Atomnye podvodnye lodki**. 2-e, perer. i dop. izd. Leningrad: Gos. soiuznoe izd-vo sudostroit. promyshl., 1963. 230 p.

_____. **Boevye atomnye korabli stran NATO**. 1959. 51 p.

Cherkashin, Nikolai. **"Khirosima" vsplyvaet v polden': geroi i zhertvy pervogo sovetskogo strategicheskogo podvodnogo kreisera-raketonostsa K-19**. Moskva: "Andreevskii flag", 1993. (Vakhtennyi zhurnal. Avariinaia trevoga). 80 p.

_____. **Unesennye bezdnoi: gibel' "Kurska": khronika, versii, sud'by**. Moskva: Kollektsiia "Sovershenno sekretno", 2001. 318 p.

Chernavin, V. N. **Atomnyi podvodnyi--: flot v sud'be Rossii: razmyshleniia posle shtormov i pokhodov**. Moskva: "Andreevskii flag", 1997. (Rossiiskaia morskaia biblioteka). 470 p.

Droblenkov, Viktor Feoktistovich and Vladimir Nikolaevich Gerasimov. **Ugroza iz glubiny**. 1966. 306 p.

Giltsov, Lev, Nicolaï Mormoul and Leonid Ossipenko. **La dramatique histoire des sous-marins nucléaires soviétiques: des exploits, des échecs, et des catastrophes cachées pendant trente ans**. Paris: R. Laffont, c1992. (Collection "Vécu"). 357 p.

Kellogg, Sanoma Lee and Elizabeth J. Kirk, eds. **Reducing wastes from decommissioned nuclear submarines in the Russian Northwest: political, technical, and economic aspects of international cooperation: proceedings from the NATO Advanced Research Workshop "Recycling, Remediation, and**

Restoration Strategies for Contaminated Civilian and Military Sites in the Arctic Far North," Kirkenes, Norway, 24 to 28 June, 1996. Washington, DC: American Association for the Advancement of Science, 1997. 221 p.

Kholostov, Dmitrii Ivanovich. **Sredstva korablevozhdeniɛa podvodnykh atomokhodov**. Moskva, Voen. izd-vo, 1967. 210 p.

Kirk, Elizabeth J., ed. **Decommissioned submarines in the Russian Northwest: assessing and eliminating risks (Proceedings of NATO Advanced Research Workshop on Recycling, Remediation and Restoration Strategies for Contaminated Civilian and Military Sites in the Arctic Far North (1996: Kirkenes, Norway).** Dordrecht; Boston: Kluwer Academic Publishers, c1997. (NATO ASI series. Partnership sub-series 2, Environment; 32). 179 p.
NPS/DKL Location: GENERAL TD898.14.R8 D43 1997

Kostev, Georgii Georgievich. **Nuclear safety challenges in the operation and dismantlement of Russian nuclear submarines**. Moscow: Committee for Critical Technologies and Non-Proliferation, 1997. 142 p.
English translation of Problemy bezopasnosti pri ekspluatatsii i utilizatsii atomnykh podvodnykh lodok.

_____. **Problemy bezopasnosti pri ekspluatatsii i utilizatsii atomnykh podvodnykh lodok**. Moskva: Kom-t po kriticheskim tekhnologiiam i nerasprostraneniiu, 1997. 138 p.

Krupnick, Charles. **Decommissioned Russian nuclear submarines and international cooperation**. Jefferson, NC: McFarland, 2001. 260 p.
NPS/DKL Location: GENERAL TD898.13.R8 K78 2001

Kurushin, M. **Podvodnaia lodka "Kursk": rozhdenie, zhizn´, versii gibeli, podrobnosti**. Moskva: "Olimp": Izd-vo AST, 2000. 157 p.

Kuznetsov, Vladimir Aleksandrovich. **Komponovka i raschet reaktorov sudovykh iadernykh ustanovok. Utverzhdeno v kachestve ucheb. posobiia dlia sudomekhanicheskikh fakul´tetov vysshikh inzhenernykh morskikh uchilishch. i Odesskogo in-ta inzhenerov morskogo flota**. Moskva, Transport, 1966. 218 p.

_____. **Sudovye iadernye reaktory: osnovy teorii i ekspluatatsii**. Ucheb. izd. Leningrad: "Sudostroenie", 1988. 262 p.

Lazarev, Nikolai Mikhailovich. **Pervye sovetskie atomnye podvodnye lodki i ikh voennaia priemka**. Sankt-Peterburg: Sankt-Peterburgskoe morskoe biuro mashinostroeniia "Malakhit", 1996. (Podvodnoe korablestroenie--proshloe, nastoiashchee, budushchee; vyp. 7). 184 p.

_____. **Pervye sovetskie atomnye podvodnye lodki: u istokov atomnogo voennogo korablestroeniia: trudnosti i uspekhi, liudi, sobytiia, fakty**. Izd. 2. Moskva: Paleia, 1997. 227 p.

Maziuk, Igor'. **IAdernaia ruletka Kremlia: drama atomnykh submarin SSSR**. Sevastopol': Izd-vo "Flot Ukrainy", 2000. 218 p.

Mezhenkov, Vladimir. **Atomnaia podlodka "Kursk": khronika gibeli**. Moskva Izdatel'skii dom "Pushkinskaia ploshchad'", 2000. 328 p.

Moore, Robert. **A time to die: the untold story of the Kursk tragedy**. Crown Publishers, c2002. 271 p.

Mormul', Nikolai. **Atomnye, unikal'nye, strategicheskie: zapiski ispytatelia atomnykh podvodnykh lodok**. Murmansk: Izdatel'skii Dom "999", 1997. 265 p.

Osipenko, Leonid. **Podvodniki: zapiski komandira pervoi atomnoi podvodnoi lodki**. Volgograd: Komitet po pechati i informatsii, 1997. 158 p.

_____, Lev Zhil'tsov, Nikolai Mormul'. **Atomnaia podvodnaia epopeia: podvigi, neudachi, katastrofy**. Moskva: AO "Borges", 1994. 350 p.

Savichev, Gennadii Aleksandrovich. **Pod vodoi vokrug Zemli**. 1967. 78 p.

_____. **Sluzhba morskaia**. Moskva: DOSAAF, 1977. 110 p.

Sengling, Bettina and Johannes Voswinkel. **Die Kursk; Tauchfahrt in den Tod**. Deutsche V.-A., Stgt., 2001. 287 p.

Shamanov, Nikolai Pavlovich and Gennadii Efimovich Romantsov. **Sudovye iadernye energeticheskie reaktory: osnovy teorii i rascheta**. Leningrad: "Sudostroenie", 1984. 230 p.

Shigin, V. V. **APRK "Kursk": Posleslovie k tragedii**. Moskva: OLMA-Press, 2002. 447 p.

_____. **Taina ischeznuvshei submariny: Zapiski ochevidtsa spasatel'noi operatsii APRK "Kursk"**. Moskva: OLMA-Press, 2001. 416 p.

Snell, Benjamin Aaron. **Dismantling Russia's Northern Fleet Nuclear Submarines environmental and proliferation risks**. Monterey, CA: Naval Postgraduate School; Springfield, VA: Available from National Technical Information Service, 2000. (ADA378654). 74 p.
Thesis (M.A. in National Securty Affairs), Naval Postgraduate School, 2000.
Abstract: *This thesis examines the 1986 Chernobyl accident and its consequences as the basis for an analysis of the possible dimensions of the nuclear catastrophes that could occur during the dismantlement process of Russia's Northern Fleet nuclear submarines. It assesses the potential demographic, ecological,*

and economic consequences of a nuclear accident. Given the systemic problems at Russian nuclear facilities, the risks of a catastrophic event in the poorly maintained and operated submarine yards housing over 100 operating nuclear reactors are significant. A major nuclear accident at these facilities could cause damage to the environment of global proportions. This thesis considers the potential environmental impact of a nuclear accident during the nuclear submarine dismantlement process and discusses the environmental damage that has already occurred as a result of Soviet and Russian practices. This thesis also evaluates the risk of diversion of nuclear materials to proliferators or terrorists. Lastly, this thesis examines how the United States, the European Union, and perhaps others could assist Russia in reducing the environmental and proliferation risks in this dismantlement process.
NPS/DKL Location: THESIS S64443
Electronic access: http://library.nps.navy.mil/uhtbin/hyperion-image/00Jun_Snell.pdf
Electronic access: http://handle.dtic.mil/100.2/ADA378654

Sorokin, Anatolii Ivanovich. **My s atomnykh**. Moskva, Izd-vo DOSAAF, [1968]. 112 p.

_____. **My s atomnykh**. 2-e dop. i pererab. izd. Moskva, Izd-vo DOSAAF, 1972. (Za chest' i slavu Rodiny). 253 p.

Stoecker, Sally W. **Life aboard a Soviet destroyer and a Soviet submarine**. Santa Monica, CA: Rand Corp., [1983]. (P-6910). 5 p.

Sviatov, Georgii Ivanovich. **Atomnye podvodnye lodki**. Moskva, Voenizdat, 1969. (Nauchno-populiarnaia biblioteka). 176 p.

Timofeev, Riurik Aleksandrovich. **K Severnomu poliusu na pervoi atomnoi**. Sankt-Peterburg: Sankt-Peterburgskoe morskoe biuro mashinostroeniia "Malakhit", 1995. (Podvodnoe korablestroenie--Proshloe, nastoiashchee, budushchee; vyp. 4). 86 p.

Zapol´skii, Anatolii Aleksandrovich. **Rakety startuiut s moria**. Sankt-Peterburg: Sankt-Peterburgskoe morskoe biuro mashinostroeniia "Malakhit", 1994. (Podvodnoe korablestroenie, proshloe, nastoiashchee, budushchee; vyp. 2). 151 p.

_____. **Strategicheskim raketonostsam--byt'!** Sankt-Peterburg: SPMBM "Malakhit", 1998. (Podvodnoe korablestroenie, proshloe, nastoiashchee, budushchee; vyp. 11). 194 p.

SUBMARINES -- SWEDISH -- GENERAL

Bergström, Lars & Klas Åma, eds. **Ubåtsfrågan: en kritisk granskning av den svenska nutidshistoriens viktigaste säkerhetspolitiska dilemma.** Uppsala: Verdandi, 1999. (Verdandi-debatt; 88). 199 p.

Ellerström, Hans. **Swedish submarines: Swedish export submarines**. 5. ed. [Malmö]: [Kockums AB], [1997]. 40 p.

SUBMARINES -- TURKISH -- GENERAL

Metel, Raþit. **Türk denizaltýcýlýk tarihi**. Ýstanbul: Deniz Kuvvetleri Komutanlýðý, 1960-1980. 1 v.

SUBMARINES -- YUGOSLAV -- GENERAL

de Majnik, John. **Diary of a submariner: the first hand story of the daring escape of a Yugoslav submarine and her crew from the Germans during World War II**. Inglewood W.A., Australia: Asgard Press, 1996. 93 p.

PERIODICAL ARTICLES

SUBMARINES -- GENERAL

Abele, Bradford L. "The Loss of USS GRUNION," **Submarine Review** 94-97 July 2002.

Ackley, Richard T. "Submarines for the 21st Century?" **Submarine Review** 57-60 April 1993.

Adams, Allan. "Leveraging submarine power in the 21st century," **Submarine Review** 16-24 October 1998.

"Advance in Size of Submarines," **U.S. Naval Institute Proceedings** 38(2):762 June 1912.

Alden, John D. "Hell Ship of Convoy MATA-30; U.S. POWs and U.S. Submarines," **Submarine Review** 83-90 October 1999.

_____. "Loss of Grunion possibly explained," **Submarine Review** 72-74 April 1998..

_____. "Ultra Revisited," **Submarine Review** 37-43 April 1992.

_____. "Unfaced Challenge: Submarine Versus Free World," **U.S. Naval Institute Proceedings** 90(4):26 April 1964.

_____. "The ups and downs of Electric Boat," **U.S. Naval Institute Proceedings** 125(7):64-67 July 1999.

Alves, Dora. "Submarines in East Asia," **Submarine Review** 58-66 October 1995.

Amidon, C. Philip. "The virtual submarine," **Naval Forces** 20(3):54-55 Special Issue 1999.

"Anchors aweigh for submarine data," **Mechanical Engineering** 120(11):18 November 1998.

Anderson, Jon. "A Changing World Beneath," **Navy Times** 10 April 4, 1994.

Andrade, Ernest J. "Submarine policy in the United States Navy, 1919-1941," **Military Affairs** 35(2):50-56 1971.

Annati, Massimo. "AIP (Air-Independent Propulsion) systems: A solution for everybody?" **Military Technology** 20(11):107-108+ November 1996.

_____. "Trends and requirements in torpedo development," **Military Technology** 22(11):44-49 November 1998.

_____. "Underwater special operations craft," **Military Technology** 20(3):85-89 March 1996.

Antonov, A. M. "Birth of Red November," **U.S. Naval Institute Proceedings** 121(12):79-81 December 1995.

"Argument for the Submarine," **U.S. Naval Institute Proceedings** 45(10):1783 October 1919.

Armstrong, Harold J. "Project Magnum - Taking Submarine Design and Operations into the 21th Century," **Submarine Review** 96-105 January 1999.

Aronson, Robert. "SSGN: a "second career" for the boomer force," **Undersea Warfare** 2(2):19-23 Winter 1999.

"Asia-Pacific submarine survey," **Asian Defence Journal** 48-51 October 1999.

Ayer, David. "Underwater spies," **The Washington Monthly** 30(12):46-47 December 1998.

Baciocco, A.J. "Mission Requirements: The *PULL* on Technology [Submarine Technology Symposium 16 May 2000]," **Submarine Review** 51-57 July 2000.

Backhaus, Roland G. "We can be ready in peacetime," **U. S. Naval Institute Proceedings** 127(7):88-89 July 2001.

Bacon, Roger F. "Submarine Warfare -- It's A-Changing," **U.S. Naval Institute Proceedings** 118(6):52-54 June 1992.

Baker, A. D. III. "Combat fleets," **U.S. Naval Institute Proceedings** 124(10):109 October 1998.

Ballard, Robert D. "Lurking in the deep-sea terrain," **U. S. Naval Institute Proceedings** 124(5):60-63 May 1998.

Barney, James R. and John J. Zerr. "NSSN--new attack submarine," **Program Manager** 25(2):38-41 March-April 1996.

Barron, Claude. "The operator is part of the system," **Submarine Review** 30-47 January 2001.

Barry, John M. "Undersea Warfare: upgrading the SQQ-89 sonar suite to meet littoral warfare requirements," **Surface Warfare** 23:17-19 March/April 1998.

"Battling the Subs - World War II," **All Hands** 59-63 April 1958.

Baumgartner, Michael J. "Stay Engaged through Innovation," **Submarine Review** 65-71 July 1993.

Beach, Edward L. "A centennial salute to the U.S. submarine force," **Sea Power** 43(7):33-40 July 2000.

_____. "Radar and submarines in World War II," **Defense Electronics** 11:48+ October 1979.

Beach, Paul. "Arctic challenge: under the polar ice cap," **Undersea Warfare** 3(2):5-7 Winter 2001.

Beadnell, Nick. "A submariner's perspective on operational effectiveness and resource constraints," **RUSI** 140(3):27-32 June 1995.

Beaudan, Eric. "The Hunt for U.S. Subs," **American Legion Magazine** 131:18-19+ December 1991.

Belke, Thomas J. "Roll of drums," **U.S. Naval Institute Proceedings** 109(4):58-64 April 1983.

Bell, Kelly. "Gray Wolves and the Ides of March," **World War II** 7:47-52 March 1993.

Benson, Robert. "Torpedo chaser," **All Hands** (980):32-35 December 1998.

Berg, John. "Viking submarine fights for the future," **Janes Defense Weekly** 31:38 May 5, 1999.

Bieg, V.N. "The Submarine and the Future," **U.S. Naval Institute Proceedings** 41(1):151 July/August 1915.

Blackett, Jeffrey. "Determination of a submarine's hostile intent," **RUSI** 139(6):32-34 December 1994.

Blazar, Ernest. "Converted submarines could bolster U.S. power projection," **Defense News** 14:19 March 29, 1999.

_____. "Is the submarine era over yet? Nyet!" **Navy Times** 44(40):17 July 10, 1995.

_____. "Keeping the Sub Force Levels Afloat," **Submarine Review** 74-78 July 1999.

_____. "Last gasp of the Seawolf? Perhaps," **Navy Times** 44(35):4 June 5, 1995.

_____. "Navy, CIA at odds over sub threat," **Navy Times** 44(38):4 June 26, 1995.

_____. "Navy sub plans draw fire," **Navy Times** 26 September 18, 1995.

_____. "Pentagon Backs Subs' Post-Cold War Missions," **Submarine Review** 7-11 April 1999.

_____. "A 'new dimension' in warfighting capabilities--the tactical boomer: Trading Tridents for Tomahawks," **Sea Power** 42(7):37-40 July 1999.

_____. "A new Seawolf is raising howls," **Navy Times** (Marine Corps Edition) 20 April 3, 1995.

Bloomquist, Dick L. "Air-Independent Submarine Propulsion--A Historical Perspective from Walter to Stirling," **Submarine Review** 74-80 July 1993.

Bodnar, John W. and Rebecca Dengler. "The Emergence of a Command Network," **Naval War College Review** 49(4):93-107 Autumn 1996.
http://www.nwc.navy.mil/press/review/1996/autumn/comm%2Da96.htm

Bonsignore, Ezio. "The coming SE Asia submarine race," **Military Technology** 76-81 March 1996.

_____. "Hey, psst--wanna build a submarine?" **Military Technology** 25:7 May 2001.

Borik, Frank C. "The Silent Service is On the Air: How Advanced Communnications Will Revolutionize Submarine Warfare," **Submarine Review** 102-108 July 1999.

_____. "Sub Tzu and the art of submarine warfare," **U.S. Naval Institute Proceedings** 121(11):64-72 November 1995.

Bowling, R.A. "The Mark 14 Torpedo Tribulations," **Submarine Review** 67-71 July 2002.

Bowman, Frank L. "Force level: remarks to corporate benefactors," **Submarine Review** 6-32 April 2001.

_____. "The Future Days [NSL Annual Symposium 16 June 2000]," **Submarine Review** 25-38 July 2000.

_____. "Mobile targets from under the sea: new missions in a new security environment," **Submarine Review** 4-17 January 2000.

_____. "NSL annual symposium remarks," **Submarine Review** 25-38 July 2000.

_____. "Opening Remarks at Annual NSL Symposium, **Submarine Review** 5-13 July 2001.

_____. "The Pull For Submarine Technology: Get Real [Remarks, Submarine Technology Symposium 2002]," **Submarine Review** 5-13 July 2002.

_____ "Remarks at Corporate Benefactors Day," **Submarine Review** 13-25 April 2002.

_____. "Remarks at UDT Hawaii 2001," **Submarine Review** 6-11 January 2002.

_____. "Submarines in the new world order," **Undersea Warfare** 1(3):2-8 Spring 1999.

_____. "The Virginia class submarine: the right ship for the future," **Submarine Review** 5-11 April 2000.

Boyle, Richard. "Attack Submarine Design: Let's Wake: Up and Win," **Submarine Review** 85-89 April 1999.

_____. "Bound in shallows and miseries?" **U.S. Naval Institute Proceedings** 122(10):52-55 October 1996.

_____. "Emergence of Offensive U-Boats during the Great War," **Submarine Review** 57-64 April 1992.

_____. "A Submarine Bibliography," **Submarine Review** 92-97 January 1993.

_____. "U.S. submarine design in the 20th century," **Submarine Review** 89-103 October 2000.

_____. "Wanklyn versus GARIBALDI, 28 July 1941," **Submarine Review** 99-104 October 1996.

Boyne, Peter B. "In The Beginning...There Was Special Projects!" **Submarine Review** 116-118 April 2002.

Bradley, Mark A. "Why they called the Scorpion 'Scrapiron,'" **U.S. Naval Institute Proceedings** 124(7):30-37 July 1998.

Bramlett, W.T. "The Nuclear Submarine: Riding High," **U.S. Naval Institute Proceedings** 101(2):55 February 1975.

Breemer, Jan S. "Where Are the Submarines?" **U.S. Naval Institute Proceedings** 119(1):37-42 January 1993.

Brekke, Bjorn. "Submarines today: Flotilla Commanders comment (I)," **Naval Forces** 20:38+ 1999.

Brigger, Clark V. "A hostile sub is a joint problem," **U.S. Naval Institute Proceedings** 126(7):50-53 July 2000.

Brighton, D. R. et al. "The Use of Fuel Cells to Enhance the Underwater Performance of Conventional Diesel Electric Submarines," **Journal of Power Sources** (51):375-389 1994.

Broad, William J. "A Tale of Daring American Submarine Espionage," **New York Times** [Late Edition (East Coast)] 1 November 8, 1998.

Brooks, Linton F. "Pricing Ourselves Out of the Market: The Attack Submarine Program," **Naval War College Review** 32(5):2–17 September/October 1979.

_____. "START, START II, and the Submarine Force," **Submarine Review** 25-33 October 1993.

Brower, J. Michael. "The enemy (below), the brass above," **U.S. Naval Institute Proceedings** 126(6):33 June 2000.

Brown, David A. "Navy pushes onward for Guam sub basing," **Navy Times** 50:21 February 12, 2001.

_____. "Officials sound alarm on sub shortage," **Navy Times** 49:20 July 10, 2000.

_____. "Opening the hatch to women on subs," **Navy Times** 48:21 June 28, 1999.

_____. "Sub force in flux," **Navy Times** 49:22 August 21, 2000.

Brown, Steve. "Submarine Imagery: Looking Ahead to the Next Hundred Years," **Submarine Review** 54-58 April 2002.

Buchanan, T. R. "Submarine information technology: it begins with the backbone," **Submarine Review** 81-89 January 2000.

Buff, Joseph J. "ASDS: one minisub, many roles," **Submarine Review** 48-59 January 2000.

_____. "Diesel-AIPs: Low Displacement as a Weakness," **Submarine Review** 87-95 January 1999.

_____. "Hydrothermal Vent Plumes as Acoustic Lenses," **Submarine Review** 48-60 October 1999.

_____. "Looking forward--submarines in 2050," **Submarine Review** 82-90 July 1998; "Part two," **Submarine Review** 54-63 October 1998.

Burmester, Andreas. "Production and co-production in submarine building [Special Issue]," **Naval Forces** 53-55 2001.

Burns, R. W. "Impact of technology on the defeat of the U-boat September 1939-May 1943," **IEE Proceedings, Science, Measurement and Technology** 141:343 September 1994.

Burns, Richard F. "Constructing submarines--digitally," **Sea Technology** 39(7):55-58 July 1998.

_____. "Undersea warfare--national, international trends," **Sea Technology** 35:10 November 1994.

Bush, James T. "Changing sub mission means saving money," **Navy Times** (Marine Corps Edition) 31 August 25, 1997.

Bushnell, Dennis M. "Future [2020-2030] Strategic Technology Issues, A Presentation at the Submarine Technology Symposium May 12, 1999," **Submarine Review** 58-66 July 1999.

_____. "The shape of things to come?" **Undersea Warfare** 3(2):2-4 Winter 2001.

Butler, John D. "Coming of Age: The SSGN Concept," **Submarine Review** 26-36 April 2002.

_____. "New Directions in Submarine Technology," **Submarine Review** 34-44 October 2002.

Butterworth, Cameron. "Fast catamaran sparks international naval interest," **Asia-Pacific Defence Reporter** 26(7):32-33 December/January 2001.

Byron, John L. "A minority view on Greeneville," **Submarine Review** 113-116 October 2001.

Canto, Wilfred Jr. "Virtual reality under the sea," **Naval Forces** 20(3):S22-S24 1999.

Caldwell, H. H. "DARTER and DACE at Leyte Gulf," **Submarine Review** 63-68 July 2001.

Caldwell, Harry. "Trip South [Operation High Jump]," **Submarine Review** 78-88 October 2001.

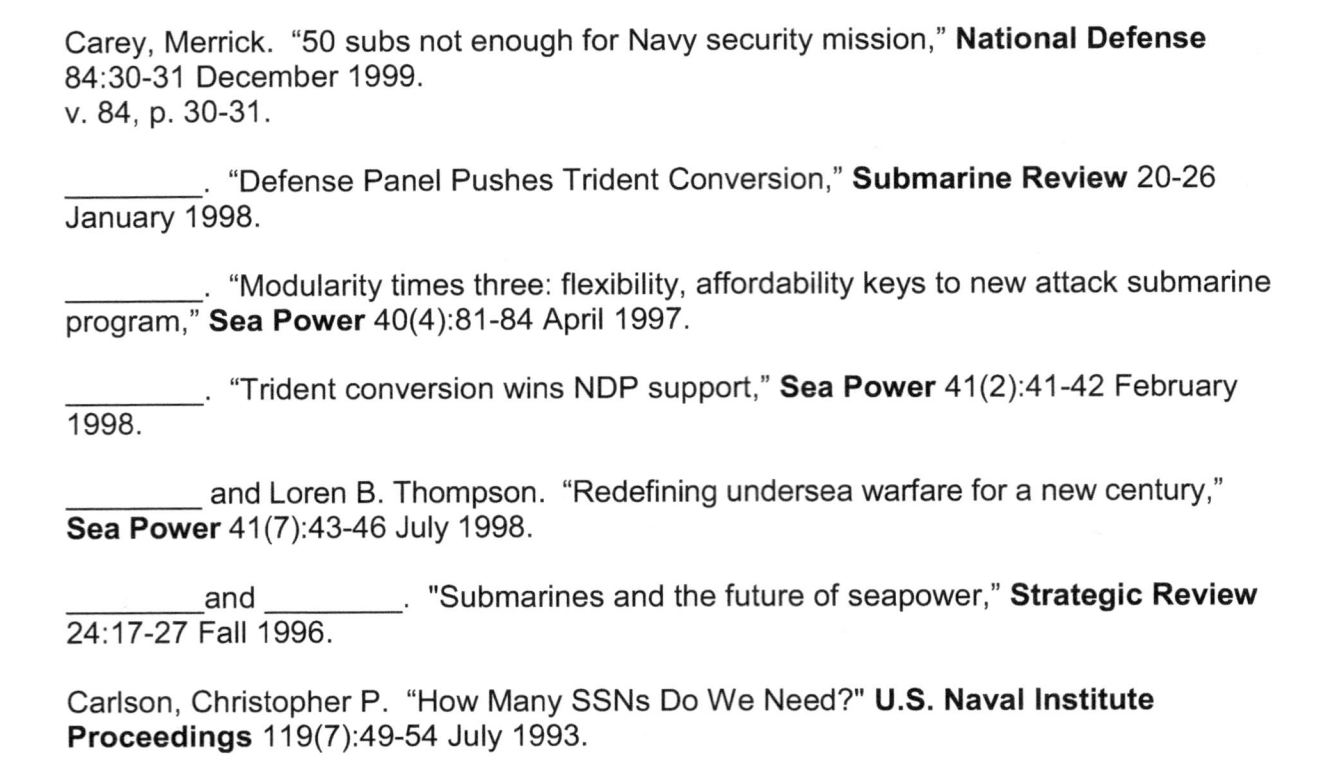

Carey, Merrick. "50 subs not enough for Navy security mission," **National Defense** 84:30-31 December 1999.
v. 84, p. 30-31.

_____. "Defense Panel Pushes Trident Conversion," **Submarine Review** 20-26 January 1998.

_____. "Modularity times three: flexibility, affordability keys to new attack submarine program," **Sea Power** 40(4):81-84 April 1997.

_____. "Trident conversion wins NDP support," **Sea Power** 41(2):41-42 February 1998.

_____ and Loren B. Thompson. "Redefining undersea warfare for a new century," **Sea Power** 41(7):43-46 July 1998.

_____ and _____. "Submarines and the future of seapower," **Strategic Review** 24:17-27 Fall 1996.

Carlson, Christopher P. "How Many SSNs Do We Need?" **U.S. Naval Institute Proceedings** 119(7):49-54 July 1993.

Carr, Barry. "Diving in the Lake - Retrofitting Conventional Submarines," **NATO's 16 Nations** 38:31-35 1993.

Carullo, Anthony. "The strategic planning process and its impact on submarine design," **Submarine Review** 94-98 April 2000.

Cavaiola, Lawrence. "Address to the Submarine Technology Symposium (May 15, 1997," **Submarine Review** 16-26 July 1997.

Chapman, Richard. "Submarines in a New Security Environment," **Submarine Review** 62-69 October 1993.

Cheatham, James T. "Atlantic Turkey Shoot," **World War II** 6:30-36 March 1992.

Chiles, H. G., Jr. "Submarines, Technology, and the Post-Cold War Era," **Submarine Review** 10-17 October 1993.

Clinton, William J. "Proclamation 7363--100th Anniversary of the U.S. Navy Submarine Force, 2000," **Weekly Compilation of Presidential Documents** 36(41):2450-2451 October 16, 2000.

Collins, James E. "ILS -- essential to submarine warfighting," **Submarine Review** 54-59 July 1991.

Compton-Hall, Richard. "First disaster," **Submarine Review** 81-86 January 2001.

_____. "Holland's Hollands--an Irish tale," **U.S. Naval Institute Proceedings** 117(2):59-63 February 1991.

_____. "Submarines, Seagulls, and Sea Lions," **Submarine Review** 140-145 April 2001.

Cook, William J. and Robert Kaylor. "How to Stop a Russian 'Surge,'" **U.S. News & World Report** 102(23):43 June 15, 1987.

Cooper, Brian H. "The threat: What are the credible threats to Australia?" **Asia-Pacific Defence Reporter** 26(7):16+ December/January 2001.

Cooper, Dale B. "Run Silent, Run Deep," **Soldier of Fortune** 19:48-53+ February 1994.

Coquinot, J. P. "Submarines on loose leash. Countermeasures?" **Armada International** 21(2):40-42+ Apr-May 1997.

Corless, Josh. "Numbers count as submarine commitments stretch USN," **Jane's Navy International** 105(2):18-24 March 2000.

_____. "Problems beneath the waves," **Jane's Navy International** 104(5):10 June 1, 1999.

_____. "The silent service gets vocal," **Jane's Navy International** 105(1):25-29 January/February 2000.

_____. "Virtual reality extends to USW," **Jane's Navy International** 104(5):3 June 1, 1999.

Corvaian, Daniel. "Submarines today: Flotilla Commanders comment," **Naval Forces** 20:40-46 1999.

Cosentino, Michele. "Back to the future," **U.S. Naval Institute Proceedings** 123(3):44-47 March 1997.

Cote, Owen R. "Attacking Mobile Targets From Under the Sea [Submarine Technology Symposium 16 May 2000]," **Submarine Review** 69-78 July 2000.

Courter, Jim and Loren B. Thompson. "Arsenal under the sea," **Sea Power** 40(6):41-44 June 1997.

_____ and _____. "Boomer reborn," **U.S. Naval Institute Proceedings** 123(11):51-53 November 1997.

_____and _____. "The next submarine -- and the one after that," **Sea Power** 38(6):23-26 June 1995.

_____, _____ and Scott C. Truver. "The U.S. submarine industrial base: unique to defense and essential to security," **Strategic Review** 23:7-14 Spring 1995.

Cox, Ken. "Horses and boats: thoughts on the U.S. Submarine Force in the 21st century," **Submarine Review** 122-127 October 1998.

Crimmins, Jim. "Mine warfare and submarines," **U.S. Naval Institute Proceedings** 120(10):80-81 October 1994.

Crowe, William J. Jr. "The Early Days [Navy League Symposium 16 June 2000 banquet address]," **Submarine Review** 9-19 July 2000.

Curtis, Ian. "Submarines and small powers finally marry," **Military Technology** 24:8 November/December 1996.

Dalton, John H. and T. E. Lloyd. "The case for and against a third Seawolf," **Navy Times** (Marine Corps Edition) 31 April 3, 1995.

Danzig, Richard. "SecNav Address to the Submarine Community," **Submarine Review** 4-17 July 1999.

Daubin, F.A. "The Fleet Submarine," **U.S. Naval Institute Proceedings** 42(6):1815 November/December 1916.

Davis, Jacquelyn K. "Submarine's role in the twenty-first century," **Sea Power** 40(7):35-37 July 1997.

Davis, John P. "USS Jimmy Carter (SSN-23) expanding future SSN missions," **Undersea Warfare** 2(1):16-18 Fall 1999.

Davis, Malcolm R. "Briefing: Asian navies," **Jane's Defence Weekly** 35(4):22-27 January 24, 2001.

de Lionis, Andres. "Allure of AIP (air-independent propulsion) beckons the navies of developing states," **Jane's Intelligence Review** 10(2):39-41 February 1998.

Dealey, Sam. "The Best Patrol-HARDER'S FIFTH," **Submarine Review** 15-39 October 2001.

Desharnais, Dennis F. "Full-spectrum USW test & evaluation tracking range facilities," **Naval Forces** 20(3):S45-S47 1999.

"Design," **Navy International** 95:453-457+ December 1990.

"DIGITAL FLUB: The Bug in the System Is a Fish," **The Washington Post** [FINAL Edition]:F.05 November 23, 1998.

Dikkenberg, John. "Diesel electric submarines: a European perspective," **Asia-Pacific Defence Reporter** 27:22-23 October/November 2000.

_____. "Regional Submarines," **Asia-Pacific Defence Reporter** 28(4)18-21 May 2002.

"'Doing it all, doing it well, doing it everyday' [Interview with Vice Adm. John J. Grossenbacher, Commander, Submarine Force, U.S. Atlantic Fleet]," **Sea Power** 44(7):9-14 July 2001.

Donnelly, John. "An asymmetric threat 'below the layer': A quiet revolution in undersea warfare," **Sea Power** 42(7):42-45 July 1999.

Dorber, Frank G. "Man of God built a futuristic weapon of war--then lost it beneath the waves," **Military History** 11(4):8+ October 1994.

Dornan, Robert K. "Roles and Missions for a Post-Cold War U.S. Submarine Force," **Submarine Review** 11-16 January 1993.

dos Santos Guimaraes, Leonam. "Should we fear Third World nuclear submarines?" **U.S. Naval Institute Proceedings** 126(3):52-55 March 2000.

_____. "Undersea warfare: balancing affordability and advanced technology," **Sea Technology** 37:11 January 1996.

Douglas, Lawrence H. "The Submarine and the Washington Conference of 1921," **Naval War College Review** 26(5):86-100 March/April 1974.

Douglas, Martin. "Submarines for the Third World," **Naval Forces** 14:20-22+ 1993.

Douglass, John W. "Navy undersea acquisition: a changing environment," **Sea Technology** 39:20 January 1998.

_____. "Undersea warfare: balancing affordability and advanced technology," **Sea Technology** 37:11 January 1996.

Downing, John. "Cold War weapon or global deterrent?" **Jane's Navy International** 103(9):23-27 September 1998.

Downing, Wayne A. "Thinking outside the box," **Submarine Review** 6-12 January 1996.

Doyle, Michael K. "The U.S. Navy and War Plan Orange, 1933–1945: Making Necessity a Virtue," **Naval War College Review** 33(3):49-63 May/June 1980.

Drake, Hal. "'Attack the U.S. With a 5-inch Gun!'" **Navy** 11:30-31+ June 1968.

Dulas, Mike. "The battlegroup commander's most unused asset: the submarine," **Submarine Review** 98-103 October 1995.

Duncan, Cecil and David Slusher. "Portable recompression chamber," **NAVSEA Journal** 28:9-16 June 1979.

Duncan, Francis. "No slap in the face," **U.S. Naval Institute Proceedings** 126(4):44-47 April 2000.

Dunk, Graeme. "Do we need a Southeast Asian water space management regime?" **Asian Defence Journal** 12-13 May 1995.

Dwyer, John B. "Silent running: submarine special operations," **Vietnam** 9:42-48 April 1997.

Dwyer, Dennis M. "A Heritage of Excellence and New Challenges to Meet: Strategic Systems Programs State of the Program 2002," **Submarine Review** 47-58 October 2002.

_____. "A rich heritage of innovation," **Sea Power** 43(10):41-43 October 2000.

Eames, Michael. "Foreseen technology and its impact on naval capabilities," **Canadian Defence Quarterly** 21(3):13-18 December 1991.

Eccles, Thomas J. "Navy's latest stealth test submarine: The large-scale vehicle Cutthroat," **Sea Power** 42(7):41 July 1999.

Edwards, R.S. "On the Fleet Submarine," **U.S. Naval Institute Proceedings** 42(6):1955 November/December 1916.

"Electric Super-Submarines," **U.S. Naval Institute Proceedings** 95(10):1803 October 1969.

Elhefnawy, Nader. "*De-Escalation:* A Mission for Russia's Submarine Force?," **Submarine Review** 41-46 January 2002.

_____. "Submarines and Space Power," **Submarine Review** 71-77 October 2001.

_____. "Undersea Future Shock," **Submarine Review** 81-87 July 2002.

Enright, Joseph F. "The short life and sudden death of the Shinano," **American Legion Magazine** 107:14-15+ August 1979; 107:18-19+ September 1979.

Erwin, Sandra I. "Cutbacks on maintenance chores mean higher retention on submarines," **National Defense** 85(561):12 August 2000.

_____. "U.S. ponders sea-based missile defense," **National Defense** 84(551):25-26 October 1999.

"ESM supports new roles for submarines," **Jane's Navy International** 103(10):26-30 December 1, 1998.

Estes, G. Brian. "Trident base East: setting a large construction program in motion," **Military Engineer** 73:322-327 September/October 1981.

Exley, Gerard M. "U.S. Navy submarine communications improvements: No-longer the 'silent service,'" **Naval Forces** 19(5):88-91 1998.

"Experience of Russia in the creation of submarines with air-independent propulsion," **Asian Defence Journal** 54-57 May 1997.

Fages, Malcolm. "Forward...from under the sea," **Undersea Warfare** 1:4-8 Fall 1998.

_____. "The Submarine Force: a century of excellence, and the challenge of the future," **Undersea Warfare** 2(4):2-6 Summer 2000.

_____. "The Submarine Force and technology today," **Submarine Review** 58-66 July 2000.

Fargo, Thomas B. "Pacific Perspective [NSL Annual Symposium 16 June 2002]," **Submarine Review** 41-50 July 2000.

_____. "A Pacific Update," **Submarine Review** 14-24 July 2001.

Farrer, Mark. "China and Taiwan: an opening capability gap underwater," **Asia-Pacific Defence Reporter** 27(1):74-75 February 2001.

_____. "Electronic warfare--vital to survival," **Asia-Pacific Defence Reporter** 26(7):38-39+ December 2000/January 2001.

Feuer, A. B. "Heroic Pigeon in the Philippines," **World War II** 11:50-56 March 1997.

_____. "USS Growler's War Against Japan," **World War II** 12:42-48+ February 1998.

Finnegan, Philip and Robert Holzer. "A Move to Aid U.S. Shipyards," **Navy Times** 38 June 14, 1993.

Fiorenza, Nicholas. "Changing roles: European submarines take on new missions," **Armed Forces Journal International** 139:20+ August 2001.

_____. "Changing Roles: European Submarines Take on New Missions," **Submarine Review** 93-98 October 2001.

_____. "Keeping the colors flying: European naval programs are national but cooperative," **Armed Forces Journal International** 138(3):72-74+ October 2000.

Fleming, Bruce. "In the once-silent submarine service," **Southwest Review** 81(1):72-78 Winter 1996.

Fox, Robert F. "Build it and they will come," **U.S. Naval Institute Proceedings** 127(4):44-47 April 2001.

Foxwell, David. "Developments in naval periscopes," **Naval Forces** 11:48-52+ 1990.

_____. "Hidden raider--air independent submarine propulsion," **NATO's Sixteen Nations** 35(2):52-54+ April 1990.

_____. "Signature Reduction: Smart Materials for Active Control," **International Defense Review** 24: 1215-1217+ November 1991.

_____. "Sub proliferation sends navies diving for cover," **Jane's IDR: International Defense Review** 30:30-31+ August 1997.

_____. "Submarine defensive warfare--US to refit enhanced torpedo defenses," **International Defense Review** 24(1):1233-1234 November 1991.

_____. "Torpedo problem for US Navy and industry," **Jane's Navy International** 103(7):6 September 1, 1998.

_____. "US NAVY - DARPA uncovers next-generation submarine technologies," **Jane's Navy International** 4 July 1, 1998.

Frank, Willard C., Jr. "Multinational Naval Cooperation in the Spanish Civil War," **Naval War College Review** 42(2):72-101 Spring 1994.

Freeman, Jeffrey G. "Orion hunts again," **U.S. Naval Institute Proceedings** 125(11):60-61 November 1999.

Freitag, Walter. "Developments in submarine technology," **Naval Forces** 10-12 2001.

Frere, Toby. "Submarine warfare," **RUSI** 138(2):46-5 April 1993.

Frick, R.E. "The submarine building program," **Submarine Review** 35-44 October 1995.

Friedman, Norman. "It's what's inside that counts," **U.S. Naval Institute Proceedings** 123(2):41-44 February 1997.

_____. "The Typhoon saga ends," **U.S. Naval Institute Proceedings** 125(2):91-92 February 1999.

_____. "Up periscope, up antenna," **Journal of Electronic Defense** 24:45-50 March 2001.

_____. "World naval developments," **U.S. Naval Institute Proceedings** 125(9):121-123 September 1999.

Fuentes, Gidget. "Still no women on subs," **Navy Times** 49:22 November 8 1999.

"Future Attack Submarine stresses affordability," **Jane's Defence Weekly** 30(19):45 November 11, 1998.

"The future lies in AIP propulsion for submarines," **Asian Defence Journal** 52-54 October 1999.

Fyfe, H.C. "Submarine Boats," **U.S. Naval Institute Proceedings** 28(3):753 September 1902.

Gabriel, Michael J., Jr. "Submarine officer development: can we do better?" **Submarine Review** 119-122 January 1999.

Galdorisi, George. "USW: We can't wish it away," **U.S. Naval Institute Proceedings** 124(6):38-40 June 1998.

Garvin, Chris S. "Ethics, the Navy, and the submarine fleet," **Submarine Review** 57-61 October 1996.

Gebhardt, Laurence P. "The U.S. Needs a Commercial-Military Submarine," **U.S. Naval Institute Proceedings** 119(12):84-86 December 1993.

Geiger, Dan. "One View of Future Technology," **Submarine Review** 44-47 October 1999.

Geisenheyner, Stefan. "Communications with submerged submarines," **Armada International** 15:28+ August/September 1991.

Genalis, Paris. "U.S. submarines in the near future," **Submarine Review** 38-46 July 1999.

George, Glenn R. "The Naval Reactors program: from Nautilus to the millennium," **Nuclear News** 41(11):26 October 1998.

George, James. "The Sense of Centurion: Two Opposing Views.' **Submarine Review** 40-46 October 1993.

George, James L. "The SSN-21 Seawolf--Progress, problems and a touch of paranoia," **Navy International** 96:138-141 May 1991.

Geranios, Nicholas K. "Rigged for Stealthy Running; Small Submarines Test Silent, Test Deep in Mysterious Idaho Lake; Navy: Prototypes up to 88 feet long cruise Lake Pend Oreille, which offers extreme depth and consistent subsurface," **The Los Angeles Times** [Record Edition] 1 August 23, 1998.

"German-Norwegian U-212 Project Delayed: The Norwegian View," **Naval Forces** 14:42-44+ 1993.

Giambastiani, E. P. "The future of our submarines," **Submarine Review** 23-30 October 1996.

_____. "Silence in our wake," **U.S. Naval Institute Proceedings** 123(10):48-50 October 1997.

Giaquinto, Joseph N. et al. "The quick strike submarine," **U.S. Naval Institute Proceedings** 121(6):41-44 June 1995.

Glasser, Robert D. "Enduring Misconceptions of Strategic Stability: The Role of Nuclear Missile-carrying Submarines," **Journal of Peace Research** 29:23-37 February 1992.

Golda, E. Michael. "The Dardanelles Campaign: A Historical Analogy for Littoral Mine Warfare," **Naval War College Review** 51(3):82-96 Summer 1998.
http://www.nwc.navy.mil/press/Review/1998/summer/art6su98.htm

Goldingham, C. S. "U. S. submarines in the blockade of Japan in the 1939-45 war," **Journal Of The Royal United Service Institution** 97:87-98 February 1952.

_____. "U. S. submarines in the blockade of Japan in the 1939-45 war," **Journal Of The Royal United Service Institution** 97:212-222 May 1952.

Goldstein, Richard. "Clay Blair, 73, Navy Veteran And an Expert on Submarines [Obituary]," **New York Times** [Late Edition (East Coast)]:1.67 December 20, 1998.

Goodman, Glenn W., Jr. "Getting SEALs ashore," **Armed Forces Journal International** 22-23 January 1997.

_____. "Redesigning the Navy: Three new ships hold the key to long-term modernization of surface and submarine fleets," **Armed Forces Journal International** 135(8):28+ March 1998.

_____. "Silent running: US and Russian navies continue to put a premium on submarine quieting," **Armed Forces Journal International** 135(9):26-28 April 1998.

_____. "Undersea mobility platform," **Armed Forces Journal International** 135:52-53 May 1998.

Gordon, Mike. "Opposition surfaces ['Silent service' says no to further cuts].," **Navy Times** 49:14-16 January 17, 2000.

Gorenflo, Mark. "Retaining the Submarine Junion Officer A Modest Proposal," **Submarine Review** 76-85 April 2001.

_____. "Submarine Foreece Structure: An Exercise in Applied RADCON Math," **Submarine Review** 23-33 October 2002.

Gouge, Michael. "Some Thoughts on Propulsion," **Submarine Review** 83-84 April 1999.

Graham, M.T. "U-2513 Remembered," 83-92 **Submarine Review** July 2001.

Graves, Barbara and Edward C. Whitman. "The Virginia class: America's next submarine," **Undersea Warfare** 1(2):2-7 Winter 1998/1999.

Grazebrook, A. W. "Tough job for the ASC task force," **Asia-Pacific Defence Reporter** 26:51-52 June/July 2000.

Green, Jerry. "Submarine Warfare - New Challenges," **Submarine Review** 54-61 October 1992.

Griffin, James. "Simulation-based design for undersea warfare systems," **Naval Forces** 20(3):S28-S30 1999.

Grimes, Vincent P. "Building conventional subs in the US," **Naval Forces** 15:20-24 1994.

Grossenbacher, John J. "Address to the NDIA Clambake sub base, New London, September 2000," **Submarine Review** 13-21 January 2001.

_____. "Address to the NDIA Clambake sub base, New London, September 2001," **Submarine Review** 12-20 January 2002.

_____. "Remarks at the 2002 Submarine Birthday Ball," **Submarine Review** 115-119 July 2002.

Gruner, William P. and Henry E. Payne, III. "Submarine Maneuver Control," **U.S. Naval Institute Proceedings** 118(7):56-60 July 1992.

Guttman, Jon. "Undersea struggle for Guadalcanal," **World War II** 6:57 July 1991.

Haas, Mike. "Submarine Raiding During The Korean War," **Submarine Review** 86-96 April 2002.

Hackney, Gene. "Penetrating acoustic frequency levels," **Surface Warfare** 24:15 May/June 1999.

Haffa, Robert P., Jr. and James H. Patton, Jr. "Analogues of stealth: Submarines and aircraft," **Comparative Strategy** 10:257-271 July/September 1991.

Hall, Jim. "War in the littorals," **Surface Warfare** 24:8-10 May/June 1999.

Hall, Robert D. Jr. "The inventor of the first practical submarine began his work in a secret Irish attempt to sink the Royal Navy," **Military History** 17(4):20-22, 28+ October 2000.

Halloran, Richard. "Silent service: the mission of U.S. fast-attack submarines has fundamentally changed," **Asia-Pacific Defense Forum** 22:16-21 Winter 1997-1998.

Hamadyk, Donald M. "Over the Horizon" A View of Submarine Developments," **Submarine Review** 21-33 January 2002.

Hamblen, William. "Next generation stealth submarines," **Sea Technology** 39(11):59 November 1998.

Handler, Joshua. "Waging Submarine Warfare," **Bulletin of the Atomic Scientists** 43(7):40 September 1987.

Hanley, John T. "Implications of the changing nature of conflict for the submarine force," **Naval War College Review** 46(4):9-28 Autumn 1993.

Hanna, Robert G. "Air Independent Propulsion for SSNs," **Submarine Review** 66-68 January 2000.

Hansen, Butch. "A new SSBN operating cycle for Kings Bay," **Undersea Warfare** 2(1):9-11 Fall 1999.

Hansen, Richard P. "Submarine warfare as an instrument of policy," **Submarine Review** 43-61 April 1991.

Harboe-Hansen, Hans. "Viking--the future Nordic submarine?" **Naval Forces** 19(5):30-32+ 1998.

"Harnessing technology is the key to sub force's strength," **C4I News** 1 July 29, 1998.

Harral, Brooks J. "Submarine power--the final arbiter," **Submarine Review** 31-36 July 1990.

_____. "Unseen persuaders," **Naval History** 4(3):10-13 Summer 1990.

Hart, Kenneth. "The Silent Service Must Communicate," **U.S. Naval Institute Proceedings** 123(2):75-77 February 1997.

Hartsfield, J. Carl. "Made in America: from Flintlocks to Tomahawks and beyond," **Submarine Review** 74-80 April 2000.

Hasslinger, Karl. "Junior officers design the Submarine Force for the next hundred years," **Undersea Warfare** 2(4):7-10 Summer 2000.

_____ and Kevin Mooney. Selling the Submarine Force," **Submarine Review** 43-48 April 1999.

Hausherr, Burckart and Reinhard Dinse. "Production progress of class 212A [Special Issue].," **Naval Forces** 24-26+ 2001.

Heffron, John S. "The Virginia attack boat brings new potency to the fleet," **Sea Power** 47(7):27 July 2003.

Henry, Mark. "Submarine External Rotary Rack Weapons Systems," **Submarine Review** 60-65 January 2000.

Hessman, James D. and Edward J. Walsh. "Strong voice for the silent service [Interview with Rear Adm. Thomas D. Ryan, Dirctor of Submarine Warfare]," **Sea Power** 36(6):9+ June 1993.

_____, _____ and Gordon I. Peterson. "The U.S. submarine force today--operational demands grow as numbers fall," **Sea Power** 42(7):9-14 July 1999.

_____, _____ and Vincent C. Thomas, Jr. "A voice for the silent service [Interview with Vice Adm. Roger F. Bacon]," **Sea Power** 34(7):48-49+ July 1991.

Hewish, Mark. "Snooping above the waves: sensors that increase a submarine's effectiveness against surface ships, aircraft and land targets are in increasing demand," **Jane's International Defense Review: IDR** 34:48-49+ August 2001.

_____ and Joris Janssen Lok. "Ready for action: integrated submarine combat systems," **Jane's International Defense Review: IDR** 32:42-47 February 1999.

Hindinger, John R. "Emphasize tactical training!" **U.S. Naval Institute Proceedings** 123(10):38-39 October 1997.

Hinge, Alan. "Developments in Minewarfare and Mine Countermeasures in the Western Pacific," **Journal of the Australian Naval Institute** 19:23-28 August 1993.

Hoekley, Rhonda et al. "The Monte Carlo approach to mine warfare," **Sea Power** 44(2):44-46 February 2001.

Holderness, V. F. "Relaunch the non-nuclear boats," **U.S. Naval Institute Proceedings** 121(61):45-46 June 1995.

Holland, Cecelia and Ray Hillman. "Hapless voyage of H-3," MHQ: The Quarterly Journal of Military History 12:70-75 Spring 2000.

Holland, Jerry. "Preparing for war: now or later?" **Submarine Review** 69-80 October 2000.

_____. "Submarine Command and Control in the New World Order," **Submarine Review** 46-53 October 1992.

Holland, William J. Jr. "100 years of submarines," **U.S. Naval Institute Proceedings** 126(1):71-73 January 2000.

_____. "Acquiescence is Agreement: Reflections on Submarine Roles and Missions From the Submarine Technology Symposium," **Submarine Review** 43-46 July 2001.

_____. "ISSUE: RELATIVE VULNERABILITY: A Fleet to Fight in the Littorals," **Submarine Review** 33-44 April 2001.

_____. "SSN: The Queen of the Seas," **Naval War College Review** 44(2):113-118 Spring 1991.

_____. "Uncertainty Efftects in the Submarine Fire Control Problem," **Submarine Review** 49-54 April 1999.

Holzer, Robert. "DARPA, U.S. Navy renew sub tech ties," **Defense News** 13:10 September 14-20 1998.

_____. "DoD Orders Navy to Conduct Study of Non-Nuclear Sub," **Defense News** 7:3+ February 10, 1992.

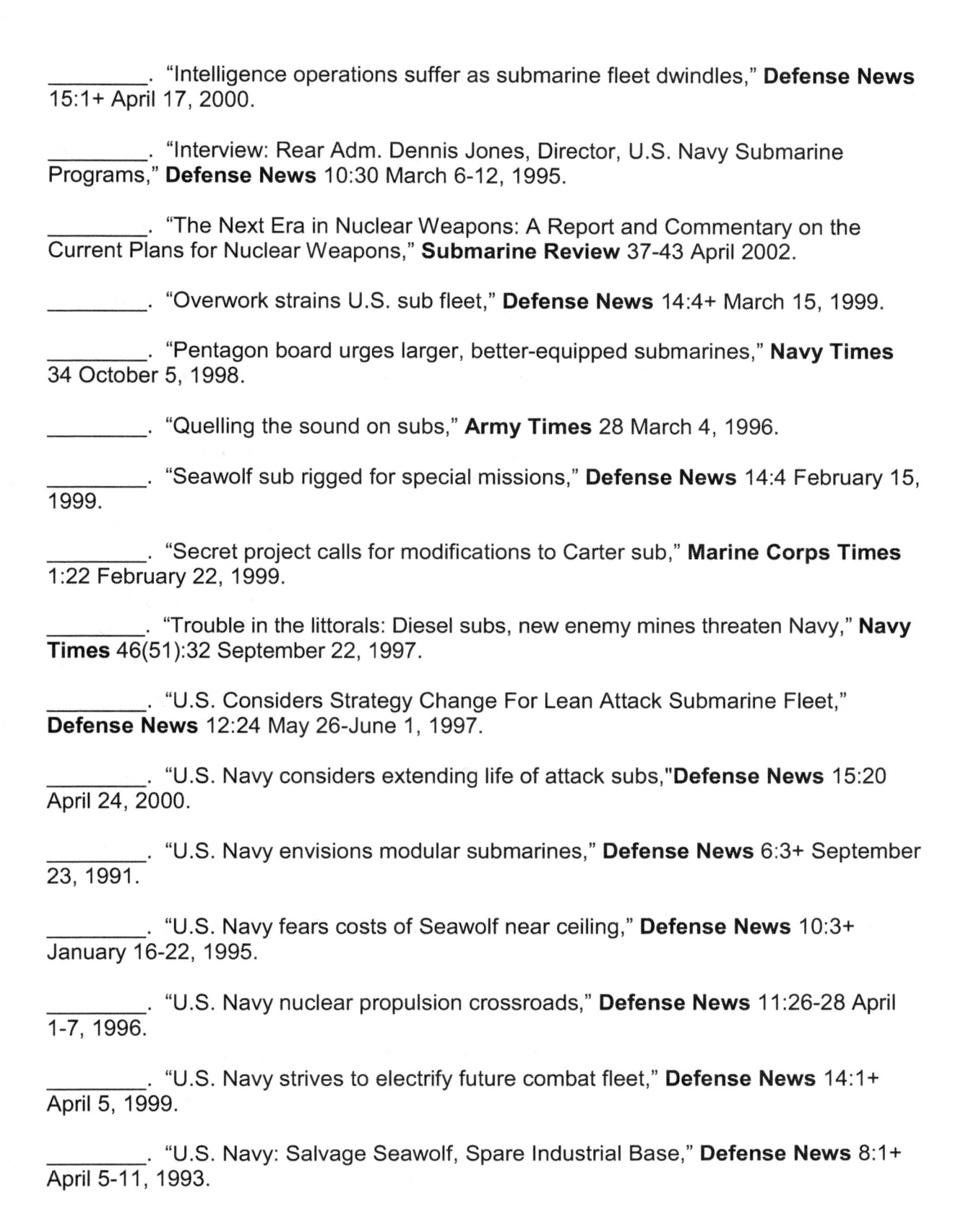

_____. "Intelligence operations suffer as submarine fleet dwindles," **Defense News** 15:1+ April 17, 2000.

_____. "Interview: Rear Adm. Dennis Jones, Director, U.S. Navy Submarine Programs," **Defense News** 10:30 March 6-12, 1995.

_____. "The Next Era in Nuclear Weapons: A Report and Commentary on the Current Plans for Nuclear Weapons," **Submarine Review** 37-43 April 2002.

_____. "Overwork strains U.S. sub fleet," **Defense News** 14:4+ March 15, 1999.

_____. "Pentagon board urges larger, better-equipped submarines," **Navy Times** 34 October 5, 1998.

_____. "Quelling the sound on subs," **Army Times** 28 March 4, 1996.

_____. "Seawolf sub rigged for special missions," **Defense News** 14:4 February 15, 1999.

_____. "Secret project calls for modifications to Carter sub," **Marine Corps Times** 1:22 February 22, 1999.

_____. "Trouble in the littorals: Diesel subs, new enemy mines threaten Navy," **Navy Times** 46(51):32 September 22, 1997.

_____. "U.S. Considers Strategy Change For Lean Attack Submarine Fleet," **Defense News** 12:24 May 26-June 1, 1997.

_____. "U.S. Navy considers extending life of attack subs," **Defense News** 15:20 April 24, 2000.

_____. "U.S. Navy envisions modular submarines," **Defense News** 6:3+ September 23, 1991.

_____. "U.S. Navy fears costs of Seawolf near ceiling," **Defense News** 10:3+ January 16-22, 1995.

_____. "U.S. Navy nuclear propulsion crossroads," **Defense News** 11:26-28 April 1-7, 1996.

_____. "U.S. Navy strives to electrify future combat fleet," **Defense News** 14:1+ April 5, 1999.

_____. "U.S. Navy: Salvage Seawolf, Spare Industrial Base," **Defense News** 8:1+ April 5-11, 1993.

_____. "U.S. submariners urge halt to dwindling fleet," **Navy Times** 48:26 July 5, 1999.

_____. "U.S. submarines urge halt to dwindling fleet," **Defense News** 14:6 June 28, 1999.

_____. "UAVs Someday Could Be Launched From A Sub," **Navy Times** (Marine Corps Edition) 31 November 3, 1997.

_____. "Undersea battles: has Navy lost its edge?" **Navy Times** (Marine Corps Edition) 26 May 18, 1998.

_____. "Undersea Warfare needs focus to win funding," **Defense News** 13:1+ May 11-17, 1998.

_____. "Utility of subs rises as targeting grows more precise," **Defense News** 15:17 April 10, 2000.

_____. "With limited subs, U.S. Navy looks westward to Guam," **Defense News** 15:4 August 28, 2000.

_____ and David Silverberg. "Congress, U.S. Navy Face off over Diesel Sub Exports: Service Argues Vital Technology May Leak Out," **Defense News** 7:9 August 10-16, 1992.

Honaker, Neil Travis. "False assumptions, wistful dreams," **U.S. Naval Institute Proceedings** 124(1):70-73 January 1998.

Hoon, Shim Jae. "Kim the cool," **Far Eastern Economic Review** 161(28):16 July 9, 1998.

_____. "Submarine shocker," **Far Eastern Economic Review** 159(40):18 October 3, 1996.

Horne, Chuck. "Sub the nation needs," **U.S. Naval Institute Proceedings** 121(3):14 March 1995.

Householder, Patrick. "War ready," **All Hands** 996:20-21 April 2000.

Houley, W. P. "2015," **U.S. Naval Institute Proceedings** 119(10):49-52 October 1993.

Howard, Peter. "Rolls-Royce details new division aims," **Jane's Navy International** 105(5):37 June 2000.

Howe, John A. "Wolfpack: Measure and Counter," **Naval War College Review** 23(8):61-65 April 1971.

Hulina, Victor E. "Our Submarine Navy in Transition: A Traditional OEM's Perspective," **Submarine Review** 81-85 April 2002.

Hunter, Jack. "FAB, The First Submarine Towed Array?" **Submarine Review** 129-131 January 2000.

"The Ideal Submarine," **U.S. Naval Institute Proceedings** 43(2):400 February 1917.

"In the near future we will deal within international communities with the subject of submarine rescue [interview with Yuri Kormilitsin, General Designer of conventional submarines in the Rubin Design Bureau].," **Naval Forces** 21:34-36+ 2000.

"Intelligence rises for US Pacific subs," **Jane's Defence Weekly** 1 December 16, 1998.

"Interview with CINCPACFLT, Admiral Archie Clemins," **Undersea Warfare** 1(4):2-5 Summer 1999.

Isby, David C. "UNITED STATES - SSN combat sub Lexington; system upgrades," **Jane's Defence Upgrades** 2 November 13, 1998.

_____. "UNITED STATES - SSBN to SSGN conversion studied," **Jane's Defence Upgrades** 3 July 3, 1998.

Ivarsson, Ola and Hans Pommer. "Air-independent propulsion systems for submarines," **Naval Forces** 56-59 2001.

Jacobsen, Klaus. "Submarine combat systems--a users perspective," **Naval Forces** 19:8-10+ 1998.

"JANAC Submarine Credits Revised," **Submarine Review** 44-46 January 2000.

"JDW country survey: United Kingdom," **Jane's Defence Weekly** 16(18):809+ November 2, 1991.

Joergensen, Tim Sloth. "U.S. Navy Operations in Littoral Waters: 2000 and Beyond," **Naval War College Review** 51(2):20-29 Spring 1998.
http://www.nwc.navy.mil/press/review/1998/spring/art2%2Dsp8.htm

Jones, D. A. "Submarine force plans and programs: preparing for the challenges of the 21st Century," **Submarine Review** 26-33 October 1995.

_____. "U.S. Navy submarine force: where we are and where we are going," **Submarine Review** 15-22 April 1995.

Kaminski, Paul G. "Affordability in the submarine force," **Defense Issues** 11(65):1-5 June 16, 1996.

http://www.defenselink.mil/speeches/1996/s19960616-kaminski.html

Karniol, Robert. "Interview with Rear Admiral Al Konetzni, Commander Submarine Force, US Pacific Fleet," **Janes Defense Weekly** 31:32 January 27, 1999.

Kauchak, Marty. "Submarines remain in the Hill's scope," **Armed Forces Journal International** 139:6-8 August 2001.

_____. "Transformation in their scope [exclusive interview with Admiral Frank L. "Skip" Bowman, Director, Naval Reactors].," **Armed Forces Journal International** 139:34-37 November 2001.

Kaufhold, Edmund E. "More on submarine LANS," **Submarine Review** 131-133 July 2000.

Kaufman, Steve and Robert Kaufman. "The silent service," **Retired Officer** 51:34-41 February 1995.

Kearney, T.A. "The Submarine Its Purpose and Development," **U.S. Naval Institute Proceedings** 41(4):1239 July/August 1915.

Keil, Robin. "Submarines in Southeast Asia," **Naval Forces** 17:16+ 1996.

Keeler, R Norris. "Regulus-the forgotten weapon," **Signal** 53(1):90 September 1998.

Kelle, Karl-Heinz. "One hundred years of submarines and now?" **Naval Forces** 21(6):18-21 2000.

Keller, Stephen H. "Clearing The Way For Coalition Warfare," **Sea Power** 40(12):50-52 December 1997.

Kellogg, Ned. "Operation Hardtack as a Submerged Target: Life Aboard a Diesel Submarine in the 1950s," **Submarine Review** 135-149 October 2002.

Kelly, C Brian. "The 1917 sinking of the liner Laconia helped push the United States into World War I," **Military History** 15(1):74 April 1998.

Kemp, Ian. "Country survey: The United Kingdom," **Jane's Defence Weekly** 30(17):19-21+ October 28, 1998.

_____ and Hans J. Andersson. "Country briefing: Sweden," **Jane's Defence Weekly** 31(24):18-20+ June 16, 1999.

Kennedy, Floyd D., Jr. "Navy Fears Empty Flight Decks - Enough Carriers, but Not Enough Aircraft," **National Defense** 36-37 April 1992.

Kennedy, Harold. "Virginia-class boats portend new ship-building approach," **National Defense** 84:28-31 May 2000.

Kenny, Alejandro. "Today's submarine forces," **Naval Forces** 21:82-83 2000.

Kenyon, Henry S. "Goalkeeper blocks incoming submarine shots: prototype underwater interceptor increases warship survivability," **Signal** 55:63-65 March 2001.

Kharitonov, N. "Striking from under the water," **Soviet Military Review** 41-43 June 1978.

Khlopkin, N.S. "Development of Marine Nuclear Power Plants for Submarines and Surface Ships," **Submarine Review** 63-68 October 1999.

Kim, Duk-Ki. "Cooperative Maritime Security in Northeast Asia," **Naval War College Review** 52(1):53-77 Winter 1999.
http://www.nwc.navy.mil/press/review/1999/winter/art3%2Dw99.htm

"A Kiosk-Hydro for Submarines," **U.S. Naval Institute Proceedings** 40(6):3826 November/December 1914.

Kirby, John. "The size of the fight," **All Hands** 36-41 July 1999.

Kirschenbaum, Susan S. "Lessons Learned and Mis-conceptions: Eighteen Years of Submarine Decision Research," **Submarine Review** 44-50 July 2002.

Knox, James. "Silent service - submarine proliferation in Southeast Asia," **Harvard International Review** 19:54-55+ Winter 1996/1997.

Koch, Andrew. "Four USN SSBNs may convert to conventional role," **Janes Defense Weekly** 31:10 February 17, 1999.

_____. "START II and costs doom Trident conversions," **Janes Defense Weekly** 33:8 April 26, 2000.

_____. "US Navy sets sights on electric attack submarine," **Janes Defense Weekly** 34:8 July 26, 2000.

_____. "USN approves submarine missile concept," **Janes Defense Weekly** 31:18 March 10, 1999.

_____. "USN must boost SSN fleet or risk 'not meeting warfighting needs'," **Janes Defense Weekly** 31:6 March 24, 1999.

Konetzni, Al. "How many subs do we need?" **U.S. Naval Institute Proceedings** 126(11):56-57 November 2000.

Koonce, Robert A. "Bridging the Gap [retention]," **Submarine Review** 36-40 April 1999.

Kostiuk, Michael. "Removal Of The Nuclear Strike Option From United States Attack Submarines," **Submarine Review** 85-90 January 1998.

Kowenhoven, William H. and Frederick J. Harris. "NSSN (New Attack Submarine): A 21st century design," **U.S. Naval Institute Proceedings** 123(61):35-38 June 1997.

Kreck, Thomas. "Energy from AIP-systems: the key to future submarine propulsion," **Naval Forces** 21:12-14+ 2000.

Kreh, William R. "Submarine Development Group One - The Navy's Unique Command With an In-Depth Mission," **Navy** 13:27-29 October 1970.

Kreisher, Otto. "Innovations under the sea," **Sea Power** 44(7):42-44 July 2001.

_____. "The trinity torpedo: Mk54 hybrid blends the best of three," **Sea Power** 42(7):47-48 July 1999.

Kristof, Nicholas. "The State of Periscope Technology," **Submarine Review** 51-57 October 2001.

Krug, Rainer. "The future role of undersea warfare," **Naval Forces** 21(2):18-23 2000.

Kuhlmann, Dietrich. "Submarine strike comes of age," **Undersea Warfare** 2(3):7-8 Spring 2000.

Kurak, Steve. "Multi-level security networks on submarines," **Submarine Review** 134-139 July 2000.

Kurdin, Igor and Wayne Grasdock. "Loss of a Yankee SSBN," **Submarine Review** 56-57 October 2000.

Kutner, Joshua A. "SEAL teams seek efficient maritime mobility systems," **National Defense** 83:24-25 February 1999.

LaVo, Carl. "Commanding Officer Breaking Down [some sub skippers could not take the pressure early in WWII]," **Naval History** 6(3):29-34 Fall 1992.

Lacroix, Frank W. "Answering the challenges," **U.S. Naval Institute Proceedings** 127(7):29 July 2001.

Lake, Julian. "Case for the diesel-electric submarine," **U.S. Naval Institute Proceedings** 121(6):63 June 1995.

Lambeth, Benjamin S. "Why Submariners Should Talk to Fighter Pilots," **Submarine Review** 36-43 October 1999.

Largess, Richard P. "The Origin of Albacore," **Submarine Review** 63-73 January 1999.

Largess, Robert P. "The Origin of the Towed Array [An Interview with Marvin Lasky]," **Submarine Review** 47-67 January 2002.

_____. "USS TRITON: The Ultimate Submersible," **Submarine Review** 101-107 January 1994; **Submarine Review** 41-46 April 1994.

"The lasting legacy: nuclear submarine disposal," **Jane's Navy International** 103(1):12-20 January/February 1998.

Lautenschlager, Karl. "The Submarine in Naval Warfare, 1901-2001," **International Security** 11(3):94-140 Winter 1986/1987.

Lavery, Jason. "Finnish-German submarine cooperation 1923-35," **Scandinavian Studies** 71(4):393-418 Winter 1999.

Lawson, Richard. "Interview with COMSUBPAC Rear Admiral Henry C. McKinney, USN," **Submarine Review** 95-104 October 1993.

Layton, Edwin T. Capt. "Rendezvous in reverse [Japanese second attack on Pearl Harbor]," **U.S. Naval Institute Proceedings** 79(5):478-485 May 1953.

Lecroix, F.W. "Opportunity Calls for Focus," **Submarine Review** 47-53 April 2002.

Legro, Jeffrey W. "Military culture and inadvertent escalation in World War II," **International Security** 18(4):108-142 Spring 1994.

Leidig, Charles J. "Joint Vision 2010 -- a submarine's guide," **Submarine Review** 104-111 July 1997.

_____. "Submarine force multipliers," **Submarine Review** 22-29 January 2001.

Lemkin, Bruce. "The new leader of the pack: The USS SEAWOLF," **U.S. Naval Institute Proceedings** 117(6):42-45 June 1991.

Leopold, George. "Summit Agreement May Lead to Revaluation of U.S. Forces," **Defense News** 4 June 22-28, 1992.

Lepotier Capt. "Submarines in the Battle of Leyte Gulf," **Military Review** 30:102-108 December 1950.

LeSueur, Stephen C. "Navy's Mine Warfare Plan Envisions New MCM Command & Control Support Ship," **Inside Defense Electronics** 5:7-8 December 20, 1991.

Levine, Emil. "Who helped the Barb?" **Naval History** 15(3):44-37 June 2001.

Levy, James. "Ready or Not? The Home Fleet at the Outset of World War II," **Naval War College Review** 52(4):90-108 Autumn 1999.
http://www.nwc.navy.mil/press/review/1999/autumn/art5%2Da99.htm

Lewis, David. "DD-21: Another Seawolf?" **U.S. Naval Institute Proceedings** 127(8):54-57 August 2001.

Lewis, Oliver. "Impact of network centric warfare on submarine operations," **Submarine Review** 109-111 January 1999.

Lieberman, Joseph. "The Reasons Why," **U.S. Naval Institute Proceedings** 118 (6):55-58 June 1992.

Lillie, Bill. "The Undersea Platform for Future Maritime Dominance," **Submarine Review** 3-57 July 1999.

Lisiewicz, John S. "Unmanned undersea vehicles," **Naval Forces** 20:75 1999.

"Lockheed Martin delivers new USW system to Navy," **Sea Power** 42(8):2 August 1999.

Lodmell, Joseph. "It only takes one," **U.S. Naval Institute Proceedings** 112(12):30-33 December 1996.

Loesener, Rolf. "Integrated Submarine Combat Systems: a systematic approach to integrate sensors, communications, weapons and C2 systems," **Naval Forces** 22:53-57 2001.

Lok, Joris Janssen. "Conventional submarines: at the forefront of naval developments," **Jane's Defence '96** 126-131 1996.

_____. "Fincantieri awaits go-ahead for work on Improved Sauro submarines," **Jane's International Defense Review: IDR** 31:67 August 1998.

_____. "Mini submarines and special forces pose maximum threat," **Jane's International Defense Review** 63 June 1, 1998.

_____. "New challenges force change on Royal Navy," **Jane's Defence Weekly** 28(9):41-42+ September 3, 1997.

_____. "Regional submarine programme stalled," **Janes Defense Weekly** 29:33+ February 18, 1998.

_____. "Setting a Collision Course for Nuclear Catastrophe," **Janes Defense Weekly** 18:16 March 13, 1993.

_____. "STN Atlas seeks international partners for naval ventures," **Janes Defense Weekly** 29:16-17 April 1, 1998.

_____. "Strong Signals from SDC," **Jane's Defense Weekly** 18:38-39 September 12, 1992.

Lyman, Mel. "Crimson Tide: They Got It All Wrong," **Submarine Review** 30-35 April 1999.

Lyne, Jen Scott. "The submarine: Outdated menace or force multiplier?" **U.S. Naval Institute Proceedings** 125(7):90-91 July 1999.

Lyon, Waldo K. "Submarine Combat in the Ice," **U.S. Naval Institute Proceedings** 118(2):33-40 February 1992.

Machtley, Ronald K. "Yes, submarines are still needed!" **Sea Power** 35(3):26-28 March 1992.

Macri, Fred. "Sun Tzu and the art of submarine warfare," **Submarine Review** 81-84 October 1997.

Madsen, Kaj Toft. "Fighting the beast: Nonnuclear subs & littoral warfare," **U.S. Naval Institute Proceedings** 122(8):28-30 August 1996.

Maier, Wolfgang. "Autonomous underwater vehicles: the upcoming solution for mine countermeasures," **Naval Forces** 21:76-80 2000.

Maiolo, Joseph A. "Deception and intelligence failure: Anglo-German preparations for U-boat warfare in the 1930s," **Journal of Strategic Studies** 22:55-76 December 1999.

Mann, Paul. "Sub accident sparks proliferation worries," **Aviation Week & Space Technology** 153:32-33 August 21, 2000.

Martin, M. E. "Disruptive technology in Undersea Warfare for the 21st century: Part one," **Submarine Review** 48-61 April 2000; "Part Two," **Submarine Review** 107-122 July 2000.

Martini, Ron. "A Submarine Internet Site," **Submarine Review** 123-126 July 2001.

Martinovich, Steven. "USS Indianapolis captain cleared," **American History** 36(5):8-9 December 2001.

Marty, Mark M. " Joint Professional Military Education: a New Paradigm for Submarine Junior Officers," **Submarine Review** 113-116 January 1999.

Mason, L.G. "Why Submarines?" **Canadian Defence Quarterly** 21:21-23 June 1992.

Matthews, William. "Future sub: more than a 'torpedo boat,'" **Navy Times** 48:19 March 29, 1999.

_____. "Rebirth of sorts in works for the Trident," **Navy Times** 14:24 March 22, 1999.

_____. "Trident subs could enjoy rebirth," **Defense News** 14:24 April 5, 1999.

_____ and Bradley Peniston. "Sub fleet taking a dive: Is the Navy wasting its best asset?" **Navy Times** 48(24):16-18 March 22, 1999.

Maurer, Dr. Maurer and Lawrence J. Paszek. "Origin of the Laconia Order," **Air University Review** 15:26-37 March/April 1964.

_____ and _____. "Origin of the Laconia Order," **Royal United Service Institution Journal** 109:338-344 November 1964.

Mazumdar, Mrityunjoy. "Global conventional submarine programmes (first of two parts),"; **Naval Forces** 21(4):18-22 2000; "Part II of Two," **Naval Forces** 21(5):36-41 2000.

McCue, Gary. "The Centennial connection: USS Holland and Electric Boat," **Sea Power** 41(2):49 February 1999.

McDonald, C.A.K. "Real story of Scorpion?" **U.S. Naval Institute Proceedings** 125(6):28-33 June 1999.

McGee, E. D. "To sink and swim: the USS FLIER," **Submarine Review** 94-98 October 1996.

_____ and Peter J. Hanway. "Haul down the subs!" **U.S. Naval Institute Proceedings** 125(12):67-68 December 1999.

McIlvaine, Brian. "Improving submarine warfighting endurance," **Submarine Review** 96-103 January 2001.

McLennan, R. A. "Avoiding the Undersea Surprise," **U.S. Naval Institute Proceedings** 120(5):122+ May 1994.

McMichael, William H. "Should four Trident subs be scrapped or rebuilt? [To comply with treaty regulations, the Navy needs to reduce its ballistic nuclear missile submarine fleet by four boats], " **Navy Times** 50:10-12 December 25, 2000.

_____. "What if it were our sub?" **Marine Corps Times** 2:24-25 September 4, 2000.

McMillian, Craig E. "Submarine C4ISR antenna systems," **Naval Forces** 20:64-65 1999.

Meacham, J. A. "The Mine Countermeasures Ship," **U.S. Naval Institute Proceedings** 94(4):128-129 April 1968.

Meadows, Sandra I. "Navy casts multimission sub for pollution free dominance," **National Defense** 81:28-29 March 1997.

Merrill, John. "April 1900: inventor-builder John P. Holland delivers first U.S. submarine--part one," **Submarine Review** 91-98 July 1998.; "Part two," **Submarine Review** 64-69 October 1998.

_____. "Convoy: The Forgotten years 1919-1939 Part I," **Submarine Review** 53-61 January 1999; "Part II Between the Wars," **Submarine Review** 63-69 April 1999.

_____. "Looking Around: Short History of Periscopes-Part I," **Submarine Review** 68-88 January 2002; Part II," **Submarine Review** 59-78 April 2002.

_____. "Slide rule strategy begins: World War II operations research," **Submarine Review** 106-117 January 2000.

_____. "Submarine Bells to Sonar & Radar Submarine Signal Company (1901-1946): Part I," **Submarine Review** 85-113 October 2002.

Mies, Richard W. "NATO and partner submarines in a changing geopolitical environment," **NATO's Sixteen Nations** 63-65 (SACLANT Special Issue) 1998.

_____. "Remarks to the NSL Annual Symposium June 5, 1997 [current state of the U.S. Submarine Force]," **Submarine Review** 27-39 July 1997.

_____. "The SSBN in national security," **Undersea Warfare** 2(1):2-6 Fall 1999.

Miller, David. "Conventional submarines 1990," **Defense & Diplomacy** 8(4):10-12+ April 1990).

_____. "An Inside Look at the US Submarine Community," **Naval Forces** 14:613-617 1993.

_____. "The Silent Menace: Diesel-Electric Submarines in 1993," **International Defense Review** 26:613-617 August 1993.

_____. "Submarines in the Gulf," **Military Technology** 17:42-45 June 1993.

Minnick, Wendell and Robert Karniol. "Russians talk subs with Taipei," **Janes Defense Weekly** 35:35 June 20, 2001.

Momiyama, Thomas S. "All and nothing," **Air & Space Smithsonian** 16(4):22-3 October/November 2001.

Morris, Richard Knowles. "Who built those subs?" **Naval History** 12(5):31-34 September/October 1998.

Morris, William. "Undersea dominance in the littoral and open ocean," **Surface Warfare** 24:11-14 May/June 1999.

Morrison, David C. "New subs and where to build them," **National Journal** 27:1952-1953 July 29, 1995.

Mrityunjoy, Mazumdar. "Global conventional submarine programmes," **Naval Forces** 21:18-20+ 2000.

Muir, Malcolm, Jr. "A Stillborn System: The Sub-Launched Cruise Missile," **MHQ** 6:80-83 Winter 1994.

"The Multitubular Submarine," **U.S. Naval Institute Proceedings** 43(6):1297 June 1917.

Munsey, Christopher. "Sub officers weigh in on fate of Waddle: many feel sympathy, respect for former Greenville commander," **Navy Times** 50:8 May 7, 2001.

_____. "The world is watching," **Navy Times** 50:8 March 5, 2001.

Murdock, Paul. "SSNs (nuclear attack submarines) aren't enough," **U.S. Naval Institute Proceedings** 122(2):48-51 February 1996.

Murphy, Leo. "To catch the quiet ones," **U.S. Naval Institute Proceedings** 123(7):59-62 July 1997.

Nanos, G. P. "Strategic Systems Update," **Submarine Review** 12-17 April 1997.

Naval War College CASE STUDIES Part II: Strategic Thought and Practice in the Modern Security Environment; VI. THE FIRST WORLD WAR, 1914-1918: TOTAL WAR AND STRATEGIC ASSESSMENT.
http://www.nwc.navy.mil/cncscasestudies/cases/case06.htm

Naval War College CASE STUDIES Part II: Strategic Thought and Practice in the Modern Security Environment; VII. THE FIRST WORLD WAR, 1914-1918: TOTAL WAR AND STRATEGIC ASSESSMENT.
http://www.nwc.navy.mil/cnwcasestudies/cases/case07.htm

"Navy Captain, Submariner Albert `Ace' Burley Dies at 77," **The Washington Post** [FINAL Edition] B06, November 9, 1998.

"Neff System of Submarine Propulsion," **U.S. Naval Institute Proceedings** 41(6):2065 November/December 1915.

Nelan, Bruce W. "The spies from the sea," **Time** 148(16):44-45 September 30, 1996.

"New interests spur force development," **Jane's Defence Weekly** 25(21):23-26 May 22, 1996.

"The new security environment," **Submarine Review** 27-33 January 1995.

Newman, Richard J. "Submarine salesmanship: how did the submarine community get to the front of the requirements queue?" **Air Force Magazine** 84:60-64 January 2001.

_____. "Tales from the sea floor," **U.S. News & World Report** 125(20):44 November 23, 1998.

Newton, George B. "Submarine Arctic operations requirements, challenges, progress," **Submarine Review** 90-105 October 1991.

_____. "Don't Forget the Arctic," **Submarine Review** 91-100 April 2001.

Nicholls, David. "WWII Submarine Activity in Australia," **Submarine Review** 90-93 July 2002.

Nisbett, Shawn T. "The Impact of Modern Radar Technology on Submarine Tactical Employment," **Submarine Review** 69-78 January 2000.

Nitschke, Stefan. "From the hunted to the hunter," **Naval Forces** 22:18+ 2001.

Noble, Dennis L. and Truman R. Strobridge. "Winter of decision," **National Defense** 68:47-50 February 1984.; 68:57-62 March 1984.

Norris, William L. "The Coming Threat," **Submarine Review** 70-73 April 2001.

_____. "National missile defense and strategic forces," **Submarine Review** 28-31 January 2000.

Norton, Douglas M. "The Open Secret: The U.S. Navy in the Battle of the Atlantic, April-December 1941," **Naval War College Review** 26(4):63-83 January/February 1974.

"Nuclear capability is scaled back," **Jane's Defence Weekly** 30(17):1 October 28, 1998.

"Nuclear submarine disposal and recycling," **Submarine Review** 8-24 April 1998.

O'Connell, Jerome A. "Radar and the U-Boat," **U.S. Naval Institute Proceedings** 89(9):53-65 September 1963.

O'Rourke, Ron. "Luncheon Address to Subtech Symposium," **Submarine Review** 11-25 October 1999.

Offley, Ed. "Lost in the Deep: Scorpion's Tale Hidden in Web of Contradiction," **Navy Times** 12-15 November 22, 1993.

Oliver, Daniel T. "Awards Luncheon Address," **Submarine Review** 19-27 July 1999.

"Old friends and enemies: diesel subs around the world," **Journal of Electronic Defense** 24:38 March 2001.

"One on one with Adm. Frank Bowman, Director of U.S. Navy Nuclear Propulsion," **Defense News** 15:62 September 18, 2000.

Ostlund, Mike. "FINDING UNCLE BILL: Suprising New Information Uncovered About USS GUDGEON's Loss in 1944," **Submarine Review** 142-148 October 2001.

_____. "Correction to 'Finding Uncle Bill'," **Submarine Review** 136-137 April 2002.

"Our submarine century [Special insert]," **Undersea Warfare** 2(3):1-15 Spring 2000.

Owens, William A. "The View from OPNAV," **Submarine Review** 18-26 July 1993.

Packer, D.L. "The Network Centric Naval Officer," **Submarine Review** 69-82 October 1999.

Padgett, John B., III. "The Pull For Submarine Technology: Needed: Technology to Support Pacific OPLANs [Remarks, Submarine Technology Symposium 2002]," **Submarine Review** 14-19 July 2002.

"Parker 'Baby' Submarine," **U.S. Naval Institute Proceedings** 41(6):2067 November/December 1915.

Parry, Mike. "Virginia can be a Streetfighter," **U.S. Naval Institute Proceedings** 126(6):30-32 June 2000.

Parsley, Kenneth P. "Naval inshore undersea warfare contributes to RIMPAC 98 training," **The Officer** 74(10):49-50 November 1998.

Paterson, James. "Progress in submarine periscopes," **Asia-Pacific Defence Reporter** 25(8):44 January 2000.

Patton, James H., Jr. "Quanitative Submarine Communications Requirements," **Submarine Review** 65-70 October 2001.

_____. "The Rise and Fall of the Submarine Force - Again," **Submarine Review** 31-36 April 1993.

_____. "Stealth is a zero-sum game: A submariner's view of the advanced tactical fighter," **Military Review** 5:4-17 Spring 1991.

_____. "Stealth, sea control, and air superiority," **Airpower Journal** 7(1):52-56 Spring 1993.

_____. "The submarine force in the twenty-first century," **National Security Studies Quarterly** 5:61-76 Autumn 1999.

_____. "Submarines, stealth, and silver bullets: low-rate production for high leverage platforms," **Comparative Strategy** 14:59-79 January/March 1995.

_____. "Trident can fire more than nukes," **U.S. Naval Institute Proceedings** 124(8):36-38 August 1998.

Pearson, Jeff. "Have We Crossed the Line?" **Submarine Review** 97-99 April 1999.

Peele, Reynolds B. "Combat power projection 'forward...from (under) the sea.'" **Marine Corps Gazette** 79:12-15 June 1995.

Pekelney, Richard and Terry Lindell. "A Short History of the ARMA Gyrocompass," **Submarine Review** 100-106 October 1999.

Pell, Claiborne. "America's Future Under Sea Power," **Naval War College Review** 21(1):47-52 October 1968.

Pengelley, Rupert. "Grappling for submarine supremacy [1996 Edition]," **Jane's International Defense Review** 48 July 1, 1996.

Peniston, Bradley. "5-foot airlock opens sub to new missions," **Navy Times** 49(8):22 November 29, 1999.

_____. "Old subs make way for the new Jimmy Carter," **Navy Times** 48:8 February 15, 1999.

_____. "Submariners speak out about the future," **Navy Times** 48:20 June 14, 1999.

_____. "Swedish subs serve as model to U.S. fleet: Women have been part of some crews for 10 years," **Navy Times** 48(39):24 July 5, 1999.

"Pentagon steers new course on attack-submarine research," **Jane's Defence Weekly** 30(9):1 September 2, 1998.

Peppe, P. Kevin. "Attack submarines should attack! Attack! Attack!" **U.S. Naval Institute Proceedings** 118(9):62-64 September 1991.

_____. "Rethinking tomorrow's attack submarine force," **Submarine Review** 51-62 January 1995.

_____. "SSNs: Supporting the battle group?" **U.S. Naval Institute Proceedings** 123(5):40-43 May 1997.

_____. "Submarines in the Littorals," **U.S. Naval Institute Proceedings** 119(7):46-48 July 1993.

_____. "Victory and Perhaps Defeat," **Submarine Review** 55-61 October 1993.

Perry, Doug. "NR-1-within visual sight of the bottom," **Undersea Warfare** 1(4):18-21 Summer 1999.

Petzoldt, Travis M. "The Arabian Gulf as a Model for Littoral USW," **Submarine Review** 47-62 July 2001.

Pexton, Patrick. "Attack subs will stay at both San Diego and Pearl," **Navy Times** (Marine Corps Edition) 17 January 30, 1995.

Pirie, Robert B. "Acting SecNav's Address at Corporate Benefactors Day," **Submarine Review** 47-53 April 2001.

Pirie, Robin. "In Memory of Scorpion," **Submarine Review** 35-39 January 1999.

Polmar, Norman. "Different angle of attack," **U.S. Naval Institute Proceedings** 123(8):87-88 August 1997.

_____. "How fast is fast?" **U.S. Naval Institute Proceedings** 122(11):87-88 November 1996.

_____. "How many spy subs...?" **U.S. Naval Institute Proceedings** 122(12):87-88 December 1996.

_____. "How many submarines?" **U.S. Naval Institute Proceedings** 124(2):87-88 February 1998.

_____. "Innovation in sub design, at last," **U.S. Naval Institute Proceedings** 125(2):87-88 February 1999.

_____. "Manning the (smaller) nuclear fleet," **U.S. Naval Institute Proceedings** 121(8):87-88 August 1995.

_____. "The name game," **U.S. Naval Institute Proceedings** 124(6):87 June 1998.

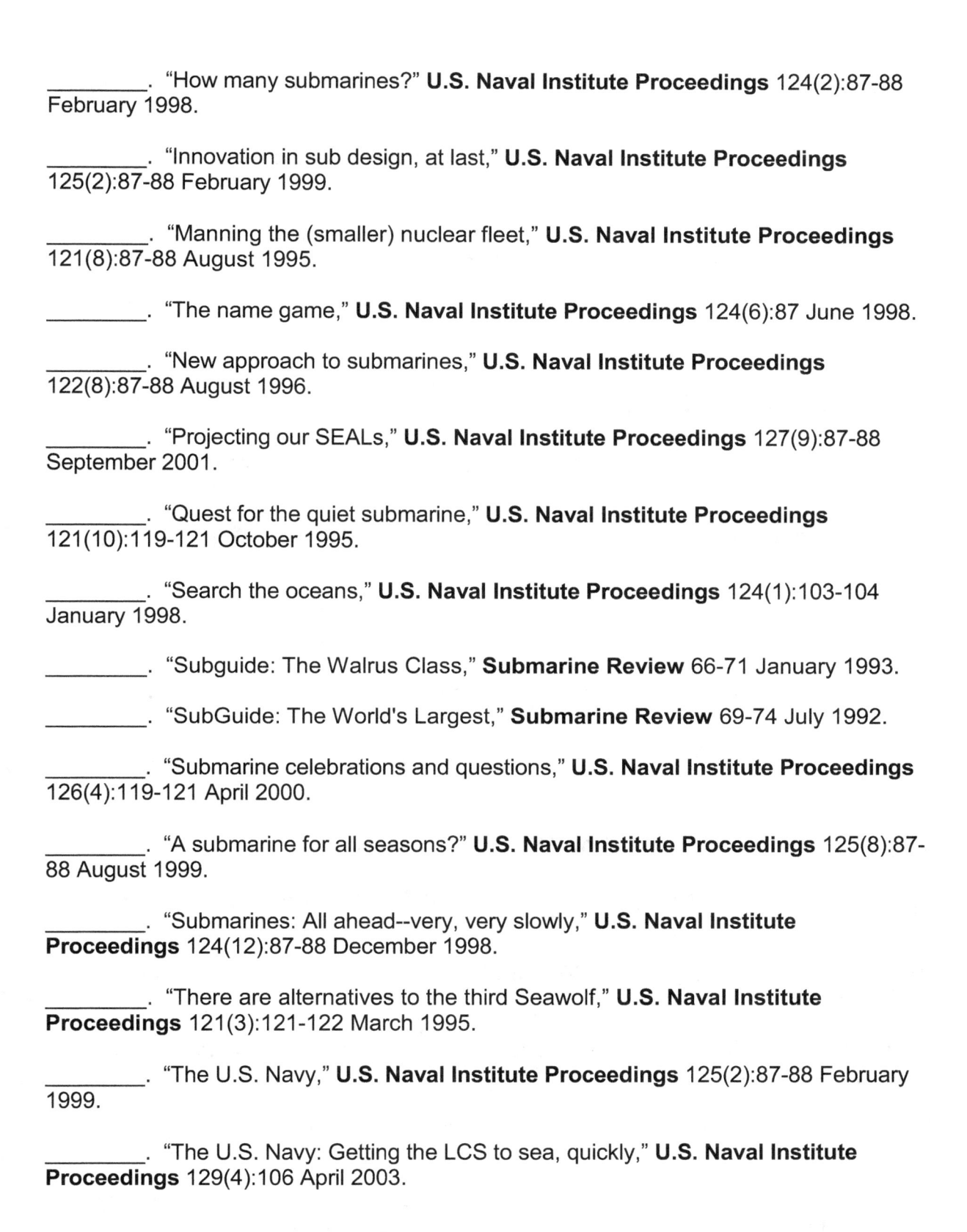

_____. "New approach to submarines," **U.S. Naval Institute Proceedings** 122(8):87-88 August 1996.

_____. "Projecting our SEALs," **U.S. Naval Institute Proceedings** 127(9):87-88 September 2001.

_____. "Quest for the quiet submarine," **U.S. Naval Institute Proceedings** 121(10):119-121 October 1995.

_____. "Search the oceans," **U.S. Naval Institute Proceedings** 124(1):103-104 January 1998.

_____. "Subguide: The Walrus Class," **Submarine Review** 66-71 January 1993.

_____. "SubGuide: The World's Largest," **Submarine Review** 69-74 July 1992.

_____. "Submarine celebrations and questions," **U.S. Naval Institute Proceedings** 126(4):119-121 April 2000.

_____. "A submarine for all seasons?" **U.S. Naval Institute Proceedings** 125(8):87-88 August 1999.

_____. "Submarines: All ahead--very, very slowly," **U.S. Naval Institute Proceedings** 124(12):87-88 December 1998.

_____. "There are alternatives to the third Seawolf," **U.S. Naval Institute Proceedings** 121(3):121-122 March 1995.

_____. "The U.S. Navy," **U.S. Naval Institute Proceedings** 125(2):87-88 February 1999.

_____. "The U.S. Navy: Getting the LCS to sea, quickly," **U.S. Naval Institute Proceedings** 129(4):106 April 2003.

_____. "U.S. Navy: Submarines in a minefield," **U.S. Naval Institute Proceedings** 120(4):120-122 April 1994.

_____. "Where have all the nukes gone?" **U.S. Naval Institute Proceedings** 124(11):87-88 November 1998.

Powis, Jonathan. "The Future of Submarine Escape and Rescue Liaison Office SMERLO," **Submarine Review** 71-74 October 2002.

Preston, Antony. "Air independent submarine propulsion," **Naval Forces** 12(1):48-50+ 1991.

_____. "Conventional submarines--programmes and markets," **Military Technology** 15:37+ March 1991.

_____. "Developing submarine technology," **Asian Defence Journal** 2:47-51 February 1993.

_____. "Developments in undersea warfare," **Asian Defence Journal** 7:37-4 July 1990.

_____. "Naval mid-term needs," **Armada International** 19(4):6-8+ August/September 1995.

_____. "New air independent submarine propulsion system," **Naval Forces** 14(2):62-64 1993.

_____. "The submarine threat to Asian navies," **Asian Defence Journal** 18+ October 1995.

_____. "Undersea warfare," **Naval Forces** 13(3):48-53 1992.

"Position of the Submarine In 1906," **U.S. Naval Institute Proceedings** 33(1):441 March 1907.

"Possibility of Dispensing with the Mother Ship," **U.S. Naval Institute Proceedings** 42(2):614 March/April 1916.

Price, Kelly D. "The Improved Los Angeles Class Fast Attack Submarine," **Submarine Review** 59-66 January 1997.

Prina, L. Edgar. "The Future of SSNs; CSIS Plays Twenty Questions," **Sea Power** 36(10):29-32+ October 1993.

Pugliese, William N. and Terrence L. Tinkel. "Build a Stealthy Submarine by 2010," **U.S. Naval Institute Proceedings** 118(2):82-83 February 1992.

Puleo, Steve. "Trapped in the depths," **American History** 36(2):46-54 June 2001.

"Rapid Development of Submarines (1911)," **U.S. Naval Institute Proceedings** 37(2):636 March/April 1911.

Razmus, Jerry. "SSBN security," **Submarine Review** 25-35 April 1996; 34-45 October 1996.

Reardon, Kevin J. "Ensuring the undersea advantage," **U.S. Naval Institute Proceedings** 113(10):66-68+ October 1987.

Reckamp, Douglas E. "Tactical Nuclear Deterrence by the Naval Reserves," **Submarine Review** 80-84 January 1999.

"Resistance of Submarine Hulls," **U.S. Naval Institute Proceedings** 41(4):1318 July/August 1915.

Rethinaraj, T S Gopi. "ATV: all at sea before it hits the water," **Jane's Intelligence Review** 10:31-35 June 1998.

Reuter, Joachim. "'Triton'--an advanced new missile system for submarines[special issue]," **Naval Forces** 66-68 2001.

Rhades, Jurgen. "Air-independent submarine propulsion revisited," **Military Technology** 14(9):47-50+ September 1990.

Rhys-Jones, Graham. "I. SSN: Queen of the Seas?" **Naval War College Review** 44(2):119-120 Spring 1991.

Roach, Terence. "Review of regional submarine forces," **Asia-Pacific Defence Reporter** 21:25-28 September/October 1995.

Roberts, Jerry. "The silent service comes of age," **Sea History** 18-19 Winter 2000/2001.

Robertson, Thomas. "Air independent propulsion: A look at what is currently on offer," **Naval Forces** 17(6):36-39 1996.

Robinson, Carence A., Jr. "Radar Detection Challenges Submarine Warfare Shroud Source," **Signal** 31-34 March 1992.

Rockwell, David L. "Submarine C2W: A game of one?" **Journal of Electronic Defense** 20(3):41-46 March 1997.

_____. "Submarine sensors come to the surface," **Journal of Electronic Defense** 24(8):56-59 August 2001.

Rodgaard, John. "Attack from below," **U.S. Naval Institute Proceedings** 126(12):64-67 December 2000.

Rodgers, W.L. "Large vs. Small Submarines," **U.S. Naval Institute Proceedings** 43(1):148 January 1917.

Roos, John G. "Weighing the options: US Navy faces tough choices in modernizing its attack-sub fleet," **Armed Forces Journal International** 138:48+ June 2001.

Rosenblatt, Richard M. "Submarine air independent propulsion and the U.S. Navy," **Submarine Review** 119-124 July 1997.

Rosenlof, Eric. "Contingency blues," **U.S. Naval Institute Proceedings** 121(1):53-57 January 1995.

Ross, Steve. "USS Boise's 1998 deployment: SSN operations in the 21st century," **Undersea Warfare** 1(2):9-11 Winter 1998/1999.

Rota, Dane L. "Deep battle: Submarines in the Pacific," **Military Review** 72(10):76-77 October 1992.

Ruff, Dave. "SUBLANT SSBN Operations," **Submarine Review** 45-50 October 2001.

Ruhe, William J. "Blowing the Japanese out of shallow water," **U.S. Naval Institute Proceedings** 115(12):60-64 December 1989.

Ryan, Paul J. "Attack Submarines Should Lead Battle Groups," **U.S. Naval Institute Proceedings** 119(2):86-88 February 1993.

Ryan, Stephen L. "Shallow threats: has the shallow water submarine threat to blue ocean navies been overrated?" **Asian Defence Journal** 14+ July 1995.

Sanchez, Jonas. "Hollywood and Submarines," **Submarine Review** 98-103 July 2002.

"Saving the Industrial Base or Saving Face?" **Naval Forces** 14:4 1993.

Saxon, Timothy D. "Anglo-Japanese Naval Cooperation, 1914-1918," **Naval War College Review** 53(1):62-92 Winter 2000.
http://www.nwc.navy.mil/press/Review/2000/winter/art3-w00.htm

Scherr, Michael R. "Undersea warfare launchers," **Naval Forces** 20(4):76-81 1999.

Schmidt, Wade H. "The Multipurpose Platform of Choice," **Submarine Review** 47-52 October 1993.

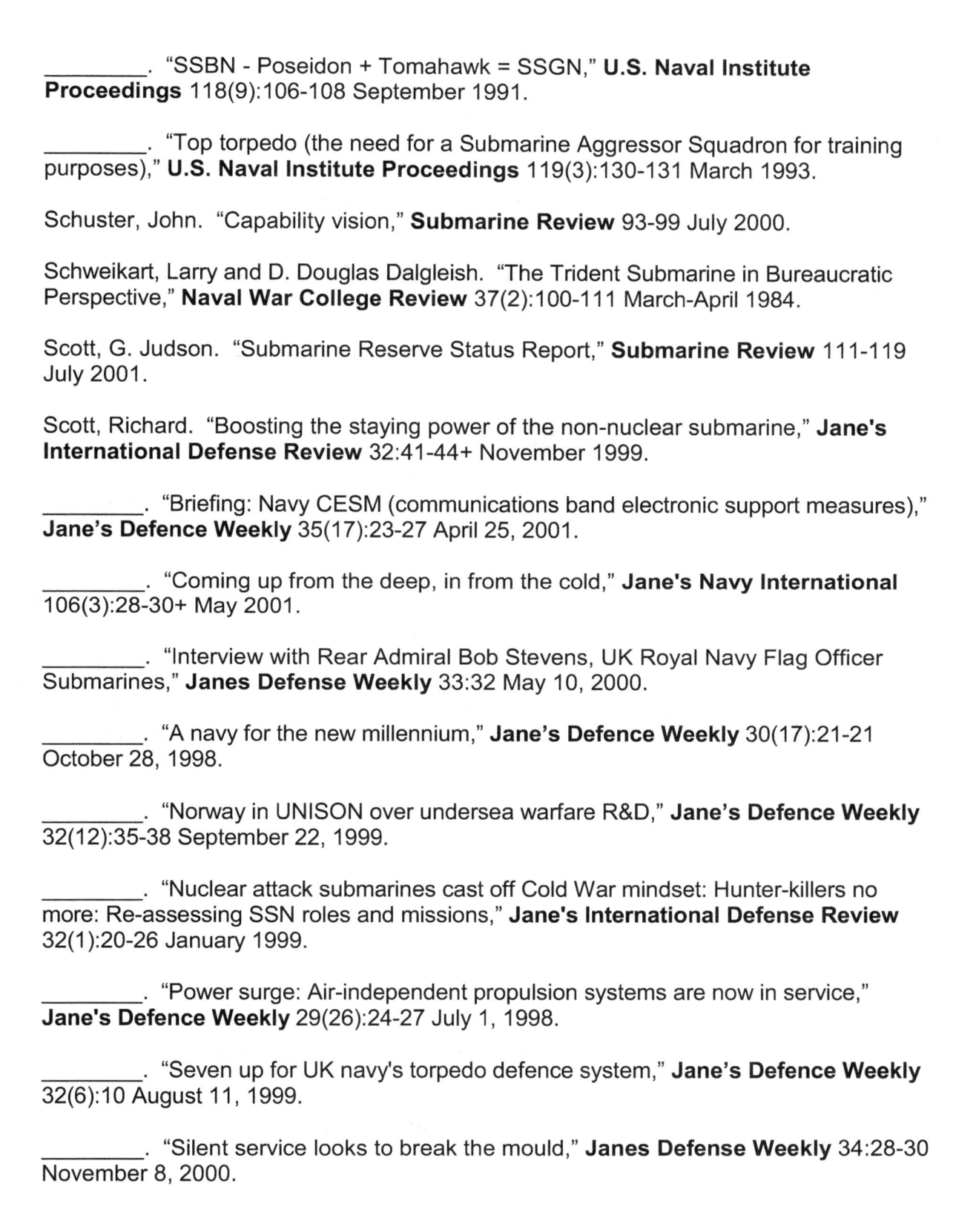

_____. "SSBN - Poseidon + Tomahawk = SSGN," **U.S. Naval Institute Proceedings** 118(9):106-108 September 1991.

_____. "Top torpedo (the need for a Submarine Aggressor Squadron for training purposes)," **U.S. Naval Institute Proceedings** 119(3):130-131 March 1993.

Schuster, John. "Capability vision," **Submarine Review** 93-99 July 2000.

Schweikart, Larry and D. Douglas Dalgleish. "The Trident Submarine in Bureaucratic Perspective," **Naval War College Review** 37(2):100-111 March-April 1984.

Scott, G. Judson. "Submarine Reserve Status Report," **Submarine Review** 111-119 July 2001.

Scott, Richard. "Boosting the staying power of the non-nuclear submarine," **Jane's International Defense Review** 32:41-44+ November 1999.

_____. "Briefing: Navy CESM (communications band electronic support measures)," **Jane's Defence Weekly** 35(17):23-27 April 25, 2001.

_____. "Coming up from the deep, in from the cold," **Jane's Navy International** 106(3):28-30+ May 2001.

_____. "Interview with Rear Admiral Bob Stevens, UK Royal Navy Flag Officer Submarines," **Janes Defense Weekly** 33:32 May 10, 2000.

_____. "A navy for the new millennium," **Jane's Defence Weekly** 30(17):21-21 October 28, 1998.

_____. "Norway in UNISON over undersea warfare R&D," **Jane's Defence Weekly** 32(12):35-38 September 22, 1999.

_____. "Nuclear attack submarines cast off Cold War mindset: Hunter-killers no more: Re-assessing SSN roles and missions," **Jane's International Defense Review** 32(1):20-26 January 1999.

_____. "Power surge: Air-independent propulsion systems are now in service," **Jane's Defence Weekly** 29(26):24-27 July 1, 1998.

_____. "Seven up for UK navy's torpedo defence system," **Jane's Defence Weekly** 32(6):10 August 11, 1999.

_____. "Silent service looks to break the mould," **Janes Defense Weekly** 34:28-30 November 8, 2000.

_____. "Submarines stay the course," **Janes Defense Weekly** 34:32-33+ October 18, 2000.

_____. "Torpedoes home in on new targets," **Asia-Pacific Defence Reporter** 27(4):30-34 June/July 2001.

_____ and Mark Hewish. "Remote hunting key to littoral waters," **Jane's International Defense Review: IDR** 32:48-54 December 1999.

Scott, Richard, et al. "Tactical Solutions For Submariners,' **Jane's Navy International** 102(3):37-42+ June 1997.

Seal, Scott. "Where's My Gunboat? The Time is now for Trident SSGN," **Submarine Review** 40-45 January 1999.

Seese, Robert J. "Nightly routine shattered: The U-boats and Navy blimps met by surprise," **World War II** 5:26-33 July 1990.

Seigle, Greg. "USN explores options to boost SSN numbers," **Janes Defense Weekly** 12:10 November 24, 1999.

Sen, Philip. "Surviving SUBSUNK," **International** 106(2):12-14+ March 2001.

Sengupta, Prasun K. "Submarine fleet build-up in Asia-Pacific," **Asian Defence Journal** 26+ August 2000.

_____. "Submarines for Asia-Pacific navies: From SSK (diesel-electric boat) to SSN/SSBN (nuclear-powered attack submarine/ballistic nuclear submarine)," **Asian Defence Journal** 2:50-52+ February 1998.

Seykowski, Rosemary. "From Sea To Land," **Surface Warfare** 22:24-25 November/December 1997.

Shadwick, Martin. "Submarines, white papers.and survival," **Canadian Defence Quarterly** 27(4):5 Summer 1998.

"Shall the Submarine Be Outlawed?" **U.S. Naval Institute Proceedings** 45(9):1625 September 1919.

Shannon, Jim. "Undersea warfare is TEAM warfare," **U.S. Naval Institute Proceedings** 122(6):48-49 June 1996.

Sharkey, J.B. "The Submarine in Netcentric Warfare," **Submarine Review** 70-73 July 1999.

Shobe, Katharine K. "The role of the human operator in the submarine," **Submarine Review** 58-64 October 2001.

Silmaris, John E. "Undersea Warfare: the challenges ahead," **Naval Forces** 20:8-13 1999.

Simons, Eric. "New diesel submarines for export," **Asia-Pacific Defence Reporter** 25(3):16-17 April/May 1999.

Singer, Jeremy. "Subs' ability to send, receive data needs update, Navy says," **Navy Times** 49:17 August 28, 2000.

_____. "U.S. submarines need better satellite communications," **Space News** 11:32 August 28, 2000.

_____. "U.S. subs need improved data transfer for new roles," **Defense News** 15:7 August 21, 2000.

Sirmalis, John E. "Undersea warfare: The challenges ahead," **Naval Forces** 20(1):8-13 1999.

Skelton, Ike. "National Military Strategy [NSL Corporate Benefactors Day February 5, 2002]," **Submarine Review** 6-12 April 2002.

Skerret, R.F. "Progress in Submarine Craft (1909)," **U.S. Naval Institute Proceedings** 35(4):328 December 1909.

"Skipper calls collision 'a burden I will carry'," **Navy Times** 50:23 March 12, 2001.

Slade, Stuart. "Submarine modernization," **Naval Forces** 19(2):28-30+ 1998.

Smith, C. Alphonso. "Battle of the Caribbean," **U.S. Naval Institute Proceedings** 80(9):976-982 September 1954.

Smith, Robert H. "The Fleet Ballistic Missile Submarine: An Irresistible Future," **Naval War College Review** 35(2):3-9 March/April 1982.

Smith, T.J. "A Logical Explanation as to the Loss of USS Scorpion (SSN 589)," **Submarine Review** 107-116 October 1999.

Smith, William D. "The submarine century," **Submarine Review** 12-14 April 2000.

_____. "Submarine developments: new weapons--new roles," **Naval Forces** 21:16-20 2000.

_____. "Submarine missions: preparing the future battlespace," **Submarine Review** 49-52 July 1999.

_____. "Superstealth submarines for the next century," **Naval Forces** 20(1):28-32 1999.

Snyder, Jim. "Undersea Warfare: The Battle Below," **Surface Warfare** 22:14-16 January/February 1997.

Sokolski, Henry D. "Nonapocalyptic Proliferation: A New Strategic Threat?" **Washington Quarterly** 17:115-127 Spring 1994.

Somes, Timothy E. "Musing on Naval Maneuver Warfare," **Naval War College Review** 51(3):122-128 Summer 1998.
http://www.nwc.navy.mil/press/review/1998/summer/sd1su98.htm

"SOSUS network now used to track whales, seaquakes," **Sea Power** 41(8):24 August 1998.

Spassky, I.D. and V.P. Semyonov. "To build a better sub," **U.S. Naval Institute Proceedings** 123(8):58-61 August 1997.

Speare, L. "Submarine Boats Past, Present, and Future," **U.S. Naval Institute Proceedings** 28(4):1000 December 1902.

Speer, Richard T. "Let pass safely the Awa Maru," **U.S. Naval Institute Proceedings** 100(4):69-76 April 1974.

Spikes, Clayton H. "Littoral warfare technology testing," **Sea Technology** 38:71 June 1997.

Stallings, John D. "'The best boat E.B. ever built' [USS Salmon]," **Naval History** 13(4):16-21 July/August 1999.

Starr, Barbara. "Jane's interview (with US Navy RADM Dennis Jones, director of submarine warfare)," **Jane's Defence Weekly** 23(22):40 June 3, 1995.

_____. "Navy wants contractors' submarine partnership," **Jane's Defence Weekly** 27:8 January 22, 1997.

_____ and John Boatman. "(US Navy special)," **Jane's Defence Weekly** 17(7):241-243+ February 15, 1992.

"Status of the Submarine (1916)," **U.S. Naval Institute Proceedings** 42(3):972 May/June 1916.

"Steam Submarines," **U.S. Naval Institute Proceedings** 45(4):647 April 1919.

Stefanovsky, Vladimir. "Their System Still Needs Victims," **U.S. Naval Institute Proceedings** 118(8):64-68 August 1992.

Steinhauer, Jules Verne. "Gangway-The Electric Revolution is Coming!" **Submarine Review** 117-121 April 2001.

_____. "Some Thoughts on Submarine Escape, Rescue and Salvage," **Submarine Review** 108-115 April 2002.

Stenberg, Pelle. "AIP--the Swedish way," **Submarine Review** 72-79 October 1996.

Stephens, Alan. "Close air support is crucial," **Asia-Pacific Defence Reporter** 26(7):40-41 December/January 2001.

Stevens, Paul S. "DD21's USW mission," **Surface Warfare** 24:26-27 May/June 1999.

Still, Bryan C. "A Perspective on Submarine Officer Retention," **Submarine Review** 86-90 April 2001.

Stirling, Y. "The Submarine," **U.S. Naval Institute Proceedings** 43(7):1371 July 1917.

Stone, Andrea. "Too cramped for comfort?" **Navy Times** 49:24 June 5, 2000.

"Sub show at Smithsonian," **All Hands** 998:12 June 2000.
http://www.mediacen.navy.mil/pubs/allhands/jun00/pg6m.htm

"Submarine Bibliography," **Submarine Review** 95-104 April 1994.

"Submarine Census, 2000," **U.S. Naval Institute Proceedings** 126(12):90+ December 2000.

"The Submarine Menace (1914)," **U.S. Naval Institute Proceedings** 40(4):1187 July/August 1914.

"Submarine mystery deepens," **Jane's Defence Weekly** 26(13):3 September 25, 1996.

"Submarine order may add to Greek-Turkish tensions," **Sea Power** 41(12):33 December 1998.

"Submarined," **Far Eastern Economic Review** 159(46):7 November 14, 1996.

"Submarines and Density of Seawater," **U.S. Naval Institute** Proceedings 41(1):223 January/February 1915.

"Submarines and Mines, 1910," **U.S. Naval Institute Proceedings** 36(4):1192 December 1910.

"Submarines and network-centric warfare," **Sea Technology** 39(9):78-79 September 1998.

"Submarines and Ramming Tactics," **U.S. Naval Institute Proceedings** 41(1):221 January/February 1915.

"Submarines as Sea-Going Vessels (1910)," **U.S. Naval Institute Proceedings** 36(3):910 September 1910.

"Submarines today: Flotilla commanders comment," **Naval Forces**, 20(6):40-46 1999.

"Subs introduced to command and weapons control systems," **Jane's International Defense Review** 20 August 1, 1998.

Suggs, Ralph E. "Remarks prepared for submarine conference," **Submarine Review** 25-27 January 2000.

Sullivan, Paul E. and Morgan A. Heavener. "Ensuring undersea superiority and access: the groundbreaking Virginia-class SSN program," **Sea Power** 44(7):36-38 July 2001.

"Sun Tzu & the art of submarine warfare," **U.S. Naval Institute Proceedings** 121(11):64-72 November 1995.

"The Super-Submarine," **U.S. Naval Institute Proceedings** 42(3):966 May/June 1916.

Sviatov, George. "American submarines from a Russian point of view," **Submarine Review** 91-98 January 1996.

_____. "Fleet of modified Seawolfs & Virginias makes sense," **U.S. Naval Institute Proceedings** 124(12):68-71 December 1998.

_____. "Perspectives of American ballistic missile submarines development," **Submarine Review** 123-128 July 2000.

_____. "Recollections of a Maverick," **Submarine Review** 91-96 January 2002.

_____. "Sinusoid of the Arms Race and American Strategy," **Submarine Review** 69-79 July 2001.

Sweetman, Bill. "The many faces of Virginia: USN's multipurpose platform will revolutionize undersea warfare," **Jane's International Defense Review** 33(12):46-51 December 2000.

_____. "Operation Barney," **U.S. Naval Institute Proceedings** 121(6):58-59 June 1995.

Swicker, Charles C. "Ballistic Missile Defense From The Sea: The Commander's Perspective," **Naval War College Review** 50:7-25 Spring 1997.

Tailyour, Patrick. "Torpedo Development," **Submarine Review** 93-99 October 1999.

Talbott, J. E. "Weapons Development, War Planning and Policy: The U.S. Navy and the Submarine, 1917–1941," **Naval War College Review** 37(3):53-71 May/June 1984.

Tangredi, Sam J. "The Good, the Bad, and the Stealthy: Surface Views of the Future of the Submarine Force (or A view from After Steering)," **Submarine Review** 116-125 October 2001.

_____. "Future Security Environment 2001-2025 -Part I, The Concensus View," **Submarine Review** 54-69 April 2001; Part II, Divergent Views, Debates and Wild Cards," **Submarine Review** 25-39 July 2001.

_____. "Globalization and Naval Operations: Seven Critical Effects," **Submarine Review** 7-18 October 2002.

_____. "SECNAVs and submarines--building the force," **Submarine Review** 38-47 January 1995.

_____. and Randall G. Bowdish. "Core of Naval Operations: Strategic and Operational Concepts of the United States Navy," **Submarine Review** 11-23 January 1999.

Terpstra, Richard P. "Oh, how offensive!" **Undersea Warfare** 2(3):18-20 Spring 2000.

Terraine, John. "Atlantic Victory: 50 Years On," **RUSI** 138(5):53-59 October 1993.

Teuteberg, Hanno. "Submarines today: Flotilla Commanders comment," **Naval Forces** 21:68-69 2000.

Thomas, Vincent C. "Realistic scenario for undersea warfare--interview with Vice Adm. Daniel L. Cooper," **Sea Power** 33(7):7+ July 1990.

Thompson, G.R. "Technology Advances and Enablers," **Submarine Review** 29-33 July 2002.

Thompson, Joseph M. "Purple submarines," **Submarine Review** 42-47 April 1996.

Thompson, Loren B. "New attack submarine: The next capital ship?" **Sea Power** 41(4):67-68+ April 1998.

Thompson, Phillip. "Attack submarines and network-centric warfare," **Submarine Review** 81-86 October 2000.

_____. "Global access and submarine relevance in the 21st century," **Submarine Review** 40-44 April 2000.

Thompson, Richard. "AN HISTORIC BLUNDER: Further Downsizing RDT&E Infrastructure," **Submarine Review** 75-80 January 2001.

_____. "Strategic Implications of the New World Order for the U.S. Submarine Force," **Submarine Review** 19-25 January 1993.

_____. "Strike Warfare in the 21st Century: Relying on the Kindness of Strangers?" **Submarine Review** 35-43 July 2002.

Tomb, J.H. "Submarines Types, Operations and Accidents," **U.S. Naval Institute Proceedings** 31(4):965 December 1905.

Toti, William J. "Fast attacks and boomers: the Smithsonian exhibit," **U.S. Naval Institute Proceedings** 126(6):37 June 2000.

Trapp, Malte. "Simulator training for submarine crews," **Naval Forces** 21:64-66 2000.

Treadwell, Terry. "Undersea aircraft carriers," **Aviation History** 9:50-56 November 1998.

Tritten, James J. "The Submarine's Role in Future Naval Warfare," **Submarine Review** 16-28 July 1992.

_____. "The Trident System: Submarines, Missiles and Strategic Doctrine," **Naval War College Review** 36(1):61-76 January/February 1983.

Truver, Scott C. "Briefing: US Navy programme review," **Jane's Defence Weekly** 35(14):22-28 April 4, 2001.

_____. "The latest, best, and most affordable," **Sea Power** 39(4):75-76+ April 1996.

_____. "Today, tomorrow, and (Navy) after next--an overview of selected U.S. Navy programs," **Sea Power** 42(10):32-34+ October 1999.

_____. "Tomorrow's fleet, pt I," **U.S. Naval Institute Proceedings** 121(7):89-94 July 1995.

_____. "Tomorrow's fleet, pt 2," **U.S. Naval Institute Proceedings** 122(8):55-60 August 1996.

_____. "Tomorrow's fleet, pt 1," **U.S. Naval Institute Proceedings** 125(1):83-88 February 1999.

_____. "Tomorrow's fleet, pt 2," **U.S. Naval Institute Proceedings** 125(2):65-68 February 1999.

_____ and Morgan A. Heavener. "'Mission success assured.'[Advanced SEAL Delivery System viewed as major force multiplier]." **Sea Power** 44(6):45-46 July 2001.

Tuohy, William. "Escape From TANG," **Submarine Review** 97-107 April 2002.

Turner, James E., Jr. "We were absolutely breaking new ground [New Attack Submarine (NSSN)]," **Sea Power** 41(7):9-13 July 1998.

_____. "A shipbuilder's perspective of logistics," **Submarine Review** 107-113 October 1996.

"Type 209/1400mod new submarines for the South African Navy," **Naval Forces** 22:95-96 2001.

"Type 212s on a Med cruise," **Military Technology** 50-53 March 1996.

"Type 214--a new submarine export design," **Military Technology** 22(12):99 December 1998.

Ude, Udo. "Trends in conventional submarine developments," **Naval Forces** 22:33-34+ 2001.

Uhlig, Frank, Jr. "A Real Revolution in Naval Affairs and What it Achieved," **Naval War College Review** 52(2):138-142 Spring 1999.
http://www.nwc.navy.mil/press/review/1999/spring/s%26d%2Dsp9.htm

"Ultra-Violet Rays for Purifying the Air In Submarines," **U.S. Naval Institute Proceedings** 38(2):754 March/April 1912.

"Underwater Guide for Submarines," **U.S. Naval Institute Proceedings** 41(4):1317 July/August 1915.

"Undersea Warfare (classified) Monterey, California, USA 7-10 APRIL," **Jane's Defence Weekly** 1 January 1, 1997.

"Undersea Warfare Conference USA 10-13 MARCH," **Jane's Defence Weekly** 1 January 1, 1997.

"UNDERSEA WARFARE TOPS USN PRIORITIES," **Jane's Defence Weekly** 27(7):7 February 19, 1997.

"Underway on nuclear power," **All Hands** (979):11 November 1998.

"Up periscope," **The Economist** 341(7986):39-40 October 5, 1996.

"The U.S. submarine force today [Interview with Rear Adm. Malcolm I. Fages]," **Sea Power** 42(7):9-14 July 1999.

"Usefulness of Merchant Submarines," **U.S. Naval Institute Proceedings** 42(5):1686 September/October 1916.

"USN aims to counter the silent threat," **Jane's Defence Weekly** 27:7 February 19, 1997.

Vanoss, Vincent M. "Recruiting, training, and leading the Submarine Force in the 21st century," **Submarine Review** 84-93 April 2000.

Vanzo, John. "Saga of U-505: A Crewman's Story," **World War II** 12:26-32+ July 1997.

Vego, Milan N. "The Role of the Attack Submarines in Soviet Naval Theory," **Naval War College Review** 36(6):48-64 November/December 1983.

Vlattas, John. "Shifting from blue to brown: pursuing the diesel submarine into the littoral," **Submarine Review** 90-96 April 1999.

Walker, Robert J., III. "The perfect shooter," **Submarine Review** 83-90 April 1997.

Walker, William B. "Here are the submarines...where are the tactics?" **U.S. Naval Institute Proceedings** 120(7):26-30 July 1994.

Wall, Robert. "Navy investigates UAV-sub teaming," **Aviation Week & Space Technology** 155:67 July 9, 2001.

Wallace, Michael D. and Charles A. Meconis. "Submarine proliferation and regional conflict," **Journal of Peace Research** 32:79-95 February 1995.

Walsh, Don. "The AIP alternative," **Sea Power** 42(12):34-37 December 1999.

_____. "Nautilus: What's in a name?" **U.S. Naval Institute Proceedings** 127(2):88 February 2001.

_____. "Psst...wanna buy a sub?" **U.S. Naval Institute Proceedings** 124(6):89 June 1998.

_____. "Sub wars off Waikiki," **U.S. Naval Institute Proceedings** 124(10):104 October 1998.

Walsh, Edward J. "Boomer conversion challenges may pay off for deep attack," **U.S. Naval Institute Proceedings** 127(9):8 September 2001.

Walsh, Jon. "A New Kind of Target Motion Analysis: The Short-Range Encounter Problem," **Submarine Review** 61-71 January 2001.

Ward, Don. "New horizons: As subs surface from Cold War, the view is friendlier but busier," **Navy Times** 43(6):14-16+ November 15, 1993.

"Warfare beneath the sea: WWII's silent service," **VFW, Veterans of Foreign Wars Magazine** 80(8):14-15 April 1993.

"WASS advances A184 torpedo," **Jane's Defence Weekly** 30(23):28 December 9, 1998.

Wathen, Jason. "COMSUBPAC future ideas initiative," **Undersea Warfare** 1(3):9-11 Spring 1999.

Watkins, James D. "The recent days: Address to the Submarine Force Centennial Birthday Ball," **Submarine Review** 20-24 July 2000.

Watson, Gary Jr. "Running too silent & too deep?" **U.S. Naval Institute Proceedings** 123(4):30-34 April 1997.

Watts, Robert C., IV. "Fast Attack Dilemma," **Submarine Review** 96-103 July 2001.

Weir, Gary E. "Deep ocean, Cold War," **Undersea Warfare** 2(3):3-6 Spring 2000.

_____. "Search for an American submarine strategy and design, 1916-1936," **Naval War College Review** 44(1):34-48 Winter 1991.

_____. "Silent defense: one hundred years of the American submarine force," **All Hands** 12-15 April 2000.

_____. "Silent defense: 1900-1940," **Undersea Warfare** 1(4):9-12 Summer 1999.

_____. "Surviving the Peace: The Advent of American Naval Oceanography, 1914-1924," **Naval War College Review** 50(4):85-103 Autumn 1997.
http://www.nwc.navy.mil/press/review/1997/autumn/art6%2Da97.htm

Weiss, Lora G. "The submarine as the ultimate asymmetric threat," **Submarine Review** 67-69 July 1999.

Welch, John K. "Electric Boat's Centennial," **Sea Power** 42(2):35 February 1999.

Welch, Richard F. "One century ago, Irish immigrant John P. Holland perfected the U.S. Navy's first practical submarine," **Military History** 16(4):16-19 October 1999.

"Welcome to the Naval Undersea Warfare Center," **Naval Forces** 20(3):S3-S4 1999.

Wells, Anthony R. "U.S. Naval Power and the Pursuit of Peace in an Era of International Terrorism and Weapons of Mass Destruction," **Submarine Review** 64-70 October 2002.

Wertheim, Eric. "Lest we forget," **U.S. Naval Institute Proceedings** 125(4):108 April 1999.

_____. "Lest we forget," **U.S. Naval Institute Proceedings** 127(6):94 June 2001.

West, Leslie. "The ASDS advantage," **Sea Power** 41(7):38-42 July 1998.

Westwood, James T. "Soviet Reaction to the U.S. Maritime Strategy," **Naval War College Review** 41(3):62-68 Summer 1988.

Wettern, Desmond. "Forgotten threat in the deep," **Defense & Diplomacy** 8(11):51-53 November/December 1990.

_____. "New light on warship design philosophy (in Soviet Union)," **Asia-Pacific Defence Reporter** 18(2):30-31 August 1991.

_____. "Threat that never was--some 'what if' questions for Western leaders," **Sea Power** 34(11):31-34 November 1991.

Whitlock, Duane L. "The Silent War against the Japanese Navy," **Naval War College Review** 48(4):43-52 Autumn 1995.

Whitman, Edward C. Rising to Victory: The Pacific Submarine Strategy in World War II Part I: Retreat and Retrenchment," **Undersea Warfare** 3(3):20-24 Spring 2001.

_____. "Rising to victory: the Pacific submarine strategy in World War II Part II: Winning through," **Undersea Warfare** 3(4):20-24 Summer 2001.

_____. "Submarine commandos: 'Carlson's Raiders' at Makin Atoll," **Undersea Warfare** 3(2):22-25 Winter 2001.

_____. "Submarine hero--George Levick Street, III," **Undersea Warfare** 2(1):27-28 Fall 1999.

_____. "Submarine hero--Howard Walter Gilmore," **Undersea Warfare** 1(4):22-23 Summer 1999.

_____. "Submarines in network centric warfare," **Sea Power** 42(7):33-36 July 1999.

Wilbur, Charles H. "Remember the San Luis!" **U.S. Naval Institute Proceedings** 122(3):86-88 March 1996.

Wilkie, Robert. "Navy 2001 Back to the Future," **Naval War College Review** 53(2):196-207 Spring 2000.
http://www.nwc.navy.mil/press/review/2000/spring/s%26d%2Dsp0.htm

Willingham, Stephen. "Navy aims to cut submarine downtime," **National Defense** 84:32-33 May 2000.

Wilson, Jim. "Concrete submarines," **Popular Mechanics** 175(12):84-85 December 1998.

_____. "Deep and deadly," **Popular Mechanics** 177(7):76-85 July 2000.

Windolph, Wolfgang. "The better AIP," **Naval Forces** 19:114-116+ 1998.

Wingo, Walter. "New subs, mines will dominate undersea warfare," **Design News** 52(25):20 December 15, 1997.

Winkelmann, Gaete. "Submarines today: Flotilla commanders comment (II)," **Naval Forces** 20(5):16-21 1999.

Winkler, David F. "The deadly Trigger," **Sea Power** 43(7):24 July 2000.

_____. "WWII submarines and the naval industrial complex," **Sea Power** 43(4):25 April 2000.

_____. "The X-1," **Sea Power** 44(7):25 July 2001.

Wolfe, Frank. "DARPA, Navy start research on future attack submarine," **Defense Daily** 1 August 31, 1998.

Wolfe, Kevan. "Australia's Defence White Paper--an overview," **Asia-Pacific Defence Reporter** 26(7):21-22 December2000/January 2001.

Wolfowitz, Paul. "Address to Submarine Technology Symposium," **Submarine Review** 28-37 July 1999.

Wolkensdorfer, Daniel J. and Taylor P. Lonsdale. "AWS and Undersea Warfare in the New World Order," **Wings of Gold** 69-70 Spring 1993.

Wood, David. "The future of deterrence: as the world's nuclear club expands, old rules no longer apply," **Navy Times** 49:28-29 September 18, 2000.

"The world market for conventional submarines: attempting a revised assessment," **Naval Forces** 21:42+ 2000.

Wright, James E. "Submarine design for the littorals," **U.S. Naval Institute Proceedings** 121(12):39-41 December 1995.

Wright, Paul Troy. "Is there a better explanation?" **U.S. Naval Institute Proceedings** 127(7):28-29 July 2001.

Wukovits, John F. "Life on the edge," **Military History** 9(1):50-57 April 1992.

Ya'ari, Yedidia "Didi." "A Case for Maneuverability," **Naval War College Review** 50(4):125-132 Autumn 1997.
http://www.nwc.navy.mil/press/review/1997/autumn/s%26d2%2Da97.htm

Yawn, Michael. "Submarine escape training--are we serious?" **U.S. Naval Institute Proceedings** 127(6):68-69 June 2001.

Young, Charles B. "New and Revolutionary Technology for Tomorrow's Navy," **Submarine Review** 6-14 October 2001.

_____. "Submarine vision," **Submarine Review** 79-90 July 2000.

Young, Donald J. "West coast war zone," **World War II** 13:27-32 July 1998.

Young, Richard. The Legal Status of Submarine Areas Beneath the High Seas," **The American Journal of International Law** 45(2):225-239 April 1951.

Zulkarnen, Isaak. "Holland's RDM teams up with PSC-NDSB in submarine offer," **Asian Defence Journal** 40 January/February 2001.

SUBMARINES BY COUNTRY

ARGENTINIA

Wixler, Keith E. "Argentina's Geopolitics and Her Revolutionary Diesel-Electric Submarines," **Naval War College Review** 42(1):86-107 Winter 1989.

AUSTRALIAN

Alves, Dora. "Australia's Collins Class Submarines," **Submarine Review** 97-101 July 1999.

_____. "RAN's Collins Class Combat System," **Submarine Review** 91-92 October 2001.

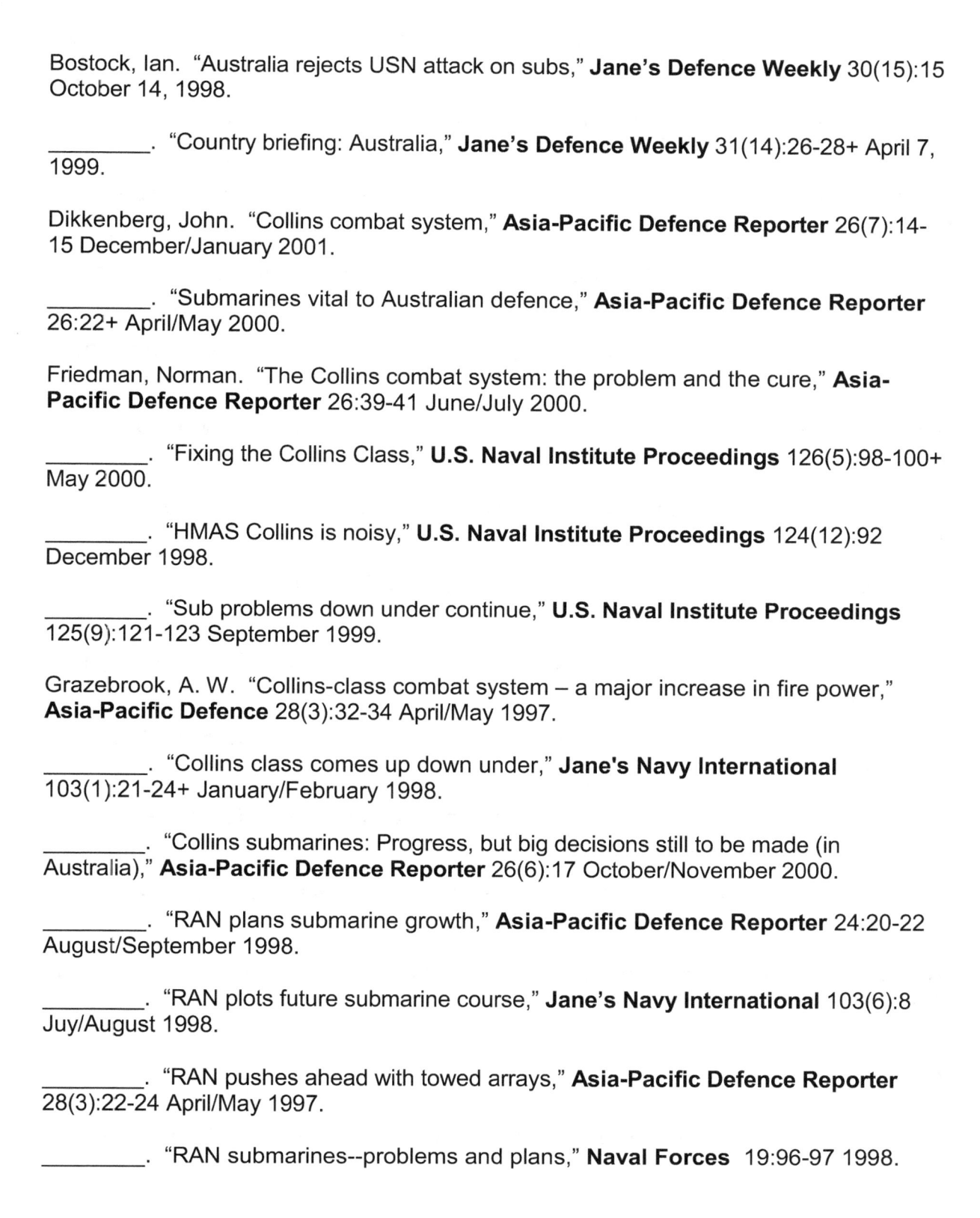

Bostock, Ian. "Australia rejects USN attack on subs," **Jane's Defence Weekly** 30(15):15 October 14, 1998.

_____. "Country briefing: Australia," **Jane's Defence Weekly** 31(14):26-28+ April 7, 1999.

Dikkenberg, John. "Collins combat system," **Asia-Pacific Defence Reporter** 26(7):14-15 December/January 2001.

_____. "Submarines vital to Australian defence," **Asia-Pacific Defence Reporter** 26:22+ April/May 2000.

Friedman, Norman. "The Collins combat system: the problem and the cure," **Asia-Pacific Defence Reporter** 26:39-41 June/July 2000.

_____. "Fixing the Collins Class," **U.S. Naval Institute Proceedings** 126(5):98-100+ May 2000.

_____. "HMAS Collins is noisy," **U.S. Naval Institute Proceedings** 124(12):92 December 1998.

_____. "Sub problems down under continue," **U.S. Naval Institute Proceedings** 125(9):121-123 September 1999.

Grazebrook, A. W. "Collins-class combat system – a major increase in fire power," **Asia-Pacific Defence** 28(3):32-34 April/May 1997.

_____. "Collins class comes up down under," **Jane's Navy International** 103(1):21-24+ January/February 1998.

_____. "Collins submarines: Progress, but big decisions still to be made (in Australia)," **Asia-Pacific Defence Reporter** 26(6):17 October/November 2000.

_____. "RAN plans submarine growth," **Asia-Pacific Defence Reporter** 24:20-22 August/September 1998.

_____. "RAN plots future submarine course," **Jane's Navy International** 103(6):8 Juy/August 1998.

_____. "RAN pushes ahead with towed arrays," **Asia-Pacific Defence Reporter** 28(3):22-24 April/May 1997.

_____. "RAN submarines--problems and plans," **Naval Forces** 19:96-97 1998.

_____. "Regional submarine forces expand," **Asia-Pacific Defence Reporter 1998 Annual Reference Edition** 24(1):28-29 1998.

_____. "US pressure in RAN submarine competition," **Asia-Pacific Defence Reporter** 27:38 August/September 2000.

Hendricks, David M. "The submarine force of the Royal Australian Navy," **Undersea Warfare** 1(4):24-25 Summer 1999.

Kerr, Julian. "First Details of New Combat System for Australian Submarines," **Asia-Pacific Defence Reporter** 28(1):9 January 2002.

Owen, Frank. "Could Australia have handled a Kursk incident?" **Asia-Pacific Defence Reporter** 27(6):36-37 October/November 2000.

Preston, Antony. "Into the Depths -- The COLLINS Class Submarines," **Military Technology** 17:58-60 December 1993.

_____. "'Serious Problems' reported with new Australian SSKs," **Sea Power** 41(12):31-32 December 1998.

"RAN Collins submarines now on track," **Asia-Pacific Defence Reporter** 26(7):12-13 December/January 2001.

Ricketts. Peter. "Collins--six years and $1bn to go," **Asia-Pacific Defence Reporter** 27:37 August/September 2000.

Roach, Terence. "Collins-class sonar suite advances," **Asia-Pacific Defence Reporter** 24(2):12 February/March 1998.

_____. "Opportunity knocks for the Collins class, **Asia-Pacific Defence Reporter** 26(6):38-39 October/November 2000.

Walsh, Don. "AE-2: Subsunk--Subfound," **U.S. Naval Institute Proceedings** 124(12):89 December 1998.

Young, Peter Lewis. "The Australian New Submarine Programme - An Update," **Asian Defence Journal** 6-7 December 1992.

AUSTRIAN

Harbron, John. "Franz Josef's forgotten U-boat captains," **History Today** 46(6):51-56 June 1996.

BRITISH

Gilbert, Nigel John. "British Submarine Operations in World War II," **U.S. Naval Institute Proceedings** 89(3):73-81 March 1963.

Kemp, Paul. "A wholly avoidable accident: the loss of HM Submarine ARTEMIS, 1 July 1971," **Submarine Review** 39-45 April 1997.

Stevens, Robert. "The British Submarine Service: past, present, and future," **RUSI Journal** 146(3):38-40 June 2001.

Tall, Jeff. "The history of the Royal Navy Submarine Service," **RUSI Journal** 146(3):41-45 June 2001.

Willett, Lee. "The Most Important Type of Warship in the World: The Royal Navy Submarine Service and Britain's Strategic Defence Review," **Submarine Review** 32-43 January 2000.

Whitman, Edward C. "Daring the Dardanelles--British submarines in the Sea of Marmara during World War I," **Undersea Warfare** 2(4):22-24 Summer 2000.

Wilson, G.A.S.C. "RN (Royal Navy) pays off conventional line," **U.S. Naval Institute Proceedings** 120(3):88-90 March 1994.

Wolfe, Frank. "British trim Trident buy, Navy says no impact on cost," **Defense Daily** 1 October 22, 1998.

CANADIAN

Crickard, Fred. "Upholders, core capabilities and multi-purpose, combat-capable forces," **Canadian Defence Quarterly** 27:36-37 Summer 1998.

Ferguson, Julie H. "Pursuing the Upholder option," **Sea Power** 40(12):46-48 December 1997.

Grant, Dale. "Canada gets a deal on upholder buy," **Naval Forces** 19:8-10 1998.

Hobson, Sharon. "Upholder-class submarines resurface in Canada," **Jane's Navy International** 106(9):26-30 November 2001.

Lynch, Thomas G. "Submarine developments in Canada--where to from here?" **Navy International** 95:448-450+ December 1990.

Maloney, Sean M. "Canadian subs protect fisheries," **U.S. Naval Institute Proceedings** 124(3):74-76 March 1998.

McLean, Doug and Doug Hales. "Why Canada needs submarines," **Canadian Defence Quarterly** 26:20+ Summer 1997.

CHINESE

Ahrari, Ehsan. "China's Naval Forces look to extend their blue-water reach [1998 Edition]," **Jane's Intelligence Review** 10(4):31-36 April 1, 1998.

Bussert, James C. "Chinese submarines quietly amass strength in Pacific," **Signal** 49:75-77 June 1995.

"Changing face of China," **Jane's Defence Weekly** 30(24):20-22+ December 16, 1998.

"China's submarine force plans its great leap forward," **Jane's Navy International** 102(2):14-15+ April 1997.

"Country briefing: People's Republic of China," **Jane's Defence Weekly** 36(2):22-27 July 11, 2001.

Dikkenberg, John. "Just how good are the Kilos?" **Asia-Pacific Defence Reporter** 28(8):16-17 November 2002.

Farrer, Mark. "Submarine force in change--the Peoples Republic of China," **Asia-Pacific Defence Reporter** 24:12-14 October/November 1998.

Kim, Duk-Ki. "The modernization of China's submarine forces," **Submarine Review** 50-58 January 1997.

Sae-Liu, Robert. "Second Song submarine vital to China's huge programme," **Janes Defense Weekly** 32:17 August 18, 1999.

DUTCH

Alden, John D. "Dutch Officers Find Their Fathers' Lost Submarines," **Submarine Review** 46-51 January 1999.

_____. "Dutch Submarines in World War II: The European Theater," **Submarine Review** 35-40 April 1994.

_____. "Dutch Submarines in World War II - The Far East," **Submarine Review** 75-81 April 1993.

Spaans, Hans. "Equipping the royal Netherlands navy," **Military Technology** 22(10):87-89 October 1998.

EGYPTIAN

Scott, Richard and Joris Janssen Lok. "Talks bring sale of Zwaardvis submarines to Egypt closer," **Jane's Defence Weekly** 30(1):18 July 8, 1998.

FRENCH

"DCN to build France's fourth new-generation SSBN," **Asia-Pacific Defence Reporter** 27(6):50 October/November 2000.

Howard, Peter. "SUBMARINES - First metal cut for Scorpene," **Jane's Navy International** 103(7):4 September 1, 1998.

Lewis, J. A. C. "France conducts missile test from Triomphant boat," **Janes Defense Weekly** 31:12 May 26, 1999.

_____. "France powers up on project for new six-strong submarine class," **Jane's Defence Weekly** 1 October 14, 1998.

GERMAN

Barnes, R.H. " German Submarine Action in World War I," **U.S. Naval Institute Proceedings** 68(10):1440 October 1942.

Bowling, R.A. "Escort-of-Convoy, Still the Only Way," **U.S. Naval Institute Proceedings** 95(12):46 December 1969; **U.S. Naval Institute Proceedings** 96(6):111 June 1970.

_____. "More About Hitler's U-Boats," **Submarine Review** 18-127 January 2001.

Boyle, Richard. "Emergence of Offensive U-Boats during the Great War," **Submarine Review** 57-64 April 1992.

Bridge, T. D. "U-boat crews sank 14.5mn tons of shipping: 70% were killed," **Army Quarterly & Defence Journal** 128:450-453 October 1998.

Cortesi, Lawrence. "Capturing the U-505," **American History Illustrated** 29(1):46 March 1994.

Davis, Gerald H. "The 'Ancona' Affair: A Case of Preventive Diplomacy," **Journal of Modern History** 38(3):267-277 September 1966.

"Development and Efficiency of German Submarine Boats (1913)," **U.S. Naval Institute Proceedings** 40(1):208 January/February 1914.

Duncan, Francis. "Deutschland--Merchant Submarine," **U.S. Naval Institute Proceedings** 91(4):68 April 1965; pt. 2 91(8):120 August 1965.

Ehle, Jurgen. "The German Navy after the Cold War and Reunification," **Naval War College Review** 51(4):63-84 Autumn 1998.
http://www.nwc.navy.mil/press/review/1998/autumn/art4%2Da98.htm

Else, J.E. Jr. "U-Boats Off Our Coasts," **U.S. Naval Institute Proceedings** 91(10):84 October 1965.

Enos, Ralph. "Demolishing U-Boat Myths [Review of 'Hitler's U-Boat War' By Clay Blair]," **Submarine Review** 133-140 April 1999.

_____. "More About Hitler's U-boats," **Submarine Review** 99-102 April 2000.

_____. "Onkel Karl and Uncle Charlie, Donitz and Lockwood: A Comparison of Style; Part I," 81-88 **Submarine Review** July 1999; "Part II," **Submarine Review** 28-35 October 1999.

"First German Submarine (1906)," **U.S. Naval Institute Proceedings** 32(4):1580 December 1906.

Fordyce, Samuel W. "The German U-boat Campaign in World War II: An Analysis," **Navy** 8:37-40+ April 1965.

Gayer, A. "German Submarine Operations, 1914-1918," **U.S. Naval Institute Proceedings** 52(4):621 April 1926; 52(8):1572 August 1926.

"German Cruising Submarines of 1917 Type," **U.S. Naval Institute Proceedings** 43(4):779 April 1917.

"German Midget Submarines," **U.S. Naval Institute Proceedings** 84(3):102 March 1958.

"German sub found in Gulf waters [cover]; Joint effort yields World War II relic," **MMS Today** 6-7 Summer 2001.
http://198.252.9.108/govper/MMSToday/www.mms.gov/ooc/newweb/publications/Summer01.pdf

"The German Submarine Cruiser (1918)," **U.S. Naval Institute Proceedings** 44(8):1918 August 1918.

German submarine design and construction--Past, present and future," **Maritime Defence** 16: 281-289 September 1991.

"German Submarine Designs (1918)," **U.S. Naval Institute Proceedings** 44(10):2363 October 1918.

"German Submarine Development, 1906-1914," **U.S. Naval Institute Proceedings** 40(6):1796 November/December 1914.

"German Submarine Fleet at Beginning of 1914," **U.S. Naval Institute Proceedings** 40(2):510 March/April 1914.

"German Submarine Flotillas In 1912," **U.S. Naval Institute Proceedings** 38(2):756 June 1912.

"German Submarine Flotillas In 1913," **U.S. Naval Institute Proceedings** 39(2):845 June 1913.

"German Submarines 1907," **U.S. Naval Institute Proceedings** 33(3):1294 September 1907.

"German Submarines 1908," **U.S. Naval Institute Proceedings** 34(2):704 June 1908.

"German Submarines 1914," **U.S. Naval Institute Proceedings** 40(4):1183 July/August 1914.

"German Submarines 1916," **U.S. Naval Institute Proceedings** 42(4):1276 July/August 1916.

"German U-3 Sinks," **U.S. Naval Institute Proceedings** 38(2):757 June 1912.

Glennon, A.N. "Weapon That Came Too Late," **U.S. Naval Institute Proceedings** 87(3):85 March 1961.

Graham, M.T. "U-2513 Remembered," 83-92 **Submarine Review** July 2001.

Grant, R.M. "How Many U-Boats Have Been Sunk?" **U.S. Naval Institute Proceedings** 465, 1598.

_____. "Known Sunk--German Submarine War Losses, 1914-1918," **U.S. Naval Institute Proceedings** 64(1):66 January 1938.

Jacobsen, Klaus. "The German Submarine Force: now and tomorrow," **Jane's Navy International** 106:10-15 June 2001.

Keil, Robin. "Taking the challenge [interview with Captain M. Borchert, Commander Submarine Flotilla, German Navy].," **Naval Forces** 18:10+ 1997.

Kronke, Carl. "Outlook for German submarines," **Military Technology** 18(4):55+ April 1994.

"Krupp-Germania Submersibles (in 1909)," **U.S. Naval Institute Proceedings** 35(4):1308 December 1909.

Kurzak, K.H. "German U-Boat Construction," **U.S. Naval Institute Proceedings** 81(4):375 April 1955.

Langdon, R.M. "Live Men Do Tell Tales (U-582 and Peleus)," **U.S. Naval Institute Proceedings** 587, 17.

"Larger German Submarines (1918)," **U.S. Naval Institute Proceedings** 44(4):370 February 1918.

Layman, R.D. "U-Boat With Wings," **U.S. Naval Institute Proceedings** 94(4):54 April 1968; 94(8):114 August 1986; 94(10):110 October 1968; 95(2):110 February 1969.

Long, Wellington. "Cruise of the U-53," **U.S. Naval Institute Proceedings** 92(10):86 October 1966.

Lundeberg, Philip K. "The German Naval Critique of the U-Boat Campaign, 1915-1918," **Military Affairs** 27(3):105-118 Fall 1963.

_____. "German Naval Literature of World War II – a bibliographical survey," **U.S. Naval Institute Proceedings** 82(1):95-106 January 1956.

MacIntyre, Donald. "Three Aces-Trumped!" **U.S. Naval Institute Proceedings** 82(9):923 September 1956.

Merrill, James M. "Submarine scare, 1918," **Military Affairs** 17(4):181-190 Winter 1953.

"Midget," **U.S. Naval Institute Proceedings** 72(2):221 February 1946.

Mulligan, Timothy P. "German U-Boat Crews in World War II: Sociology of an Elite," **The Journal of Military History** 56(2):261 April 1992.

_____. "U-Boats Destroyed: German Submarine Losses in the World Wars," **The Journal of Military History** 62(3):651-652 July 1998.

O'Connell, J.A. "Radar and the U-Boat," **U.S. Naval Institute Proceedings** 89(9):53 September 1963; 90(8):123 August 1964.

O'Connor, Jerome M. "Into the gray wolves' den [Germany's U-boat bunkers on the Bay of Biscay]," **Naval History** 14(3):18-25 June 2000.

Polmar, Norman. "More About Hitler's U-boats," **Submarine Review** 142-143 July 2000.

Price, Jennifer M. "German U-Boat Losses During World War II: Details of Destruction," **Sea Power** 41(12):53-54 December 1998.

Ritterhoff, Jurgen. "Class 214--a new class of air-independent submarines," **Naval Forces** 19(5):94-98+ 1998.

_____. "German submarine developments: recent achievements and future trends," **Naval Forces** 21:28-30 2000.

Robertson, W.W. "Graphical Analysis of U-Boat Activities, 1917-18," **U.S. Naval Institute Proceedings** 68(5):672 May 1942.

Rohwer, Jürgen. "Literaturbericht über den Einsatz von U-booten im Zweiten Weltkrieg," **Wehrwissenschaftliche Rundschau Jg.** 6:152-157 1956.

Saeger, Hans. "German submarine technology," **Submarine Review** 61-71 October 1996.

_____. "The International Status of the German Submarine Industry," **Military Technology** 18:62-63+ April 1994.

Salvarezza, Michael and Christopher P. Weaver. "The last German U-boat to sink an American ship in World War II was herself sent to the bottom," **World War II** 13:20+ February 1999.

Sarty, Roger. "The limits of Ultra: the Schnorkel U-boat offensive against North America, November 1944-January 1945," **Intelligence and National Security** 12:44-68 April 1997.

Schuster, Carl O. "German submarine operations in the Black Sea during World War II," **Strategy & Tactics** 14-16 July/August 1988.

Sides, Ann B. "When submarine UB-123 attacked the ferry Leinster, it torpedoed Germany's last hope for a 'soft peace' in 1918," **Military History** 15(4):24-26 October 1998.

Sieche, E. F. "German Human Torpedoes & Midget Submarines," **Warship** no. 14 April 1980.

Smith, C.A. "Battle of the Caribbean," **U.S. Naval Institute Proceedings** 80(9):976 September 1954.

Smith, R.W. "Q-Ship--Cause and Effect," **U.S. Naval Institute Proceedings** 79(5):533 May 1953; 79(8):899-900 August 1953.

Spilman, C.H. "German Submarine War," **U.S. Naval Institute Proceedings** 75(2):683 February 1949; 75(9):1107 September 1949; 539, 83; 544, 763.

"Submarine launch heralds the next generation of German Navy boats," **Jane's International Defense Review** 24 September 1, 1998.

Sweetman, Bill "The U-Boat Peril Overcome," **U.S. Naval Institute Proceedings** 119(6):30-31 June 1993.

Syrett, David. "Communications intelligence and the sinking of the U-1062: 30 September 1944," **The Journal of Military History** 58(4):685 October 1994.

Todd, Tom. "The 'Mad Bavarian' built Germany's first U-boat in 1851, but his greatest challenge was to find a navy that was interested," **Military History** 15(4):12-14 October 1998.

"U-boat found in Gulf of Mexico," **Naval History** 15(5):61 October 2001.

Waters, J.M. Jr. "Stay Tough," **U.S. Naval Institute Proceedings** 92(12):95 December 1966; 93(4):108 April 1967.

Wilson, Graham. "The last U-boat [U-573]," **Journal of the Australian Naval Institute** 25:32-34 July/September 1999.

GREEK

Dinse, Reinhard and Bodo Bohrmann. "The class 214 submarine programme for the Hellenic Navy [Special Issue]," **Naval Forces** 30-31 2001.

"Greece ponders second stage of submarine retrofits," **Jane's Defence Upgrades**:3 October 16, 1998.

INDIAN

Bedi, Rahul. "India presses ahead with SSN to boost Navy's nuclear profile," **Janes Defense Weekly** 30:26 July 22, 1998.

_____. "India suffers hardship amid hardware buys," **Jane's Defence Weekly** 29(21):28-29 May 27, 1998.

Farrer, Mark. "India moving to dominate Indian Ocean," **Asia-Pacific Defence Reporter** 28(5):34-35 June 2002.

_____. "Indian submarine capability a growth industry," **Defence Reporter: Australia & Asia-Pacific** 27:60-62 May 2001.

_____. "Nuclear submarine for India--the enigmatic ATV (Advanced Technology Vessel)," **Asia-Pacific Defence Reporter** 24(5):16 August/September 1998.

Joshi, Manoj. "India's nuclear submarine plans," **Asia-Pacific Defence Reporter** 21:52 March/April 1995.

Kainikara, Sanu. "India is emerging as a regional military power," **Asia-Pacific Defence Reporter** 26(7):24-25 December 2000/January 2001.

_____. "Indian Navy sailing into the 21st century," **Asia-Pacific Defence Reporter** 27(1):54+ February 2001.

Koch, Andrew. "Nuclear-powered submarines: India's strategic trump card," **Jane's Intelligence Review** 10:29-31 June 1998.

Mak, J. N. "The Indian Navy: Friend or Foe?" **Asia Pacific Defence Review** 10-21 April 1994.

Raghuvanshi, Vivek. "Indian Navy reaches nuclear power milestone," **Defense News** 16:18 November 5-11, 2001.

Sakhuja, Vijay. "Sea based deterrence and Indian security," **Strategic Analysis** 15:21-32 April 2001.

INDONESIAN

"Indonesian Navy reorganises," **Asia-Pacific Defence Reporter** 26(7):45-46 December 2000/January 2001.

IRANIAN

Dunn, Michael C. "The Iranian Submarines: A New Naval Arms Race?" **Washington Report on Middle East Affairs** 40+ December 1992/January 1993.

ISRAELI

Baumgartner, Henry. "The sub that vanished; Israeli submarine INS Dakar," **Mechanical Engineering-CIME** 121(8):56-58 August 1, 1999.

Charalambous, Charlie. "Discovery of Israeli submarine Dakar "more exciting" than Titanic," **Agence France Presse**, May 30, 1999.

Dettweiler, Thomas K., Steve Abdalla and David Brown. "Bringing the INS Dakar 'Home'," **Sea Technology** 42(4):21-26 April 2001.

Diamond, Howard. "New U.S.-Israeli strategeic dialogue announced; Israel acquires new submarine," **Arms Control Today** 29:26 July/August 1999.

Hirschberg, Peter. "The New Hunt for The Dakar," **The Jerusalem Report** 18 September 10, 1992.

"Israel Offers Cash Reward For Information On Lost Sub," **Seattle Post-Intelligencer** A2 January 28, 1998.

Johnson, Donna. "Role of image processing in the search for the I.N.S. Dakar," **MTS/IEEE - Riding the Crest into the 21st Century**, September 13-September 16 1999, Seattle, WA, USA, p 234-236.

Kesary, Michael and Ifiach Fogelson. "The Israeli submarine DOLPHIN," **Naval Forces** 19:61-79 1998.

Kinzer, Stephen. "Israel Is Advertising for Clues to a Sub Lost in 60's," **New York Times** [Late Edition (East Coast)]:1.20 October 11, 1998.

Segal, Naomi. "Hope of finding lost sub dashed," **Ethnic NewsWatch** 5 April 21, 1996.

Susser, Leslie. "Brief Encounter: with Mike Eldar, on the finding of the Dakar submarine," **The Jerusalem Report** p.10 June 21, 1999.
Last year the government blocked publication of Eldar's book on the Dakar, and the police questioned him on charges of aggravated espionage for allegedly seeking to publish classified material. Now that the Dakar has been found, Eldar hopes that the government and military will be more open with the families of the lost seamen and with the public, and will lift the ban on his book.

"Thirty-year search renewed for lost Israeli submarine," **Sea Power** 41(12):32-33 December 1998.

Wilson, Graham. "The loss and discovery of the Israeli submarine DAKAR," **Journal of the Australian Naval Institute** 26:30-32 January/March 2000.

ITALIAN

"Italian Submarine Mothership, 1916," **U.S. Naval Institute Proceedings** 42(3):968 May/June 1916.

"Italian Submarines, 1912," **U.S. Naval Institute Proceedings** 38(3):1136 September 1912.

"Italian Submarines, 1913," **U.S. Naval Institute Proceedings** 39(4):1755 December 1913.

"Italy's Progress In Submarine Navigation (1908)," **U.S. Naval Institute Proceedings** 34(2):719 June 1908.

"ITALY - Submarine modernisation imminent," **Jane's Defence Upgrades** 3 July 17, 1998.

Valpolini, Paolo. "Italian Navy steps up its fleet integration," **Jane's Defence Weekly** 27(14):19-23 April 9, 1997.

JAPANESE

Boyd, Carl. "American naval intelligence of Japanese submarine operations early in the Pacific war," **The Journal of Military History** 53(2):169-189 April 1989.

_____. "U.S. Navy Radio intelligence during the Second World War and the sinking of the Japanese Submarine I-52," **The Journal of Military History** 63(2):339-354 April 1999.

Frye, Stephen L. "Japanese submarine operations in the Pacific theater," **Military Review** 73(8):64-65 August 1993.

Gerry, Donald D. "Japanese submarine operational forces in World War II," **Submarine Review** 21-23 January 1997.

Milford, Frederick J. "Torpedoes of the Imperial Japanese Navy-Part One: Through 1918," **Submarine Review** 90-105 January 2000; Part II; Heavyweight Torpedoes 1918-1945," **Submarine Review** 51-66 July 2002.

KOREAN

Beaver, Paul. "Korean submarine incident makes demands on diplomacy," **Jane's Defence Weekly** 29(26):5 July 1, 1998.

Belke, Thomas J. "Incident at Kangnung: North Korea's ill-fated submarine incursion," **Submarine Review** 18-28 April 1997.

Bermudez, Joseph S. Jr. "Midget submarine infiltration upsets South Korea troubled waters," **Jane's International Defense Review** 24 August 1, 1998.

_____. "Submarine was on mission to spy on South," **Jane's Defence Weekly** 4 September 25, 1996.

Dyhouse, Tim. "Sub spies provoke massive manhunt in South Korea," **VFW, Veterans of Foreign Wars Magazine** 84(3):11 November 1996.

"The big picture: South Korea," **Life** 21(9):22 August 1998.

Karniol, Robert. "Country briefing: Republic of Korea," **Jane's Defence Weekly** 34(25):19-26 December 20, 2000.

_____. "Manhunt launched after N Korean boat hits rocks," **Jane's Defence Weekly** 26(13):4 September 25, 1996.

_____. "Vietnam buys submarines from North Korea," **Janes Defense Weekly** 30:14 December 9, 1998.

Omestad, Thomas. "Murder-suicide in North Korean sub," **U.S. News & World Report** 125(1):50 July 6, 1998.

Scott, Richard. "Capture of North Korean submarine reveals new capabilities [1]," **Jane's Defence Weekly** 29(26):4 July 1, 1998.

Scott, Richard and Paul Beaver. "North Korea - No escape for midget submarine crew," **Jane's Navy International** 103(6):11 July/August 1998.

"South Korea raises captured midget submarine," **Sea Power** 41(9):34 September 1998.

"Submarine tactics: Why North Korean threats work," **Far Eastern Economic Review** 159(40):5 October 3, 1996.

"Vietnam buys submarines from North Korea," **Jane's Defence Weekly** 30(23):14 December 9, 1998.

MALAYSIAN

Grazebrook, A. W. "Malaysia considering Agosta submarine purchase," **Asia-Pacific Defence Reporter** 25:24 January 2000.

PAKISTANI

de Lionis, Andres. "Pakistan's Naval Special Service Group," **Jane's Intelligence Review** 6(3):136-137 March 1994.

Haider, Salahuddin. "Breakthrough in Karachi Blast Case," **Global News Wire - Asia Africa Intelligence Wire - Gulf News**, May 15, 2002.
More than 20 people were injured in the blast, including 12 other French citizens employed by the Direction des Constructions Navales, a defence ministry shipbuilder involved in a joint France-Pakistan submarine project.

Preston, Antony. "First Pakistani Agosta SSK rolled out at DCN Cherbourg," **Sea Power** 41(10):29 October 1998.

Scott, Richard. "Agosta 90B rolls out for Pakistan," **Janes Defense Weekly** 30:16 August 19, 1998.

POLISH

Holdanowicz, Grzegorz. "Poland seeks to update its submarine force,"**Jane's International Defense Review: IDR** 33:4 August 2000.

RUSSIAN/SOVIET

Beaver, Paul. "Russians boost submarine capabilities while other navies lag in modernization," **Sea Power** 40(1):45-50 January 1997.

Butler, Vera and Alexey Muraviev. "Russia's new SSN high technology challenges," **Asia-Pacific Defence Reporter** 26:42-43+ February/March 2000.

Crowe, Roy W. "Technology vs. training: Soviet submarines in World War Two," **Submarine Review** 49-52 October 1995.

Curran, Daniel A. "Aging Russian nuclear submarine problems," **Submarine Review** 74-77 October 1997.

Dick, Charles J. "Country briefing: Russia," **Jane's Defence Weekly** 34(5):19-26 August 2, 2000.

Dronov, B. F. and B. A. Barbanel. "Beluga: Soviet Project 1710 submarine-laboratory," **U.S. Naval Institute Proceedings** 125(6):72-76 June 1999.

Friedman, Norman. "New theories on the Kursk disaster," **U.S. Naval Institute Proceedings** 127(3):4+ March 2001.

Galeotti, Mark. "The Kursk in context," **Jane's Intelligence Review** 12:8-9 October 2000.

Garthoff, Raymond L. "Cuba between the Superpowers: Handing the Cienfuegos Crisis," **International Security** 8(1):44-46 Summer 1983.

Gerken, John D. "The Challenges of Radioactive Waste in the Russian Navy," **Submarine Review** 73-82 April 1999.

Gorbunov, Alexander V. "The submarine fleet of Russia: its past, present and future," **Submarine Review** 8-20 October 1996.

Grazebrook, A.W. "Russia's Rubin presses ahead with new submarines," **Asia-Pacific Defence Reporter** 25(8):39 January 2000.

Handler, Joshua. "Russia's Pacific Fleet - Submarine Bases and Facilities," **Jane's Intelligence Review** 6:166-171 April 1994.

_____. "Submarine Safety - The Soviet/Russian Record," **Jane's Intelligence Review** 4:328-331 July 1992.

Haney, Patrick J. "Soccer Fields and Submarines in Cuba: The Politics of Problem Definition," **Naval War College Review** 50(4):67-84 Autumn 1997.
http://www.nwc.navy.mil/press/review/1997/autumn/art5%2Da97.htm

Hoffman, David. "Reactor Blast Shows Danger Of Aging Subs," **The Washington Post** [FINAL Edition]:A22 November 16, 1998.

_____. "Rotting Nuclear Subs Pose Threat in Russia; Moscow Lacks Funds for Disposal,"; **The Washington Post** [FINAL Edition]:A1 November 16, 1998.

Jordan, John. "The 'Kilo' Class Submarine," **Jane's Intelligence Review** 4:427-431 September 1992.

Kraus, George K. Jr. "Reducing the Russian Submarine Construction Base," **Submarine Review** 33-35 January 1993.

Kurnikov, L.A. and A. N. Mushnikov. "Red Banner Baltic Fleet Submarines in the Great Patriotic War, 1941-1945," Selected Articles from **USSR Naval Digest** 30-37 November 1967.

Kuteinikov, Anatoly V. "Malachite subs post proud tradition," **U.S. Naval Institute Proceedings** 124(4):52-56 April 1998.

Markov, David. "More details surface of Rubin's 'Kilo' plans," **Jane's Intelligence Review** 9:209-215 May 1997.

Moltz, James Clay. "Russian nuclear submarine dismantlement and the naval fuel cycle," **Nonproliferation Review** 7:76-87 Spring 2000.

_____and Tamara C. Robinson. "Dismantling Russia's nuclear subs: new challenges to non-proliferation," **Arms Control Today** 29:10-15 June 1999.

Montgomery, George. "Burial at sea: the Komsomolets disaster," **Studies in Intelligence** 38:43-51 1995.

Polmar, Norman. "The first Soviet nuclear submarines," **Submarine Review** 20-26 January 1991.

_____. "The Kursk salvage plan," **U.S. Naval Institute Proceedings** 127(6):88 June 2001.

_____. "What if Kursk had been ours?" **U.S. Naval Institute Proceedings** 126(10):90-92 October 2000.

Preston, Antony. "The Soviet Navy Today," **Asian Defence Journal** 86-87+ February 1992.

Prins, I. R. and A. Carel. "Salvaging the 'Kursk'," **Naval Forces** 22:60+ July 2001.

"Russia develops new generation export submarines," **Asia-Pacific Defence Reporter** 23:32 June/July 1997.

"Russian subs 'not suffering' says US Navy," **Janes Defense Weekly** 27:5 April 16, 1997.

Schofield, Michael. "The Kursk's uneasy legacy," **U.S. Naval Institute Proceedings** 127(3):74-75 March 2001.

Sviatov, George. "Akula class Russian nuclear attack submarines," **Submarine Review** 60-73 October 1997.

_____. "Death of Kursk," **Submarine Review** 47-52 October 2000.

_____. "First Soviet nuclear submarine," **Submarine Review** 118-126 January 2000.

_____. "The New Akula Class Russian Submarine GEPAD Commissioned in Severodvinsk," **Submarine Review** 75-82 October 2002.

_____. "Perspectives of Russian Ballistic Submarine Development," **Submarine Review** 58-62 April 1999.

_____. "Severodvinsk Class Russian Nuclear Attack Sububmarine," **Submarine Review** 74-79 January 1999.

Wahlbäck, Krister. "Submarine Incursions in Swedish Waters, 1980-1992: A Comment on the Report of the Latest Official Investigation and the Debate It Brought About," *in* **The Parallel History Project on NATO and the Warsaw Pact** 30 October 2002. http://www.isn.ethz.ch/php/research/AreaStudies/Wahlback.htm

Walsh, Don. "Soviet sub penetrates Sydney Harbor!" **U.S. Naval Institute Proceedings** 124(4):105 April 1998.

"Whisky on the rocks," http://www.compunews.com/s139/sp2.htm

SINGAPORE

Farrer, Mark. "Instant capability? Singapore buys a submarine squadron," **Asia-Pacific Defence Reporter** 24(2):8 February/March 1998.

"Sjohunden renamed RSS Chieftain," **Asia-Pacific Defence Reporter** 27(4):24 June/July 2001.

SOUTH AFRICAN

Heitman, Helmoed-Romer. "Contracts secure South Africa's submarine programme," **Jane's Navy International** 105:34-36 October 2000.

Lescriner, Peter. "Submarines of class 209 type 1400 MOD for the South African Navy[special issue]," **Naval Forces** 32-33 2001.

"South Africa forms submarine cluster group," **Sea Power** 41(10):31 October 1998.

"Type 209/1400mod new submarines for the South African Navy," **Naval Forces** 22:95-96 2001.

SWEDISH

Hallstrom, Fredrik. "Sweden's Gotlands are ready for fleet operations," **U.S. Naval Institute Proceedings** 124(3):58-60 March 1998.

Lindholm, Sverker. "Gotland-class submarines--a new breed," **Sea Technology** 39(11):25-31 November 1998.

Preston, Antony. "Sweden launches second AIP (air independent propulsion) submarine," **U.S. Naval Institute Proceedings** 122(5):113 May 1996.

Tornberg, Claes E. "Swedish Future Surface Ships and Submarines," **Naval War College Review** 45(1):54-60 Winter 1992.

TAIWANESE

Dikkenberg, John. "The US-Taiwan submarine dilemma," **Asia-Pacific Defence Reporter** 28(3):41-42 March/April 2002.

Holzer, Robert. "European veto may force Taiwan to build new subs," **Defense News** 16:1+ April 30, 2001.

Meconis, Charles A. and G. Jacobs. "Submarines for Taiwan?" **Asia-Pacific Defence Reporter** 27(5):24-26 August/September 2001.

Opall, Barbara. "Taipei considers compact subs," **Defense News** 12:1+ September 8-14, 1997.

Sherman, Jason. "Pentagon seeks European sub designs for Taiwan," **Defense News** 16:6 October 22-28, 2001.

_____. "Taipei envisions domestic sub," **Defense News** 16:1+ August 27-September 2, 2001.

TURKISH

Sariibrahimoglu, Lale. "Four more submarines for Turkey," **Jane's Defence Weekly** 30(4):3 July 29, 1998.

Streetly, Martin. "Sealion For Turkey?" **Journal of Electronic Defense** 21(5):18 May 1998.

"Turkey develops blue water force to control trade routes," **Jane's Defence Weekly** 30(7):1 August 19, 1998.

"Turkey issues bids for new SSKs, torpedoes," **Sea Power** 41(12):33 December 1998.

ANTISUBMARINE WARFARE (ASW)

Abraham, Douglas A. "Active sonar detection in shallow water," **Naval Forces** 18(6):74-75 1997.

Acton, Jerry. "LAMPS MK III Acoustic Target Tracker," **Vertiflite** 29(5):30-33 July/August 1983.

Adelt, Wilfried H. "TYPE A V/STOL: One Aircraft for All Support Missions?" **Journal of Aircraft** 20(6):508-515 June 1983.

"Advanced technology for surface combatants," **Naval Forces** 18(1):S30-S32 1997.

Alden, J.D. "Unfaced Challenge: Submarine Versus Free World," **U.S. Naval Institute Proceedings** 90(4):26 April 1964.

Alexander, R.J. "Can We Really Afford Surface ASW Ships? (ProfNote)," **U.S. Naval Institute Proceedings** 100(8):107 August 1974.

_____. "Oceanographic Predictions Through ASWEPS (ProfNote)," **U.S. Naval Institute Proceedings** 88(2):145 February 1962.

"America Defends against U-Boats," **Commandant's Bulletin** 26-29 April 1992.

Anderson, T. "UK Virtual Ship - the Way Forward?" **Naval Engineers Journal** 112(1):53-58 2000.

Andrews, Frank. "Guerrilla Warfare at Sea (ProfNote)," **U.S. Naval Institute Proceedings** 87(8):129 August 1961.

Annati, Massimo. "Mission packages for maritime patrol aircraft," **Military Technology** 23(9):65-73 September 1999.

"Antisubmarine Warfare," **U.S. Naval Institute Proceedings** 86(3):109 March 1960.

Antoniak, Charles E. "Measure of Effectiveness for Sensors and Strategies," **Naval Research Logistics Quarterly** 23(2):283-295 June 1976.

Arrigan, John M. and David M. Shao. "Antisubmarine Warfare Simulation on a Minicomputer," **Winter Simulation Conference Proceedings** (Atlanta, Ga, USA) v.1 p.53-58 1981.

"ASW-New Concepts," **U.S. Naval Institute Proceedings** 88(6):104 June 1962; 89(3):106-107 March 1963.

Baggenstoss, Paul. "Active Sonar Processing for Shallow Water ASW," **Sea Technology** 35(11):59+ November 1994.

Bancroft, David. "Light-weight hybrid torpedo: MK 54 MOD 0," **Naval Forces** 20(3):S26-S27 1999.

Barrette, J. Rene and John Courtenay Lewis. "Ice-penetrating Sonobuoy System Breaks the High Arctic Barrier," **Sea Technology** 29(10):63-67 October 1988.

Barron, Claude. "The Operator is Part of the System," **Submarine Review** 30-47 January 2001.

"Battling the Subs - World War II," **All Hands** 59-63 April 1958.

Beaudan, Eric. "Antisub warfare -- Changing course," **Defense & Diplomacy** 9:52+ July/August 1991.

Benedict, John R. Jr. "Future Undersea Warfare Perspectives," **Johns Hopkins APL Technical Digest** 21(2):269-279 April 2000.

Berrou, Jean-Louis and Ronald A. Wagstaff. "Non-linear Iterative Technique for High-resolution Spectral Estimation," **ASSP Spectrum Estimation Workshop 2** (IEEE Acoustics, Speech, and Signal Processing Society, Tampa, FL, USA) p.139-145 1983.

Bishop, Charles. "Oceanography and Naval Warfare," **U.S. Naval Institute Proceedings** 85(5):75 May 1959.

Black, E.F. "ASW Role of Latin American Navies (C&D)," **U.S. Naval Institute Proceedings** 92(12):127 December 1966.

Bogart, Carl D. et al. "A Heuristic Algorithm for Position Estimation," **Proceedings of the 1997 IEE Colloquium on Target Tracking and Data Fusion** (London, UK) IEE Colloquium (Digest) n.253 p.13/1-13/4 1996.

Bogdan, William R. "Life Cycle Support of Navy Airborne Antisubmarine Warfare Tactical Software," **Journal of Nuclear Materials** (2nd: IEEE Comput Society International Computer Software & Applictions Conference, Chicago, IL, USA) p.499-503 1978.

Booda, Larry L. "ASW: An Historical Perspective," **Sea Technology** 31(11):10-15 November 1990.

_____. "Mine Warfare Moves Ahead, Plays ASW Role," **Sea Technology** 25(11):10-11,13,15 November 1984.

_____. "SEASAT Research Could Make Ocean 'Transparent' for ASW," **Sea Technology** 24(11):10-14 November 1983.

Booth, Ronald. "Sea Lance Weapon Development - System and Naval Engineering Aspects of the Capsule," **Naval Engineers Journal** 100(3):204-214 May 1988.

Bouzouane, Abdenour et al. "Toward Virtual Role Playing Based on Intelligent Agents," **Applied Artificial Intelligence** 13(8):777-794 1999.

Boyles, C. Allan and Geraldine W. Joice. "Comparison of Three Acoustic Transmission Loss Models with Experimental Data," **Johns Hopkins APL Technical Digest** 3(1):67-76 January/March 1982.

Braybrook, Roy. "Wings over water," **Armada International** 22(2):8-18 April/May 1998.

Bramlett, W.T. II. "ASW: Some Surface Views.. (ProfNote)," **U.S. Naval Institute Proceedings** 102(4):99 April 1976.

Brandenburg, R.L. "Destroyer Command: Critical ASW Subsystem," **U.S. Naval Institute Proceedings** 90(7):36 July 1964.

Brewer, G. Daniel. "Nonpetroleum-fueled Military Aircraft," **Transactions of the American Nuclear Society** (Winter Meeting - American Nuclear Society, Washington, DC, USA) 43:27-28 1982.

Brown, Peter and Trevor Kirby-Smith. "Operational Field Trials of GPS Equipped Sonobuoys," **Proceedings of the International Technical Meeting of the Satellite Division of the Institute of Navigation**, ION GPS-96 (9th: Kansas City, MO, USA) 2:1553-1561 1996.

Brown, G.V. "Arctic ASW," **U.S. Naval Institute Proceedings** 88(3):54 March 1962.

Bryson, Lindsay. "Procurement of a Warship," **Naval Architect** 21-51 January 1985.

Burchell, Wade. "Reserve helos get Magic Lantern," **U.S. Naval Institute Proceedings** 123(2):74-75 February 1997.

Burns, Richard F. "Undersea Warfare: National, International Trends," **Sea Technology** 35(11):10-11+ November 1994.

Bussert, J.C. "Soviet ASW (ProfNote)," **U.S. Naval Institute Proceedings** 98(8):112 August 1972.

Carr, Jess W. "VS-57 and the sinking of Japanese submarine I-17," **Naval Aviation News** 14-15 September/October 2001.

Cavanaugh, R.B. "ASW Effort (ProfNote)," **U.S. Naval Institute Proceedings** 90(2):140 February 1964.

Chatham, Ralph E., Matthew A. Nelson and Enson Chang. "Results from the DARPA and ONR Synthetic Aperture Sonar Programs," **Detection and Remediation Technologies for Mines and Minelike Targets V** (Orlando, FL, USA) - Proceedings of SPIE 4038:422-430 2000.

"CNO directs improvements in antisubmarine warfare," **Sea Technology** 39(9):78 September 1998.

Conner, George. "A commentary: James Wirth's article on a proposed theater missle defense strategy based on USN antisubmarine warfare (ASW) tactics and Ronald Kurth's comments on Wirth's ideas," **Airpower Journal** 11(1):97-98 Spring 1997.

Coogan, J.T. Jr. "The 'CV:' Capable Vigilance or Continued Vulnerability (ProfNote)," **U.S. Naval Institute Proceedings** 98(3):117 March 1972.

Copley, J.D. "The Q-Ships," **Army Quarterly and Defence Journal** 121(4):458-462 1991.

Cotaras, F.D. "Integrated sonar systems for shallow-water ASW," **Sea Technology** 41(11):10-16 November 2000.

Cote, Owen and Harvey Sapolsky. "**Antisubmarine Warfare after the Cold War** [MIT Security Studies Conference]," June 1997.
http://web.mit.edu/ssp/Publications/confseries/ASW/ASW_Report.html

Crawford, Jeffrey D. and Fletcher L. Sawyer. "ASW Measures of Effectiveness," **Vitro Technical Journal** 9(1):12-19 Winter 1991.

Cummings, Timothy K. "Geosat: Navy Applications of Satellite Altimetry Growing," **Sea Technology** 29(11):39-43 November 1988.

Curtis, J.J.M. "Quarterback for the Helo ASW Team (ProfNote)," **U.S. Naval Institute Proceedings** 100(9):119 September 1974.

Dandridge, Anthony and Gary B.Cogdell. "Fiber Optic Sensors: Performance, Reliability, Smallness," **Sea Technology** 35(5):31-33+ May 1994.

Danis, A.L. "Offensive ASW: Fundamental to Defense," **U.S. Naval Institute Proceedings** 83(6):583 June 1957.

Davis, Charles M. and Robert E. Einzig. "Fiber-optic Sensors Offer Unbridled Potential," **Sea Technology** 27(11):19-24 November 1986.

Dombroff, Seymour. "FADAP," **U.S. Naval Institute Proceedings** 92(8):70 August 1966; 92(11):126 November 1966; 93(2):117 February 1967.

_____. "New Antisubmarine Weapon Systems (ProfNote)," **U.S. Naval Institute Proceedings** 87(7):125 July 1961.

Doney, Art and Steve Deal. "Bring back ASW--now!" **U.S. Naval Institute Proceedings** 125(3):102-104 March 1999.

Donoghue, Patrick J., Preben Jensen and Robert M. Peabody. "Hardware and Software Integration Facility (HSIF) for SH-60F CV-HELO," **Proceedings - IEEE/AIAA Digital Avionics Systems Conference** (7th: Fort Worth, TX, USA), 212-218 1986.

Dowell, John A. "ASW Helicopter Mission Avionics," **IEEE Aerospace and Electronic Systems Magazine** 3(7):10-17 July 1988.

Eagle, James N. "Optimal Search for a Moving Target When the Search Path is Constrained," **Operations Research** 32(5):1107-1115 September/October 1984.

Elliot, J. O. "Ocean Sciences Division EM DASH 1973," **Report of NRL Progress** 1-15 November 1973.

Emery, T.R.M. "Fleet ASW Training in Port (ProfNote)," **U.S. Naval Institute Proceedings** 86(7):146 July 1963.

Finnegan, Philip. "Gulf Nations Boost Antisub Forces," **Defense News** 8:1+ March 1-7, 1993.

Fitzgerald, James R., Raymond J Christian and Robert C Manke. "Can IT work for ASW?" **U.S. Naval Institute Proceedings** 125(6):38-41 June 1999.

_____. "NetworkCentric Antisubmarine Warfare," **U.S. Naval Institute Proceedings** 124(9):92-95 September 1998.

Flaherty, C.F. "ASW-Are We Missing the Boat? (C&D)," **U.S. Naval Institute Proceedings** 92(2):112 February 1966.

Fraser, D.A. "Present at the creation: recollections of the very first anti-submarine operations using the Leigh light," **Canadian Defence Quarterly** 9:43-46 Spring 1980.

"France Plans Its Future Fleet," **Warship Technology** 10-12 July/August 2001.

"French withdraw from joint ASW program," **Sea Power** 41(8):32-33 August 1998.

"From Trimaran to TRISWACH - Seakeeping Benefits with New Hullform," **Warship Technology** 19-20 July/August 2001.

Frothingham, Thomas G. "The Submarine Situation," **Current History: A Monthly Magazine of the New York Times** 6(3):245-249 1917.

Gadsden, C. J. "Experimental Command System for Force Level Anti-submarine Warfare," **IEE Colloquium on Large Scale and Hierarchical Systems**, Colloquium (Digest) 1986/125:1-2 1986.

Gallagher, Barrett. "Searching for Subs in the Atlantic (Pict.)," **U.S. Naval Institute Proceedings** 88(7):98 July 1962.

"GAO cites deficiency in ASW proficiency ['Defense Acquisitions: Evaluation of Navy's Anti-Submarine Warfare Assessment GAO/NSIAD-99-85 ']," **Sea Power** 42(10):22-23 October 1999.

Glenn, F. A., A.L. Zaklad and R.J. Wherry, Jr. "Human Operator Simulation in the Cognitive Domain," **Proceedings of the Human Factors Society Annual Meeting** (26th: Seattle, WA, USA) 964-968 1982.

Glennon, A.N. "An Approach to ASW," **U.S. Naval Institute Proceedings** 90(9):48 September 1964.

Golden, C. T. and E.E. Spear. "Sea Lance Capsule," **International SAMPE Symposium and Exhibition: Advanced Materials Technology** (32nd: Anaheim, CA, USA) p.685-697 1987.

Gorshkov. S.G. "Navies in War and in Peace, Part 4," **U.S. Naval Institute Proceedings** 100(4):46 April 1974.

Gotla, P. "Magnetometer in Oceanology," **Oceanology International** 5(2):36-39 February 1970.

Graham, David M. "GaAsING Up for Quicker ASW Signal Processing Speeds," **Sea Technology** 28(10):26-28 October 1987.

_____. "GE Ocean Systems Meeting Demanding ASW Requirements," **Sea Technology** 31(11):31 November 1990.

Greer, E. J. "Electrical Power Engineering in Modern Surface Warships," **GEC Review** 2(3):151-157 1986.

Grenfell, E.W. "Growing Role of the Submarine," **U.S. Naval Institute Proceedings** 89(1):49 January 1963.

Guy, C. R., M.J. Williams and N.E. Gilbert. "ASW Helicopter/sonar Dynamics Mathematical ModeL," **European Rotorcraft and Powered Lift Forum** (6th: Bristol, England) p.45-61 1980.

_____, _____ and _____ "SEA KING Anti-submarine Warfare Helicopter Mathematical Model," **Mechanical Engineering Transactions** - Institution of Engineers, Australia ME7(1):23-29 April 1982.

Hadden, P.G. "Selection of Materials for a Modern Warship," **Materials & Design** 10(5):235-240 September/October 1989.

Hahs, O.A. and C.O. Wilde. "Graduate Education Program for Operational ASW (ProfNote)," **U.S. Naval Institute Proceedings** 99(2):105 February 1973.

Hammer, John L. and Wayne R. Hole. "Continuing Need for Accurate Positioning in Naval Tactics," **Proceedings - a Partnership of Marine Interests OCEANS '88** (IEEE & Marine Technology Society, Baltimore, MD, USA)1379-1383 1988.

Hardwick, C.D. "NAE Convair 580 Aeromagnetics Program," **Quarterly Bulletin of the Division of Mechanical Engineering** (4):1-16 1979.

Harper, Dale P. "The destroyer Escort England was one of the U.S. Navy's most prolific killers of Japanese submarines during WWII," **World War II** 11:8 1997.

Harrison, R.G. "Navy's H-60 Programs: Meeting New Challenges," **Vertiflite** 33(6):44-47 November/December 1987.

_____. "SH-3 SEA KING: Modifications for the Year 2000," **Vertiflite** 29(7):38-39 November/December 1983.

Harrison, Tony. "Recent Developments in Dipping Sonar," **Electronics & Power** 32(5):390-391 May 1986.

Hart, Michael R. "Awesome airdales," **All Hands** (983):19 March 1999.

Haydon, Peter. "Is anti-submarine warfare dead?" **Canadian Defence Quarterly** 22(5):15-23 May 1993.

Hazell, Paul A. "What's the future for ASW in NATO? [Sweden] Part I," **Sea Technology** 39(11):10 November 1998; "Part II," **Sea Technology** 39(12):55 December 1998.

"Helicopters in Antisubmarine Warfare," **U.S. Naval Institute Proceedings** 89(7):36 July 1963.

Heppe, R.R., L.E. Channel and C.W. Cook. "S-3A EM DASH A New Dimension in Airborne Sea Control," **Journal of Aircraft** 11(9):577-583 September 1974.

Hernandez, M.L. "Efficient Data Fusion for Multi-sensor Management," **IEEE Aerospace Conference Proceedings** (Big Sky, MT, USA) 5:52161-52170 2001.

Higgins, Philip E. and Augustus Constantinides. "Reliability Growth for Expendable Mobile ASW Training Target (EMATT): A 1-shot Device," **Proceedings of the Annual Reliability and Maintainability Symposium** (Orlando, FL, USA) 497-501 1991.

Holland, William J. Jr. "ASW Is Still Job One," **U.S. Naval Institute Proceedings** 118(8):30-34 August 1992.

_____. "Battling battery boats," **U.S. Naval Institute Proceedings** 123(6):30-33 June 1997.

_____. "Network centric warfare in ASW," **Naval Forces** 22(5):8-12 2001.

Holt, L.J. "Doctrine of Incremental Reduction," **U.S. Naval Institute Proceedings** 91(1):62 January 1965.

Hornhaver, H. "Standard Flex Distributed Architecture Combat System," **Naval Engineers Journal** 107(3):41-48 May 1995.

Howells, H. et al. "Large Scale Knowledge Based Systems for Airborne Decision support," **Knowledge-Based Systems, Proceedings of the SGES 1988 International Conference on Knowledge-Based Systems and Applied Artificial Intelligence (ES98)** (18th: Cambridge, UK) 12(5-6):215-222 1999.

"The 'Ideal' ASW Vehicle: A Modest Proposal (C&D)," **U.S. Naval Institute Proceedings** 102(3):126 March 1976.

Incze, Bruce Imre. "Improvements in the Tactical Understanding of the Oceanographic Battlespace," **Proceedings of the 1995 MTS/IEEE OCEANS Conference** (IEEE & Marine Technology Society, San Diego, CA, USA),1:731-740 1995.

"Intercept Search (C&D)," **U.S. Naval Institute Proceedings** 95(10):119-120 October 1969.

Jackson, Harry A., William D. Needham and Dale E.Sigman. "Bottom Bounce Array Sonar Submarine (BBASS)," **Naval Engineers Journal** 101(5):59-72 September 1989.

Jameson, James N. "CUBIC Develops ASW Sonobuoy Reference System," **Sea Technology** 24(11):33,35 November 1983.

Jarrett, Stephen M. "Using Anti-submarine Warfare Experience to Enhance Unattended Ground Sensor (UGS) Employment Tactics," **Unattended Ground Sensor Technologies and Applications IV - Proceedings of SPIE** 4743:338-345 2002.

Jellicoe, J.R. "On efficiency of small craft against submarines," **U.S. Naval Institute Proceedings** 43(5):1083 May 1917.

Jensen, W.L. "Helicopter ASW (C&D)," **U.S. Naval Institute Proceedings** 99(7):83-85 July 1973.

Jordan, Wesley E. "Moving forward in ASW," **Sea Technology** 38(1):53 January 1997.

Kazek, Michael S. "Comparison of Offshore Patrol Vessels," **Marine Technology** 22(4):351-357 October 1985.

Kehoe, James W. Jr., Kenneth S. Brower and Herbert A. Meier. "Impact of Design Practices on Ship Size and Cost," **Naval Engineers Journal** 94(2):68-86 April 1982.

Kennell, Colen G., Brian L. White and Edward N. Comstock. "Innovative Naval Designs for North Atlantic Operations," **Transactions - Society of Naval Architects and Marine Engineers** 93:261-281 1985.

Kernevez, N. and H. Glenat. "Description of a High Sensitivity CW Scalar DNP-NMR Magnetometer," (5th: Joint MMM-Intermag Conference, Pittsburgh, PA, USA) **IEEE Transactions on Magnetics** 27(6,pt.2):5402-5404 November 1991.

Kim, Duk-Ki. "The modernization of submarine forces and antisubmarine warfare capabilities in Northeast Asia," **Submarine Review** 91-103 October 1998.

Klain, David R. "Surface ship ASW: Modern technology--outdated procedures," **U.S. Naval Institute Proceedings** 123(10):54-57 October 1997.

Klevebrant, Hakan. "'MacAROV'--A Versatile Testbed for Real Applications," **Proceedings of the International Symposium on Unmanned Untethered Submersible Technology** (6th: Ellicott City, MD, USA) 1-14 1989.

Kurth, Ronald J. "A commentary: [James J. Wirth's 'A Joint Idea: An Antisubmarine Warfare Approach to Theater Missile Defense']," **Airpower Journal** 11(1):95-97 Spring 1997.

Lancaster, Jon W. and David B. Bailey. "Naval Airship Program for Sizing and Performance (NAPSAP)," **Journal of Aircraft** 18(8):677-682 August 1981.

Lancaster, Richard W. and George Baron. "Measuring ASW, Oceanographic Parameters with XCTD Profiling Systems," **Sea Technology** 25(11):18-19,21,23 November 1984.

Larsbrink, Goran. "Anti submarine warfare in shallow waters," **Naval Forces** 21(1):66-70 2000.

Lavis, David R., William W. Rogalski, Jr.and Kenneth B. Spaulding. "Promise of Advanced Naval Vehicles for NATO," **Marine Technology** 27(2):65-93 March 1990.

Lefebvre, Thierry, Alain Lemer and Francois Dispot. "ARACHYDE: A Sensor-to-situation Assessment Software Architecture for Passive Acoustic Signal Understanding," **IEEE Conference on Neural Networks for Ocean Engineering** (Washington, DC, USA) p.255-262 1991.

Lewis, D. D. "NATO ASW Situation," **U.S. Naval Institute Proceedings** 85(4):55 April 1959.

Linder, Bruce. "Future of joint ASW (antisubmarine warfare)," **U.S. Naval Institute Proceedings** 121(9):66-70 September 1995.

"Lockheed Martin Seeks To Capitalize On Millennium Gun Demonstration," **Defense Daily** [Special issue] 215(37):1 August 22, 2002.

Lundeberg, Philip K. "Undersea Warfare and Allied Strategy in World War I; Part I: to 1916," **Smithsonian Journal of History** 1(3):1-30 1966; "Part II: 1916-1918," **Smithsonian Journal of History** 1(4):49-72 1967.

Mackie, R.R. "ASW Officer 'Jack of all Trades,'" **U.S. Naval Institute Proceedings** 98(2):34 February 1972; 98(12):93 December 1972; 99(1):94 January 1973; 99(2):89 February 1973.

Manned Helicopters Slated for Navy Destroyers," **Undersea Technology** 10(11):45,56 November 1969.

Marsh, Gerald E. and Robert Piacesi. "Simplified Anti-submarine Warfare Problem Treated As a Steady State Markov Process," **Applied Physics Communications** 8(4):227-238 December 1988.

Marshall, W.G. "Airborne Destroyers," **U.S. Naval Institute Proceedings** 100(3):26 March 1974; 100(7):83-84 July 1974; 100(9):102 September 1974; 100(12):89 December 1974.

Martin, Robert L. "SACLANTCEN: Submarine Threat Neutralizer," **Sea Technology** 32(5):27 May 1991.

Matthews, Anthony D. and Victor B. Johnson. "Radar Acoustic Hybrid (RAH) Experiment," **Detection and Remediation Technologies for Mines and Minelike Targets V** (Orlando, FL, USA) - **Proceedings of SPIE** 4038:431-437 2000.

McCandless, Bruce. "Antisubmarine Patrol: Saga of a Seagull," **U.S. Naval Institute Proceedings** 83(5):515 May 1957.

McConnell, James M. "New Soviet Methods for Antisubmarine Warfare?" **Naval War College Review** 38(4):16-27 July/August 1985.

McGrath, T.D. "Defense Group ALFA," **U.S. Naval Institute Proceedings** 85(8):49 August 1959.

_____. "National Insurance Policy: ASW Coverage," **U.S. Naval Institute Proceedings** 94(5):46 May 1968.

_____. "Submarine Defense," **U.S. Naval Institute Proceedings** 87(7):37 July 1961.

McLean, Douglas M. "Confronting Technological and Tactical Change: Allied Antisubmarine Warfare in the Last Year of the Battle of the Atlantic," **Naval War College Review** 47(1):87-104 Winter 1994.

Mercer, David D. "Sledge Hammers, Lance Bombs, and Q-Ships," **U.S. Naval Institute Proceedings** 87(4):76-81 April 1961.

Merkel, Thomas B. "Military Application of the Transit Navigation Satellite System in the P-3C ASW Aircraft," **Scripta Metallurgica**, Proceedings of the Institute of Navigation, National Aerospace Meeting (Washington, DC, USA) 39-42 1973.

Merrill, John. "WWII: Japan's Disinterest in Merchant Ship Convoying," **Submarine Review** 48-57 January 2001.

Merritt, R. G. and R. L. Herschkowitz. "Variations on a Single Theme: Future Configurations and Growth of the Patrol Hydrofoil Combatant (PHM)," **Hovering Craft & Hydrofoil** 16(11-12):8-20 August/September 1977.

Miller, John F. "Acoustic Propagation Loss Calculations in a Complex Ocean Environment for Simulator-based Training," **International Conference on Simulators** (Brighton, England), IEE Conference Publication n.226, p.215-221 1983.

Milner, Marc. "The dawn of modern anti-submarine warfare: Allied responses to the U-boats, 1944-45," **RUSI** 134:61-68 Spring 1989.

"Mine Warfare Official Outlines Littoral Combat Ship Module," **Defense Daily** [Special issue] 216(18):1 October 25, 2002.

"Mission and organization [SACLANT]," **Naval Forces** 18(6):38 1997.

Morgan, E.A. "DASH Weapons System (ProfNote)," **U.S. Naval Institute Proceedings** 89(1):150 January 1963.

Morgan, John G. Jr. "Networking ASW systems: Anti-submarine warfare dominance," **Sea Technology** 39(11):19-22 November 1998.

Morse, Philip M. "Beginnings of Operations Research in the United States," **Operations Research** 34(1):10-17 January/February 1986.

Mraz, Stephen J. "Projecting Power Across the Water," **Machine Design** 61(25):76-82 December 7, 1989.

"Navy Hydrofoil is ASW Ship (Notebook), **U.S. Naval Institute Proceedings**" 89(8):136 August 1963.

"Navy Test Vessel May Yield New Warfare Concepts," **Pentagon Brief** 8 January 15, 2002.

"Navy To Contract Soon On New Undersea Power Technology Development [UUVs]," **Defense Daily** 207(21):1 August 1, 2000.

Neeser, R.W. "The Role of kite balloon in anti-submarine operations," **U.S. Naval Institute Proceedings** 44(8):1924 August 1918.

"New ASW Fighter Squadron Formed (Notebook)," **U.S. Naval Institute Proceedings** 91(11):152 November 1965.

"Network-centric ASW," **Naval Forces** 20(3):S6-S9 1999.

Noyes, Alfred. "How England Checked the Submarine War," **Current History: A Monthly Magazine of the New York Times** 4(1):11-15 1916.

Oda, Don J. and Barrie W. Barker. "Application of Color to ASW Tactical Displays," **Proceedings of the Society for Information Display** 20(1):16-27 1st Quarter 1979.

Oi, Atsushi. "Why Japan's Anti-submarine Warfare Failed," **U.S. Naval Institute Proceedings** 78(6):587-601 June 1952.

Ort, Coenraad and Frank Driessen. "Trends in underwater warfare," **Naval Forces** 23(3):81-89 2002.

Ortega, J.J. "Carrier ASW Pilot Motivation (ProfNote)," **U.S. Naval Institute Proceedings** 94(8):144 August 1968.

Overstreet, C. Michael et al. "Utilization of Special Architectures in Support of Parallel Discrete Event Simulation," **IEEE Proceedings of the SOUTHEASTCON '91** (Williamsburg, VA, USA) 2:822-826 1991.

"Pacific Naval Officials See ASW Shortage As Growing Problem," **Defense Daily** 207(13):1 July 20, 2000.

Patton, James H., Jr. "Defensive Anti-Air Warfare for SSNs," **Submarine Review** 86-93 January 1994.

Pay, Graham. "EH-101 - Capabilities and Operational Aspects From a Launch Customer's Viewpoint," **Vertiflite** 33(6):50-56 November/December 1987.

Pearson, I.D.C. "ASW vs. AAW, a Question of Direction," **U.S. Naval Institute Proceedings** 96(3):64 March 1970.

Penny, Dawn E. "Sensor Management in an ASW Data Fusion System," **Proceedings of the 1999 Sensor Fusion: Architectures, Algorithms, and Applications III** (Orlando, FL, USA) - **Proceedings of SPIE** 3719:418-429 1999.

_____ and Mark Williams. "Sequential Approach to Multi-sensor Resource Management Using Particle Filters," **Signal and Data Processing of Small Targets 2000 - Proceedings of SPIE** (Orlando, FL, USA) 4048:598-609 2000.

Polmar, Norman. "Historic aircraft: A helicopter for all seasons," **Naval History** 16(5):12-14 October 2002.

_____. "Thinking About Soviet ASW. (NavRev)," **U.S. Naval Institute Proceedings** 102(5):108 May 1976.

Posnett, William L. III, Ya-Tung Chin and Max F. Platzer. "Multimission STOVL Application of a Hybrid Powered-lift System," **Journal of Aircraft** 24(7):417-423 July 1987.

_____. "Sea-based Multimission STOVL Application of an AIBF/VT Hybrid Powered-lift System," **AIAA/AHS/ASEE Aircraft Systems, Design & Technology Meeting** (Dayton, OH, USA) AIAA Paper AIAA-86-2675 9 p. 1986.

Powers, George and James Henson. "ATE in the Field Supporting Airborne ASW Avionics Avionics P-3 STYLE," **AUTOTESTCON '83** (Fort Worth, TX, USA) p.12-17 1983.

Preston, Antony. "BAe Offer for ANZAC WIP," **Naval Architect** (AUG. SUPPL):2 1999.

_____. "Developments in sonar arrays," **Armada International** 25(2):37-42 April/May 2001.

_____. "First ASW Vessels (OldNav)," **U.S. Naval Institute Proceedings** 98(10):82 October 1972.

_____. "Non-nuclear submarine propulsion," **Armada International** 24(5):66-70 October/November 2000.

_____. "Sonars for small Navies," **Naval Forces** 18(2):16-20 1997.

_____. "Stealthy submarines and ASW," **Military Technology** 16(9):66-71 September 1992.

Price ,W.W. II "New Partnership: Helicopter and Destroyer (ProfNote)," **U.S. Naval Institute Proceedings** 97(6):99 June 1971.

Prina, L. Edgar. "Vision & roadmap," **Sea Power** 41(9):35-37 September 1998.

Purtell, Joseph. "SH-60B SEAHAWK: Ready for Fleet Delivery," **Vertiflite** 29(7):28-29 November/December 1983.

Quan, Frederic J-Y. and James C. Venon. "Optical Fiber Technology Will Enhance Navy's ASW Capabilities," **Sea Technology** 29(7):17,19-21 July 1988.

Quinn, Kevin M. "The forward-deployed decisionmaker," **Sea Power** 45(4):92-94 April 2002.

Reade, David. "World's MPA fleets: The only constant is change," **Sea Power** 43(12):37-40 December 2000.

Reichert, H.E. "LAMPS in the Gulf of Tonkin (ProfNote)," **U.S. Naval Institute Proceedings** 99(3):115 March 1973.

Reinicke, F.G. "Antisubmarine Pioneering," **U.S. Naval Institute Proceedings** 91(10):168 October 1965.

Reynolds, J. Guy. "ASW is a Top Priority--In Europe," **Submarine Review** 24-25 January 1999.

Rhea, John. "Fiberoptics Expands DesignPpotential for Weapons Systems," **Laser Focus** 23(8):94, 96, 98-101 August 1987.

Ricard, M. and M. Keegan. "Intelligent Autonomy for the Manta Test Vehicle," **OCEANS 2000** (IEEE & Marine Technology Society, Providence, RI, USA) 2:265-1271 2000.

Richter, Stephen B. and Lawrence J. Fusillo. "Helicopter Navigation Algorithms for the Placement of Sonobuoys in an Antisubmarine Warfare (ASW) Environment," **IEEE Proceedings of the National Aerospace and Electronics Conference - NAECON** (Dayton, OH, USA) 280-286 1988.

Robinson, John. "Navy Opens Anti-Submarine Warfare Office," **Defense Daily** 1 September 11, 1996.

Rodrick, Eugene J., David M. Maurer and Raymond B. Grochowski. "Modular Shipboard Helicopter Support SysteM," **Naval Engineers Journal** 100(3):293-305 May 1988.

Rondorf, Neil E. "The New IUSS Team," **Submarine Review** 72-78 July 2002.

Ruhe, W.J. "The Nuclear Submarine: Riding High," **U.S. Naval Institute Proceedings** 101(2):55 February 1975.

Ryan, D.P. "Small Boys and Whirly Birds," **U.S. Naval Institute Proceedings** 96(2):26 February 1970.

"SAIC Team Wins Anti-Submarine Warfare Support Contract," **Defense Daily** 199(98):1 August 18, 1998.

Salitter, Michael and Ulrich Weisser. "Shallow Water Warfare in Northern Europe," **U.S. Naval Institute Proceedings** 103(3):36 March 1977.

"Saudi Frigate Programme on Course," **Warship Technology** 10-11 May 2001.

Schatz, A.D. "DesRon Airedale Arriving (ProfNote)," **U.S. Naval Institute Proceedings** 103(9):115 September 1977.

Schofield, Colin. "Electronic charting for naval combat systems," **Sea Technology** 40(3):49 March 1999.

Schrader, Rudolf . "Antisubmarine warfare research: SACLANT ASW Research Center (Prof Note)," **U.S. Naval Institute Proceedings** 92(1):143 January 1966.

Schwager, J. E. "LITT - An Implementation for ASW Flight Simulation," **AIAA Aerospace Sciences Meeting** (22nd: Reno, NV, USA), AIAA Paper AIAA-84-0518 4 p. 1984.

Sear, A.W. Jr. "The 'Ideal' ASW Vehicle: A Modest Proposal (Prof Note)," **U.S. Naval Institute Proceedings** 101(1):104 January 1975.

Shriver Jack. "Developing real anti-diesel tactics," **Submarine Review** 90-95 April 1998.

Sinex, Charles H. and Robert S. Winokur. "Environmental Factors Affecting Military oOerations in the Littoral Battlespace," **Johns Hopkins APL Technical Digest** 14(2):112-124 April/June 1993.

Skipper, J.W. "ASW 'Six-Pack' (ProfNote)," **U.S. Naval Institute Proceedings** 97(12):103 December 1971.

Skorheim, Robert D. and George M. Pavey. "From Seismic Reflection to Submarine Detection," **Sea Technology** 27(11):39-40 November 1986.

Smart, Jeffrey H. "Seasonal and Spatial Variations in the Attenuation of Light in the North Atlantic Ocean," **Johns Hopkins APL Technical Digest** 14(3):231-243 July/September 1993.

Smirnov, Viktor. "Antisubmarine activity by submarines (based on World War II experience)," **Soviet Naval Digest** 18-24 April 1988.

Smith, James M.and Russell H. Logan. "AN/APS-116 Periscope-Detecting Radar," **IEEE Transactions on Aerospace and Electronic Systems** AES-16(1):66-73 January 1980.

Smith R.H. "ASW-Crucia1 Naval Challenge (NavRev)." **U.S. Naval Institute Proceedings** 98(5):126 May 1972.

Smith, William D. "Superstealth submarines for the next century," **Naval Forces** 20(1):28-32 1999.

Sottilare, Robert A. and Rodney A. Long. "Innovations in Training Simulation," **Naval Engineers Journal** 104(3):141-147 May 1992.

"Stalking the U-boat: USAAF Offensive Antisubmarine Operations in World War II," **Naval War College Review** 50(4):164-165 Autumn 1997.

Steele, G.P. "Killing Nuclear Submarines," **U.S. Naval Institute Proceedings** 86(11):45 November 1960.

Steele, Robert and Ralph Aurora. "U.S. Navy Bets on V/STOL," **IEEE Spectrum** 15(9):58-63 September 1978.

Streetly, Martin. "Russia launches new ASW suite," **Journal of Electronic Defense** 21(12):18 December 1998.

Stevenson, R. G. "H-2/H-3: Meeting the Navy's Challenge," **Vertiflite** 31(6):32-33,70 November/December 1985.

Stretton, Milt et al. "Use of Fast-time Simulation in Predicting Multioperator-system Efficiency," **Proceedings of the Human Factors Society Annual Meeting** (35th: San Francisco, CA, USA) 2:1215-1218 1991.

"Submarine-launched UUVs Swim Ahead," **Warship Technology** 6-7 May 2000.

Sun, H. et al. "Use of Low-cost GPS Receiver Technology for Sonobuoy Positioning," Navigational Technology for the 3rd Millennium, **Proceedings of the Institute of Navigation Annual Meeting** (52nd: Cambridge, MA, USA), 167-174 1996.

"Swedish YS-2000 Corvette Programm," **Naval Architect** 55 February 1996.

Tangredi, Sam J. "Anti-Submarine War and "Arms Control": An Inevitable Collision?" **Naval War College Review** 42(1):66-85 Winter 1989.

Thach, J.W. "ASW Navy of the Seventies," **U.S. Naval Institute Proceedings** 89(1):57 January 1963.

Treacher, J.D. "Some Aspects of ASW in the Royal Navy (ProfNote)," **U.S. Naval Institute Proceedings** 92(10):152 October 1966.

Trounce, M. "HMS 'ARK ROYAL' Shows Her Paces," **Marine Engineers Review** 34-35 January 1985.

Truver, Scott C. "Whither US anti-submarine warfare, now that the threat has gone away?" **Naval Forces** 11(5):8-9+ 1990.

Tucker, Jonathan B. "Cold War in the Ocean Depths," **High Technology** 5(7):29-35,37-38 July 1985.

Tyler, Gordon D. Jr. "Emergence of Low-frequency Active Acoustics As a Critical Antisubmarine Warfare Technology," **Johns Hopkins APL Technical Digest** 13(1):145-159 January/March 1992.

Van Saun, A. "Let's Fight Fire With Fire (Prof Note)," **U.S. Naval Institute Proceedings** 102(12):99 December 1976.

_____. "Let's Fight Fire With Fire (C&D)," **U.S. Naval Institute Proceedings** 103(3):99 October 1977.

Vego, Milan N. "A World of Difference: Soviet Antisubmarine Warfare in 1991," **Naval War College Review** 45(3):58–77 Summer 1992.

_____. "Submarines in Soviet ASW Doctrine and Tactics," **Naval War College Review** 36(2):2-16 March/April 1983.

Waddey, R.F. and L. Williams. "Analysis of ASWEPS Prediction," **Proceedings U.S. Navy Symposium on Military Oceanography** (5th: Panama City, FL, USA) v.1 1968.

Walters, Brian. "Training for the underwater threat," **Armada International** 25(2):43-44 April/May 2001.

"Wanted: An Effective ASW Ship," **U.S. Naval Institute Proceedings** 89(7):106 July 1963.

Ward, G.G. "Engine/airframe Interface Considerations for Future Open-ocean ASW Aircraft," **Proceedings of the National Conference on Environmental Effects on Aircraft & Propulsion Systems** (7th: Princeton, NJ, USA) 107-116 1967.

Watermann, Jurgen. "Magnetic detection techniques in antisubmarine warfare," **Naval Forces** 18(6):88-89 1997.

Waters, J.M. Jr. "Antisubmarine warfare-small ships: Little Ships With Long Arms," **U.S. Naval Institute Proceedings** 91(8):74 August 965.

Weatherup, R.A. "Defense Against Nuclear-Powered Submarines," **U.S. Naval Institute Proceedings** 85(12):71 December 1959.

Weiland, Monica Zubritzky, Boyd Cooke and Brad Peterson. "Designing and Implementing Decision Aids for a Complex eEvironment Using Goal Hierarchies," **Proceedings of the Annual Meeting of the Human Factors Society** (36th: Atlanta, GA, USA) 1(1):394-398 1992.

Wells, Deborah K. and Richard M. Wargelin. "Programs and Products of the Naval Oceanographic Office," **Marine Technology Society Journal** 19(3):18-25 Third Quarter 1985.

Wertheim, Eric. "Lest we forget," **U.S. Naval Institute Proceedings** 127(6):94 June 2001.

West-Vukovich, G. et al. "Honeywell/DND Helicopter Integrated Navigation System (HINS)," **IEEE PLANS '88 - Position Location and Navigation Symposium Record: Navigation into the 21st Century** 416-425 1988.

_____. "Honeywell/DND Helicopter Integrated Navigation System (HINS)," **IEEE Aerospace and Electronic Systems Magazine** 4(3, pt.1):18-28 March 1989.

"What Happened to ASW? (Notebook)," **U.S. Naval Institute Proceedings** 91(6):150 June 1965.

Whidden, W.V. "Antisubmarine support carriers: Future of the 'Second Segment,'" **U.S. Naval Institute Proceedings** 90(11):76 November 1964.

_____. "Future of the "Second Segment," **U.S. Naval Institute Proceedings** 90(11):76 November 1964.

Wildberger, A. Martin, Kenneth F. Loje and Soo-Young Lee. "Anti-submarine Warfare Campaign Simulation," **Proceedings of the 1985 Summer Computer Simulation Conference** (Chicago, IL, USA) 625-630 1985.

Wilkes, D.M. and H. Alnajjar. "Critical Distance and the Performance of Sensor Array Geometries," **IEE Proceedings: Radar, Sonar and Navigation** 141(6):333-336 December 1994.

Wilson, Alan D. "3rd annual submarine and anti-submarine warfare conference," **Sea Technology** 41(7):40 July 2000.

Winnefeld, J.A. and C.H. Builder. "ASW-Now Or Never," **U.S. Naval Institute Proceedings** 97(9):18 1971.

Winokur, Robert S. and David L. Bradley. "Naval Applications of Oceanography," **Marine Technology Society Journal** 19(2):5-11 Second Quarter 1985.

Wirth, James J. "A joint idea: An antisubmarine warfare approach to theater missile defense," **Airpower Journal** 11(1):86-95 Spring 1997.

Wise, James E.. "The Dawn of ASW," **U.S. Naval Institute Proceedings** 90(2):91-105 February 1964.

Wohl, Joseph et al. "Human Cognitive Performance in Antisubmarine Warfare: Situation Assessment and Data Fusion," **IEEE Transactions on Systems, Man and Cybernetics** 18(5):777-786 September/October 1988.

Wolfe, Frank. "Navy Seeks $1.2 Billion-$1.5 Billion Annually For ASW," **Defense Daily** 199(88):1 August 4, 1998.

Wolf, George W. Jr. "U.S. Navy sonobuoys--key to antisubmarine warfare," **Sea Technology** 39(11):41-44 November 1998.

Wright, S.E. Jr. "ASW and the Modern Submarine," **U.S. Naval Institute Proceedings** 99(4):62 1973.

Yankaskas, Kurt and Tom Slotwinski. "Acoustic Characteristics of T-AGOS 19 Class SWATH Ships," **Naval Engineers Journal** 107(3):95-119 May 1995.

Yu, Xiang and Shijian Zhu. "The Challenge to Reducing Radiated Noise of Submarine Due to New Development of Sonar System," **Process in Safety Science and Technology** Part A 3:750-754 2002.

SUBMARINE CHASERS AND PATROL BOATS

"Boats of the Volunteer Patrol Squadron (1916)," **U.S. Naval Institute Proceedings** 42(1):243 January/February 1916.

"Characteristics of Russian Patrol Boats (1917)," **U.S. Naval Institute Proceedings** 43(7):505 July 1917.

Davidonis, A.C. "American Naval Mission In the Adriatic, 1918-1921," **U.S. Naval Institute Proceedings** 71(1):41 January 1945.

Doughty, L. Jr. "Mistaken Attacks In World War," **U.S. Naval Institute Proceedings** 60(12):1729 December 1934.

"'Eagle' Boats," **U.S. Naval Institute Proceedings** 44(4):838 April 1918.

"Efficiency of Submarine Chasers," **U.S. Naval Institute Proceedings** 43(4):791 April 1917.

"50-Mile Motor Boats to Rout Submarines," **U.S. Naval Institute Proceedings** 41(4):1388 July/August 1915.

Furer, J.A. "The 110-ft. Submarine Chasers and Eagle Boats," **U.S. Naval Institute Proceedings** 45(5):753 May 1919.

Hazlett, E.E. Jr. "Austro-American Navy," **U.S. Naval Institute Proceedings** 66(12):1757 December 1940.

"Illustration of Ford's Standardized Submarine Chaser," **U.S. Naval Institute Proceedings** 44(4):831 April 1918.

"Motor Boats as Submarine Chasers," **U.S. Naval Institute Proceedings** 42(4):1311 July/August 1916.

"Motor Boats as Submarine Destroyers for U.S. Navy," **U.S. Naval Institute Proceedings** 42(2):585 March/April 1916.

"Motor Patrol Boats of U.S. Navy (1916)," **U.S. Naval Institute Proceedings** 42(4):1288 July/August 1916.

"Patrol Boats and Their Operations," **U.S. Naval Institute Proceedings** 43(7):1502 July 1917.

Raguet, E.G. "United States Submarine Chasers at Gibraltar November 1918," **U.S. Naval Institute Proceedings** 62(12):1703 December 1936.

"Report on U.S. Patrol Boats 1916," **U.S. Naval Institute Proceedings** 42(6):2002 November/December 1916.

"Russian Submarine Chasers (1916)," **U.S. Naval Institute Proceedings** 42(1):237 January/February 1916.

"Sims Reports Submarine Chasers Unsatisfactory," **U.S. Naval Institute Proceedings** 43(10):2311 October 1917.

Stevens, R.S. Jr. "Peter Charlie-Plea for Recognition," **U.S. Naval Institute Proceedings** 72(8):1031 August 1946.

"Submarine Chaser 209 (U.S.)," **U.S. Naval Institute Proceedings** 63(5):631 May 1937.

"Submarine Chasers for the British Navy (1915)," **U.S. Naval Institute Proceedings** 41(6):2022 November/December 1915.

"Submarine Chasers not up to Expectations," **U.S. Naval Institute Proceedings** 43(9):2104 September 1917.

"The U.S. Navy's 345 Patrol Boats (1917)," **U.S. Naval Institute Proceedings** 43(6):1251 June 1917.

Vicker, M. "PC Attacks," **U.S. Naval Institute Proceedings** 70(8):991 August 1944.

SUBMARINE WEBSITES

GENERAL

Warships Of The World (By Country's Fleet)
http://www.warships1.com/index_ships.htm

Warships on the Web
http://web.ukonline.co.uk/aj.cashmore/index.html

Submarine World Network (mostly U.S. Links)
http://www.rontini.com/

Kockums Submarines (not just Swedish anymore)
http://www.submarines.com/Submarines/submarinesmain.html

SUBMARINE WEBSITES BY COUNTRY

AMERICAN

The Submarine (Official U.S. Navy Site)
http://www.chinfo.navy.mil/navpalib/ships/submarines/

The Submarine Center
http://www.geocities.com/Pentagon/Quarters/9000/

DANFS Online: Submarines
http://www.hazegray.org/danfs/submar/

United States Navy Submarine Fleet
http://www.subnet.com/fleet/fleet.htm

AUSTRALIAN

Submarine Service Fact File
http://www.navy.gov.au/6_facts/3_submarine.htm

Discovery of the AE2 off Gallipoli in the Sea of Marmara
http://www.navy.gov.au/history/ae2/default.htm

Report on the Discovery of the AE2
http://www.navy.gov.au/history/ae2/1998report.pdf

BRITISH

BriTsub
http://www.britsub.net/

Royal Navy Submarines
http://members.iinet.net.au/~eadej/

British Submarines of World War II
http://web.ukonline.co.uk/chalcraft/wtv/ww2sm2wtv.html

Royal Navy Submarine Museum
http://www.rnsubmus.co.uk/welcome.htm

Warships.Net: The British Royal Navy, It's History, Ships and Men.
http://www.warships.net/royalnavy/rnshiptypes/submarines/

CANADIAN

Canadian Navy Victoria Class
http://www.navy.dnd.ca/mspa_fleet/vic_overview_e.asp

Canadian Navy of Yesterday and Today
http://www.hazegray.org/navhist/canada/

DUTCH

Submarines of the Royal Netherlands Navy 1906-2000
http://www.dutchsubmarines.com/

FRENCH

The French Navy's Submarines In July 1939
http://perso.wanadoo.fr/bertrand.daubigny/FSS39uk.htm

GERMAN

The U-boat War 1939-1945, with U-boat War in World War One
http://uboat.net/

The U-Boat Bases in France
http://www.uboat-bases.com/fr/index.php

ITALIAN

Submarines of the Italian Navy
http://www.sommergibili.com/

I Sommergibili Italiani
http://www.regiamarina.net/arsenals/ships_it/subs_it.htm

JAPANESE

Submarines of the Imperial Japanese Navy
http://www.combinedfleet.com/ss.htm

POLISH

THE POLISH NAVY 1918 - 1945 /A Short Story/
http://www.computerage.demon.co.uk/navy/index.html

SINGAPORE

The Riken Project
http://www.kockums.se/News/oldnews/riken.html

TURKISH

Welcome to the History of Turkish Submarines
http://www.geocities.com/Pentagon/Bunker/7704/

MIDGET SUBMARINES

Fine Scale Models
http://www.walruscarpenter.com/midget.html

WWII SUBMARINE HISTORY

The Bosun's Website

http://www.bosun.net/index.html - home

WWII SUBMARINE STATISTICS

Statistics - Submarine Fleets over the 1939-1945 Wa
http://members.tripod.com/mackenziegregory/log/Underwater/30Statistics-SubmarineFle.html